Psychological Statistics and Psychometrics Using Stata

SCOTT A. BALDWIN
Brigham Young University

A Stata Press Publication
StataCorp LLC
College Station, Texas

Copyright © 2019 StataCorp LLC
All rights reserved. First edition 2019

Published by Stata Press, 4905 Lakeway Drive, College Station, Texas 77845
Typeset in LaTeX 2_ε
Printed in the United States of America
10 9 8 7 6 5 4 3 2 1

Print ISBN-10: 1-59718-303-2
Print ISBN-13: 978-1-59718-303-1
ePub ISBN-10: 1-59718-304-0
ePub ISBN-13: 978-1-59718-304-8
Mobi ISBN-10: 1-59718-305-9
Mobi ISBN-13: 978-1-59718-305-5

Library of Congress Control Number: 2019935130

No part of this book may be reproduced, stored in a retrieval system, or transcribed, in any form or by any means—electronic, mechanical, photocopy, recording, or otherwise—without the prior written permission of StataCorp LLC.

Stata, **Stata**, Stata Press, Mata, **Mata**, and NetCourse are registered trademarks of StataCorp LLC.

Stata and Stata Press are registered trademarks with the World Intellectual Property Organization of the United Nations.

NetCourseNow is a trademark of StataCorp LLC.

LaTeX 2_ε is a trademark of the American Mathematical Society.

Psychological Statistics
and Psychometrics Using Stata

To Autumn

Contents

	List of figures	xvii
	List of tables	xxiii
	Acknowledgments	xxv
	Notation and typography	xxvii
I	**Getting oriented to Stata**	**1**
1	**Introduction**	**3**
	1.1 Structure of the book	3
	1.2 Benefits of Stata	4
	1.3 Scientific context	5
2	**Introduction to Stata**	**9**
	2.1 Point-and-click versus writing commands	9
	2.2 The Stata interface	10
	2.3 Getting data in Stata	12
	2.4 Viewing and describing data	14
	2.4.1 list, in, and if	15
	2.5 Creating new variables	17
	2.5.1 Missing data	18
	2.5.2 Labels	19
	2.6 Summarizing data	20
	2.6.1 summarize	20
	2.6.2 table and tabulate	21
	2.7 Graphing data	22
	2.7.1 Histograms	23
	2.7.2 Box plots	24

		2.7.3	Scatterplots	25
	2.8	Reproducible analysis		27
		2.8.1	Do-files	28
		2.8.2	Log files	30
		2.8.3	Project Manager	30
		2.8.4	Workflow	31
	2.9	Getting help		32
		2.9.1	Help documents	32
		2.9.2	PDF documentation	33
	2.10	Extending Stata		33
		2.10.1	Statistical Software Components	33
		2.10.2	Writing your own programs	33

II Understanding relationships between variables — 35

3 Regression with continuous predictors — 37

	3.1	Data		38
	3.2	Exploration		38
		3.2.1	Demonstration	40
			Simulation program	40
	3.3	Bivariate regression		42
		3.3.1	Lines	43
		3.3.2	Regression equation	43
		3.3.3	Estimation	45
		3.3.4	Interpretation	46
			Slope	46
			Intercept	47
		3.3.5	Residuals and predicted values	48
		3.3.6	Partitioning variance	51
		3.3.7	Confidence intervals	54
		3.3.8	Null hypothesis significance testing	60

		3.3.9	Additional methods for understanding models	63
			Using predicted scores to understand model implications	64
			Composite contrasts	73
	3.4	Conclusions		74
4	**Regression with categorical and continuous predictors**			**75**
	4.1	Data for this chapter		76
	4.2	Why categorical predictors need special care		77
	4.3	Dummy coding		78
		4.3.1	Example: Incorrect use of categorical variable	85
	4.4	Multiple predictors		86
		4.4.1	Interpretation	86
			Model fit	86
			Intercept	87
			Slopes	88
		4.4.2	Unique variance	89
	4.5	Interactions		90
		4.5.1	Categorical by continuous interactions	91
			Dichotomous by continuous interactions	91
			Polytomous by continuous interactions	101
			Joint test for interactions with polytomous variables	105
		4.5.2	Continuous by continuous interactions	107
	4.6	Summary		109
5	**t tests and one-way ANOVA**			**111**
	5.1	Data		112
	5.2	Comparing two means		112
		5.2.1	t test	114
		5.2.2	Effect size	115
	5.3	Comparing three or more means		116
		5.3.1	Analysis of variance	116

		5.3.2	Multiple comparisons	120
			Planned comparisons	122
			Direct adjustment for multiple comparisons	126
	5.4	Summary		129
6	**Factorial ANOVA**			**131**
	6.1	Data for this chapter		132
	6.2	Factorial design with two factors		134
		6.2.1	Examining and visualizing the data	134
		6.2.2	Main effects	138
			Testing the null hypothesis	139
		6.2.3	Interactions	139
		6.2.4	Partitioning the variance	140
		6.2.5	2 x 2 source table	142
		6.2.6	Using anova to estimate a factorial ANOVA	144
		6.2.7	Simple effects	146
		6.2.8	Effect size	149
	6.3	Factorial design with three factors		151
		6.3.1	Examining and visualizing the data	152
		6.3.2	Marginal means	154
		6.3.3	Main effects and interactions	155
		6.3.4	Three-way interaction	158
		6.3.5	Fitting the model with anova	159
		6.3.6	Interpreting the interaction	160
		6.3.7	A note about effect size	165
	6.4	Conclusion		166
7	**Repeated-measures models**			**167**
	7.1	Data for this chapter		169
	7.2	Basic model		172

	7.3	Using mixed to fit a repeated-measures model	175
		7.3.1 Covariance structures	176
		Compound symmetry (exchangeable)	177
		First-order autoregressive	180
		Toeplitz	183
		Unstructured	186
		7.3.2 Degrees of freedom	189
		7.3.3 Pairwise comparisons	190
	7.4	Models with multiple factors	192
	7.5	Estimating heteroskedastic residuals	197
	7.6	Summary	200
8	**Planning studies: Power and sample-size calculations**		**201**
	8.1	Foundational ideas	202
		8.1.1 Null and alternative distributions	202
		8.1.2 Simulating draws out of the null and alternative distributions	204
	8.2	Computing power manually	210
	8.3	Stata's commands	214
		8.3.1 Two-sample z test	214
		8.3.2 Two-sample t test	215
		8.3.3 Correlation	220
		8.3.4 One-way ANOVA	223
		8.3.5 Factorial ANOVA	226
	8.4	The central importance of power	229
		8.4.1 Type M and S errors	230
		Type S errors	231
		Type M errors	233
	8.5	Summary	235
9	**Multilevel models for cross-sectional data**		**237**
	9.1	Data used in this chapter	239

	9.2	Why clustered data structures matter	239
		9.2.1 Statistical issues	239
		9.2.2 Conceptual issues	242
	9.3	Basics of a multilevel model	244
		9.3.1 Partitioning sources of variance	244
		9.3.2 Random intercepts	246
		9.3.3 Estimating random intercepts	248
		9.3.4 Intraclass correlations	250
		9.3.5 Estimating cluster means	252
		Comparing pooled and unpooled means	259
		9.3.6 Adding a predictor	262
	9.4	Between-clusters and within-cluster relationships	264
		9.4.1 Partitioning variance in the predictor	265
		9.4.2 Total- versus level-specific relationships	266
		9.4.3 Exploring the between-clusters and within-cluster relationships	267
		9.4.4 Estimating the between-clusters and within-cluster effects	270
	9.5	Random slopes	273
	9.6	Summary	283
10	**Multilevel models for longitudinal data**		**285**
	10.1	Data used in this chapter	286
	10.2	Basic growth model	287
		10.2.1 Multilevel model	295
	10.3	Adding a level-2 predictor	300
	10.4	Adding a level-1 predictor	307
	10.5	Summary	310
III	**Psychometrics through the lens of factor analysis**		**313**
11	**Factor analysis: Reliability**		**315**
	11.1	What you will learn in this chapter	316
	11.2	Example data	316

11.3	Common versus unique variance	317
11.4	One-factor model	320
	11.4.1	Parts of a path model	321
	11.4.2	Where do the latent variables come from?	321
11.5	Prediction equation	323
11.6	Using sem to estimate CFA models	326
11.7	Model fit	328
	11.7.1	Computing χ^2	330
11.8	Obtaining σ_C^2 and σ_U^2	334
	11.8.1	Computing R^2 for an item	335
	11.8.2	Computing σ_C^2 and σ_U^2 for all items	338
	11.8.3	Computing reliability—ω	340
	11.8.4	Bootstrapping the standard error and 95% confidence interval for ω	341
11.9	Comparing ω with α	343
	11.9.1	Evaluating the assumption of tau-equivalence	346
	11.9.2	Parallel items	349
11.10	Correlated residuals	350
11.11	Summary	353

12 Factor analysis: Factorial validity — 355

12.1	Data for this chapter	357
12.2	Exploratory factor analysis	358
	12.2.1 Common factor model	359
	12.2.2 Extraction methods	360
	12.2.3 Interpreting loadings	362
	12.2.4 Eigenvalues	363
	12.2.5 Communality and uniqueness	364
	12.2.6 Factor analysis versus principal-component analysis	365

		12.2.7	Choosing factors and rotation	366

- 12.2.7 Choosing factors and rotation 366
  - How many factors should we extract? 366
  - Eigenvalue-greater-than-one rule 366
  - Scree plots 367
  - Parallel analysis 368
  - Orthogonal rotation—varimax 370
  - Oblique rotation—promax 374
- 12.3 Confirmatory factor analysis 377
  - 12.3.1 EFA versus CFA 377
  - 12.3.2 Estimating a CFA with sem 380
  - 12.3.3 Mean structure versus variance structure 381
  - 12.3.4 Identifying models 383
    - Imposing constraints for identification 384
    - How much information is needed to identify a model? 385
  - 12.3.5 Refitting the model with constrained latent variables 386
  - 12.3.6 Standardized solutions 389
  - 12.3.7 Global fit 392
    - RMSEA 393
    - TLI 394
    - CFI 394
    - SRMR 394
    - A summary and a caution 395
  - 12.3.8 Refining models further 395
- 12.4 Summary 396

13 Measurement invariance   397

- 13.1 Data 398
- 13.2 Measurement invariance 398
- 13.3 Measurement invariance across groups 400
  - 13.3.1 Configural invariance 400
  - 13.3.2 Metric invariance 407

	13.3.3	Scalar invariance	407
	13.3.4	Residual invariance	408
	13.3.5	Using the comparative fit index to evaluate invariance	409
13.4	Structural invariance		412
	13.4.1	Invariant factor variances	412
	13.4.2	Invariant factor means	412
13.5	Measurement invariance across time		413
	13.5.1	Configural invariance	413
		Effects coding for identification	416
		Effects-coding constraints in Stata	417
	13.5.2	Metric invariance	424
	13.5.3	Scalar invariance	425
	13.5.4	Residual invariance	426
13.6	Structural invariance		427
13.7	Summary		428

References   429

Author index   441

Subject index   445

Figures

2.1	Stata interface for Mac OS X	11
2.2	Visualization of a working directory	13
2.3	The Data Editor	15
2.4	Example histogram with normal density plot	23
2.5	Example box plot	24
2.6	Example scatterplot	26
2.7	Example scatterplot with points that vary as a function of a variable in the dataset	27
2.8	Screenshot of the Do-file Editor on a Mac	29
3.1	Histogram of `attitude`	39
3.2	Relationship between `educ` and `attitude`	42
3.3	Illustration of $y = mx + b$	43
3.4	Interpretation of b_1; Δ = difference	46
3.5	Illustration of errors in prediction	49
3.6	Partitioning the variability of a single score	53
3.7	Sampling distribution of the slope for mean-centered education when $N = 1006$	56
3.8	Sampling distribution of the slope for mean-centered education when $N = 30$	58
3.9	Illustration of 95% confidence interval coverage in the first 100 simulation replications	60
3.10	Sampling distribution of the slope under the null hypothesis that $\beta_1 = 0$. The sampling distribution is a t distribution with 1,004 degrees of freedom. The shaded areas are the critical areas of rejection for a two-tailed test and $\alpha = 0.05$.	63

3.11 `marginsplot` showing the expected relationship (including 95% confidence intervals) between `attitude` and `educ` accounting for uncertainty in model parameters 67

3.12 `marginsplot` showing the expected relationship (including 95% confidence intervals) between `attitude` and `educ` accounting for uncertainty in model parameters but with adjusted axis titles and marker labels 68

3.13 `marginsplot` showing the expected relationship (including 95% confidence intervals) between `attitude` and `educ` accounting for uncertainty in model parameters but with adjusted axis titles, marker labels, and plottypes 69

3.14 `marginsplot` showing the expected relationship (including 95% confidence intervals) between `attitude` and `educ` accounting for uncertainty in model parameters and across the entire range of `educ`; the circles are the raw data 70

3.15 Plot based on the output of `predict` rather than `margins` showing the expected relationship (including 95% confidence intervals) between `attitude` and `educ` accounting for uncertainty in model parameters and across the entire range of `educ`; the circles are the raw data 71

3.16 Plot of the expected relationship (including 95% confidence interval and prediction interval) between `attitude` and `educ` across the entire range of `educ`; the circles are the raw data 73

4.1 The impact of correlation among two predictors on R^2 90

4.2 Visualization of no interaction between average number of putts and greens in regulation category 94

4.3 Plot of the simple regression equations describing the relationship between `money_rank` and `z_putts` conditional on `reg_cat2` 98

4.4 Visualization of a polytomous by continuous interaction using golf data 104

4.5 Visualization of a continuous by continuous interaction using the golf data 109

5.1 Violin plot of the BDI scores across treatment condition 113

5.2 Means and 95% CI for each treatment 120

5.3 Relationship between type I error rate (α) and number of comparisons (m) 122

Figures

6.1	Box plot for years for the combination of attract and sex	136
6.2	Dot plot for years for the combination of attract and sex	137
6.3	Dot plot of the attract by sex interaction	145
6.4	Box plot for years for the combination of attract, sex, and crime	153
6.5	Plot of the attract by sex by crime interaction with attract on the x axis	162
6.6	Plot of the attract by sex by crime interaction with sex on the x axis	163
7.1	Number of correct PASAT answers over four trials	171
7.2	Number of correct PASAT answers over four trials and stratified by condition	193
7.3	Expected PASAT scores over the four trials and stratified by condition	197
8.1	Null and alternative distributions	203
8.2	Simulated samples from the null distribution	206
8.3	Simulated samples from the alternative distribution	207
8.4	Simulated samples from the alternative distribution with 50 participants per condition	209
8.5	Illustrating power by using z scores for both the null (z_{null}) and the alternative (z_{alt}) distributions	213
8.6	Power curve for a two-sample t test	216
8.7	Detectable difference across sample size	217
8.8	Sample sizes needed to have 80% power to detect effect sizes between $\delta = 0.2$ to $\delta = 0.5$	218
8.9	Detectable difference as a function of sample size and power	219
8.10	Sample size as function of power and effect size	220
8.11	Power to detect a population correlation of $\rho = 0.2$ as a function of sample size	221
8.12	Sample size needed to have 80% power to detect a population correlation between $\rho = 0.1$ and $\rho = 0.5$	222
8.13	Detectable population correlation as a function of sample size	223
8.14	Power as a function of sample size for a one-way ANOVA	225

8.15 Funnel plot showing the sampling distribution of a standardized mean difference as a function of power 231

8.16 Type S error rate as a function of power: the horizontal, dotted line indicates a 5% type S error rate, while the vertical, dashed line is the power level (≈ 0.1) below which the type S error rate exceeds 5% 233

8.17 Exaggeration ratio (type M error) as a function of power: the horizontal, dotted line indicates no exaggeration, while the vertical, dashed line is the power level (≈ 0.25) below which the exaggeration ratio exceeds 2 234

9.1 Residuals of the same therapist using the alliance data 241

9.2 Residual of the same therapist assuming a large within-therapist correlation 241

9.3 Illustration of the partitioning of the total residual 245

9.4 Patient and therapist data for 10 therapists and an intraclass correlation = 0.16 251

9.5 Patient and therapist data for 10 therapists and an intraclass correlation = 0.68 252

9.6 Caterpillar plot for the "no pooling" approach to computing cluster means 254

9.7 Probability of 0, 1, or 2 heads when flipping a coin twice and the probability of heads is 0.5 257

9.8 Probability of 0–1,000 heads when flipping a coin 1,000 times and the probability of heads is 0.5 258

9.9 Caterpillar plot for the partial pooling approach to computing cluster means 259

9.10 Scatterplot comparing the unpooled and pooled therapist means . . 261

9.11 Dot plot comparing the pooled and unpooled means 262

9.12 Alliance–outcome relationship from a random-intercept model . . . 263

9.13 Visualizing the between-therapists and within-therapist relationships 270

9.14 Variability in the motivate–alliance slope across therapists 276

9.15 Correlation between therapist-specific intercepts and slopes from regress 277

9.16 Comparison of the empirical Bayes intercept and the distribution of distress 281

Figures   xxi

9.17   Comparison of the empirical Bayes slope and the distribution of
       `distress` 282
9.18   Therapist-specific regression lines (empirical Bayes) and raw data . 283

10.1   Box plot for `distress` and `stress` stratified by `tx` and `time` 287
10.2   Histograms of session-to-session change 290
10.3   Spaghetti plot showing change for the first 12 participants 291
10.4   Spaghetti plot with regression line showing change for the first 12
       participants 292
10.5   Variability in the rate of change (`distress`–`time` slope) across patients 294
10.6   Scatterplot of the patient-specific intercepts and slopes for the
       `distress`–`time` relationship; the dotted line is the regression line
       of the slope predicted from the intercept 295
10.7   Expected rate of change for all participants 299
10.8   Random intercepts and random slopes for all participants 300
10.9   Visualization of a constant treatment difference 302
10.10 Visualization of a treatment difference 305
10.11 The relationship between `stress` and `distress` over time 309

11.1   Observed variance and sources of variance for the trait anxiety
       items; the value to the left of each item is the variance for the item . 320
11.2   One-factor CFA model for trait anxiety 320
11.3   Covariance among the seven trait anxiety items 322
11.4   One-factor model with Greek notation 325
11.5   One-factor CFA model with a correlation among residuals for the
       trait anxiety items 351

12.1   Scree plot for factors from factor analysis of anxiety items ... 367
12.2   Parallel analysis plot 370
12.3   Loading plot before rotation 371
12.4   Loading plot after orthogonal rotation (varimax) 373
12.5   Loading plot after oblique rotation (promax) 375
12.6   Path diagram illustrating an EFA 378

12.7	Path diagram illustrating a CFA; see sections 11.5 and 12.3.3 for a description of the Greek symbols	379
13.1	Measurement model for YBOCS across men and women	401
13.2	Measurement model for concentration items across two timepoints .	415

Tables

5.1	Comparing the effects of the unadjusted, Tukey HSD, and Scheffé p-values	129
6.1	Design of the $2 \times 2 \times 2$ jury study	133
6.2	Factorial ANOVA source table for a 2×2 design	143
6.3	Factorial ANOVA source table for a $2 \times 2 \times 2$ design	157
7.1	Test statistics and fit indices for five covariance structures applied to the PASAT data	179
7.2	Test statistics and fit indices for five covariance structures applied to the PASAT data; DF = numerator degrees of freedom	194
9.1	Examples of random factors in psychology studies	247
9.2	Total, between-clusters, and within-cluster relationships	266
11.1	The impact of weak factor loadings on internal consistency reliability	343
11.2	χ^2-difference test for the trait anxiety items	347
13.1	YBOCS measurement invariance results	405
13.2	Concentration longitudinal invariance results	421

Acknowledgments

I have a lot of people to thank. I am grateful to my students, including those that I have had a chance to mentor in their research as well as students who have taken my classes. It is exciting to see students work hard and grasp concepts that were foreign to them. What's more, teaching others truly is the best way for me to learn. Teaching requires me to master material that I thought I knew but did not always have down, and teaching requires me to think about the material from others' perspectives. I still make mistakes, of course, and sometimes think I understand something when I actually do not or fail to grasp an important nuance. Even when I try, I am not always able to see the material from others' eyes. But questions, comments, and criticisms from students encourage me to get better at statistics.

I appreciate Chuck Huber from StataCorp. His enthusiasm for this project and for Stata is palpable. When I got a little overwhelmed with the writing and coverage of the book, his supportive words and feedback made this project much more doable. I am also grateful to Michael Larson and Scott Braithwaite for being great friends. Michael was generous enough to let me use and post some of his data (see chapters 7, 11, and 12). Both Scott and Michael have listened to me talk (all right, whine) about this book for too many years. They are good sports and make me laugh a lot. Joe Olsen, another generous colleague, has read and provided feedback on numerous chapters. He caught numerous errors and helped me improve the examples. He has been there throughout my career, helping me think through thorny modeling questions and teaching me so much about structural equation modeling, psychometrics, and multilevel models. This book is better because of his feedback. Any errors that remain are definitely of my own doing.

A big thanks to my family. My kids—Jack, Carter, Thomas, James, and Lily—mean the world to me. Work can sometimes be a drag, but knowing I get to hang out with them afterward gets me through the day. I live and breath statistics at work, so it is a pleasure to do other stuff—bike, shoot hoops, go to concerts, play video games, and sketch—with them. Thanks to Autumn, my biggest support and fan. I wish I had something deep and poetic to say to her, but, as she knows after all this time, I am kind of terrible at those sorts of things. She is the best. Truly.

Notation and typography

I assume that you are somewhat familiar with Stata—that you know how to load data, create graphs, create variables, and produce descriptive statistics. If you are new to Stata, I strongly recommend that you spend time in chapter 2 getting to know the basics. When I introduce new commands, such as `regress` for performing regression, I walk you through the basics of using them. Thus, even if you have never used Stata before, you should have no trouble following the material.

This book is designed so that you can learn by doing. Read this book near your computer so that you can try the commands and reproduce the output and graphics in the book. Replication is essential to good data analysis, and there is no time like the present to get started on that. As you become familiar with the commands, you will be able to tailor them to your own research projects.

Generally, the `typewriter` font refers to Stata commands, syntax, and variables. A "dot" prompt followed by a command indicates that you can type verbatim what is displayed after the dot (in context) to replicate the results in the book.

All the data used in this book are freely available for you to download, using a net-aware Stata, from the Stata Press website, http://www.stata-press.com. In fact, when I introduce new datasets, I load them into Stata the same way that you would. For example,

```
. use http://www.stata-press.com/data/pspus/attractive.dta
```

Try it. To download the datasets and do-files to your computer, type the following commands:

```
. net from http://www.stata-press.com/data/pspus/
. net describe pspus
. net get pspus
```

Part I

Getting oriented to Stata

1 Introduction

Statistics is a major part of research in many fields, often serving as the primary method for establishing whether results support hypotheses. Indeed, it is unusual for a quantitative study in psychology to not use statistical analysis. Consequently, it's essential for psychologists to be competent users of statistics, and that is the primary aim of this book.

That is a lofty goal. Statistics is a huge, technical field. Psychologists who spend their careers studying statistics only master small portions, and most psychologists study psychological content, not statistics. Furthermore, studying a single book will not make you a competent user of statistics. Competence comes from trial and error and from applying statistical methods and ideas to your own research. Competence comes from years of building your skills. Competence comes from collaborating with more competent people and learning from them. My hope is that this book serves as a step toward competence for those starting their training and as a useful reference or development tool for those already using statistics.

1.1 Structure of the book

This book covers some foundational topics in psychological statistics and psychometrics. Topics include t tests, analysis of variance (ANOVA), regression, power analysis, multilevel models, structural equation modeling, and factor analysis. I do not provide an exhaustive treatment of each topic because many books already do that. Instead, I provide an introduction to key concepts regarding how to use these statistical models to answer research questions.

Competent statistical analysis requires the use of statistical software. Consequently, I weave the statistical content with Stata code that illustrates how to fit models and make sense of the output. Most of the code examples illustrate how to use Stata commands, such as `regress` (for regression) or `sem` (for structural equation modeling). I also illustrate how to use some of Stata's graphical commands, such as `histogram` and `twoway scatter` for scatterplots, and data management commands, such as `generate` and `egen` (to create new variables). I show how to program Stata to run simulations to help you learn a concept or as a method for understanding your models.

The book is divided into three parts. Part I provides an introduction to Stata, including the interface, loading data, do-files and Stata syntax, descriptive statistics, graphics, and help and documentation. Part II discusses regression, ANOVA, multilevel models, and statistical power. I begin with regression because it provides the foundation for the discussion of ANOVA, multilevel models, and power. Part III covers psychometrics, including a discussion of the issues of reliability and validity. In these chapters, I discuss these concepts from the perspective of factor analysis, because I believe this provides a coherent framework for thinking about these measurement concepts. Looking at these concepts from the perspective of factor analysis will also help you make connections between the psychometric concepts of reliability and validity and regression models.

I *strongly* recommend that you work through the Stata code in each chapter rather than just reading the code and seeing the results in the book. You will learn statistical concepts better by toying with code and seeing what happens, especially as you change the specification or the options. My students tell me that it is more effective to practice using Stata and to run models during class rather than just watch me use Stata or read lecture notes. Furthermore, as you get more comfortable with Stata, you will find it much easier to learn how to use other Stata commands because you will start to see the connections between the various commands.

1.2 Benefits of Stata

Statistical software is necessary, but why choose Stata? Some treat statistical software kind of like a sports team, showing undying devotion to the software and viewing criticisms of the software as personal attacks. Others are pragmatic users, simply using what is available to them or what seems useful. Some want only open source software; others want a company to back the development and certification of the software. Many use a specific software because that's what their adviser used or what was taught in their program. I started using Stata because I was a longtime Mac user. As I was finishing my graduate degree in 2003, SAS did not support Mac at all, and SPSS's Mac software was buggy and slow. Stata's Mac support was excellent, and the software did most of what I needed. I had been introduced to Stata as an undergraduate and was happy to return. I have never looked back. Stata has remained my primary statistical package.[1]

Although my reasons for choosing Stata had little to do with statistics itself, I believe Stata is an excellent choice for five reasons:

1. **Consistent syntax.** Learning Stata, like learning any programming language, is challenging. However, learning Stata is manageable because Stata's syntax is expressive and consistent. It is expressive because it is easy to understand once you have the basics down. It is consistent because command syntax does not fundamentally change from one command to another. Thus, once you get the basic structure down, it is easy to learn how to use new commands: the structure will be similar to commands you are already familiar with.

---

1. That's not to say that I do not use other software. Sometimes, collaboration requires that I do. Sometimes, I need analysis routines not available in Stata. And sometimes, I just like to learn new stuff, so I try software that is available to me.

2. **Aids the replicability of analyses**. Statistical analyses should be reproducible (Long 2009), which means that if I asked you to run an analysis a second time (or third, fourth, and so on), you will get the same answer. Likewise, if I ran your code on my computer, I would get the same results. Stata includes the ability to specify the Stata version number for your analysis. For example, you may start your analysis with the command `version 14.2`, which tells Stata to run the analysis using the code that was present for Stata version 14.2. Therefore, if something changed between version 14.2 and 15.0, you will not get different answers.[2]

3. **Comprehensive documentation**. Stata's support documentation is comprehensive and ships as part of the software. Additionally, all documentation is freely available on the web. In my opinion, Stata documentation is second-to-none in the statistical software world. The documentation provides a readable explanation of each command, including options. Furthermore, the documentation includes worked examples and discussion of output to aid your learning. Finally, the documentation provides technical details regarding the mathematical and statistical underpinnings of commands. You can learn a lot by studying the documentation.

4. **Data management**. Most analyses require data management: cleaning data, generating variables, labeling variables and values, reshaping the data into a specific format, and so on. Stata's data management commands and utilities are excellent and make data preparation straightforward and replicable. Even when I need to use another software package for analysis, I nearly always prepare and manage the data in Stata.

5. **Graphics**. Graphics are an essential part of data analysis and are often superior to tables of numbers when it comes to communicating results (Cox 2004; Gelman, Pasarica, and Dodhia 2002; Gelman 2011). Stata includes comprehensive graphical tools to aid in exploring your data and interpreting the results of your models. The graphics are customizable and can be quite beautiful. Nearly all figures in this book were created with Stata. I include the code for creating these figures so that you can reproduce the graphs. Study this code. You will appreciate how flexible Stata graphs can be, and you will learn about Stata programming.

1.3 Scientific context

As I write this introduction, psychology, and science generally, has some problems. Pressure to publish and to ensure it is something exciting and novel combined with bad methodological practices means that a lot of research is not replicable (Ioannidis 2005, 2008, 2012). Indeed, many psychological studies simply do not replicate (Open Science Collaboration 2015), leading some to call the situation a "replication crisis"

---

2. Of course, if the Stata developers caught a bug in the code that results in a different answer in version 15 than 14.2, you may want the different answer. Nevertheless, by making the version number explicit, you can ensure that the results will only change when you expect them to.

(Pashler and Harris 2012). Psychology is not alone in this, with some pointing to problems in other disciplines, such as economics, biology, and medicine (Ioannidis 2005, 2013, 2014). Do these problems mean that many (or even most) theories are not worthwhile and the scientific literature cannot be trusted? I hope it is not that bad. Regardless of the answer to that question, I think the replication crisis does suggest we ought to step back and think about why we are facing these problems.

Some argue that a major reason for these problems is the incentives that influence scientists (Baldwin 2017; Ioannidis 2014; Nelson, Simmons, and Simonsohn 2012; Simmons, Nelson, and Simonsohn 2011). For example, in universities across the world, promotion and tenure depend upon publications—hence the phrase "publish or perish". Getting your first academic job often depends upon having many publications, including some in prestigious journals. Some research positions are "soft money" jobs, which means that salary and research support comes from grant money rather than the university itself. Getting grants requires that you are productive and that previous grants worked out. Sometimes we joke that to get a grant, we have to do all the research that the grant is proposing to prove that the research will work. When your salary and reputation depend upon getting papers published and the research turning out in a specific way, the incentive is to make sure the research works out as predicted.

Given these incentives, publishing becomes the end goal of research. That is, rather than publishing being the means to communicate scientific observation, publishing and adding lines to your vita become tantamount to science itself. The book *The Compleat Academic*[3] provides advice to researchers in the early stages of their careers on things like graduate school, applying for postdocs and jobs, submitting grants, and teaching classes. The advice to graduate students states:

> The information that we need to arrive at a short list of applicants is contained in the letters of recommendation and, primarily, in the academic vita. Wise graduate students, therefore, will start at day one of their first year in a PhD program to develop a strong vita. [...] Alter your perspective so that you derive your professional self-respect entirely from what is on that document. From the start of graduate school on, throughout what we hope will be a long and productive career, you *are* your vita. (Lord 2004, 10, emphasis in original)

Given such advice, combined with the incentives for getting and keeping a job (including securing your own salary!), it is not difficult to see why publishing became equivalent to doing science.

I learned statistics in this context, as did most researchers before me. Consequently, a number of problematic research and statistical practices evolved that ultimately helped publication rates but did not improve the quality of the science. For example, consider the use of p-values to evaluate statistical significance. There are many criticisms of

---

3. The dictionary on my computer defines the word "Compleat" as "archaic spelling of complete". Leave it to academics to take the simple word *complete* and make it snooty.

1.3 Scientific context

p-values in the scientific literature (for example, Meehl [1978]). Many of the criticisms are about how p-values are used, not so much about p-values themselves. McElreath (2016) says it well:

> This audience accepts that there is something vaguely wrong about typical statistical practice in the early 21st century, dominated as it is by p-values and a confusing menagerie of testing procedures. [...] The problem in my opinion is not so much p-values as the set of odd rituals that have evolved around them, in the wilds of the sciences, as well as the exclusion of so many other useful tools. (pp. xi–xii)

In short, p-values became the primary source of evidence that a result is publishable. Consequently, the goal of analysis becomes finding a p-value that is less than 0.05. If you complete a study and find null results, there's a good chance you will not even try to publish it.

Researcher flexibility, especially with respect to design, analysis, and reporting, means that finding significant effects probably required torturing the data. More advice from *The Compleat Academic*[4] explains this well:

> To compensate for this remoteness from our participants, let us at least become familiar with the record of their behavior: the data. Examine them from every angle. Analyze the sexes separately. Make up composite indexes. If a datum suggests a new hypothesis, try to find additional evidence for it elsewhere in the data. If you see dim traces of interesting patterns, try to reorganize the data to bring them into bolder relief. If there are participants you do not like, or trials, observers, or interviewers who gave you anomalous results, drop them (temporarily). Go on a fishing expedition for something—anything—interesting. (Bem 2004, 187)

"Interesting patterns" here usually means small p-values. Such flexibility in analysis is sometimes called researcher degrees of freedom or p-hacking (Simmons, Nelson, and Simonsohn 2011) or the garden of forking paths (Gelman and Loken 2014). Examples of such analyses include running multiple experiments and only reporting results from those with significant results. If main effects are not significant, test interactions; dropping observations when doing so takes the p-value from $p = 0.09$ to $p = 0.03$. Change a continuous variable to a categorical variable because the categorical variable produces significant results. Ignore problems with estimation and fit because a result is statistically significant. Fail to look at the raw data to see what a model (which is a reduction of the data) implies about the data because the analysis is statistically significant (Simmons, Nelson, and Simonsohn 2011). I think you get the idea.

---

4. Honestly, I cannot get over that name.

Fortunately, changes are in process. For example, the Open Science Framework (https://osf.io/) provides tools for researchers to register hypotheses before seeing the data and to create a website for hosting data and analysis files. I have seen job postings where an emphasis on improving rigor and replicability of science is a job qualification. Journals are accepting registered reports, wherein studies are reviewed prior to data collection and evaluated solely on the basis of the research question and quality of the design and proposed analyses (https://cos.io/rr/).

I hope this book contributes to the positive changes. My goal is to teach statistical concepts and software in a way that helps researchers a) address their research question transparently and openly, b) better understand their data, and c) better understand the models they use and what the models imply about their theory or research area. To be sure, I teach and use p-values—they can be useful. Stata includes many tools that supplement the information provided by p-values, and as a consequence, Stata can be used in a way that improves how statistics are used in psychology specifically and in science generally. So buckle up! We have a lot of great stuff to discuss.

2 Introduction to Stata

This chapter provides a brief introduction to using Stata. My goal is to provide sufficient instruction for you to jump right in and start using Stata. Indeed, that is what you should do—jump right in. Using software is a skill, and you will not develop your Stata skills by skimming this chapter. Rather, open the book next to the computer, and work through the chapters interactively with Stata. If you are an experienced Stata user, I suggest you at least skim this chapter because you may just learn something useful.

2.1 Point-and-click versus writing commands

Before delving into a tutorial on Stata, I want to say a word about learning Stata via the point-and-click interface versus learning to write Stata commands. When I say point-and-click, I mean using Stata via its menu system with the mouse. You can do most, if not all, analyses in this book via the menus.[1] This seems appealing and approachable, especially given that many computer programs we use every day are just like that. However, relying on point-and-click in Stata has drawbacks.

First, point-and-click methods do not lend themselves to reproducibility (see section 2.8 and Long [2009]). You would have to keep a detailed record of everything you did and the order in which you did it to reproduce your data management and analysis later on. If you open the Data Editor and manually change the data, you will have to remember which observations you did that for. It would be incredibly tedious.

Second, sole reliance on point-and-click methods can make Stata or similar programs seem more mysterious than they actually are. The command syntax can seem like gibberish and something to be feared, especially as you start out. If you avoid the syntax, your belief that the commands are gibberish and hard to learn will get reinforced because you will never learn them. They will remain scary and something only the smart (or nerdy) among us can handle—your avoidance will turn the belief into a self-fulfilling prophecy. In contrast, learning the command syntax will engender confidence and self-efficacy as a data analyst.

---

1. Using the menus is great for exploring how a particular estimation routine works or even what commands are available.

Third, sole reliance on point-and-click methods can make Stata seem more limited than it really is.[2] One of Stata's benefits is its extensibility. For example, Stata lets you capture output from analyses and do further analyses with it (for example, create tables and figures). That is only possible via the command syntax. If you do not learn the command syntax, you miss out on one of the most powerful features of Stata. Cox (2015) said it well:

> Despite modern user interfaces, the heart of Stata remains the command language. What is done via menus or dialogs is, ideally, echoed as a command. The overwhelming emphasis on a command language follows from a firm belief that statistical analysis cannot be reduced to a small series of standard tasks. (p. 145)

Fourth, it is hard to get help when using just the point-and-click methods. If you email an instructor or post a question on an Internet forum asking for help, you will nearly always only receive help in the form of commands.

In sum, I strongly recommend that you learn the command language. With the exception of a few examples in this chapter, I focus exclusively on the command language. When I teach courses using Stata, I require students to do all analyses with the command language. It can be frustrating at times because computer languages are exact and inflexible. If you include extra spaces where there should not be any or capitalize a word that should be lowercase, you will get an error. However, it is this exactness that makes analyses repeatable and consistent. I bet there will be a time that you are proud of your code and what you made Stata do.

Stata's language is approachable and consistent from command to command. Consequently, the learning will come quickly once you get the basics down.

2.2 The Stata interface

Figure 2.1 is a screenshot of the Stata interface for Mac OS X. Windows and Linux interfaces will be different, as expected. However, Stata is similar across platforms, and I have not found it difficult to move between operating systems. Regardless, with a few exceptions (for example, file format options for exporting graphics), commands across platforms are identical. Further, data files created on one platform (for example, Mac) can be opened on others (for example, Windows).

Figure 2.1 shows the five main sections of the Stata interface.

1. **Command**. The Command window is where you type commands to tell Stata what you want done. Type `display 2+2` into the Command window. Output appears in the Results window. Now type `sysuse auto`. This loads a dataset called `auto` that comes with Stata and provides some information about cars.

---

2. That being said, Stata's point-and-click capabilities are massive, and the help documentation provides links to online resources for learning many of its capabilities.

2. **Results**. The Results window displays the output from commands, except for graphics. The Results window echoes the commands (that is, instructions) you provide followed by the results of the commands. If you followed my instructions above to type `display 2+2`, the Results window printed `. display 2+2` on one line and 4 on the next line. Figure 2.1 shows some additional output.

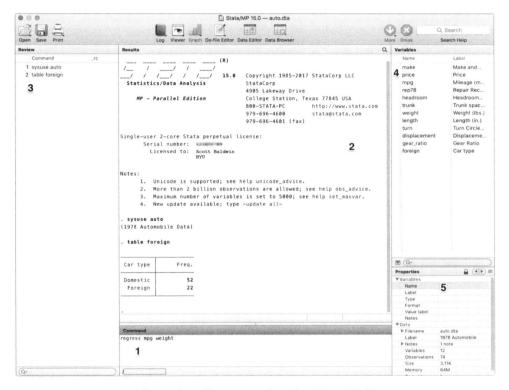

Figure 2.1. Stata interface for Mac OS X

3 **Review**. The Review window provides a list of the commands submitted to Stata and thus serves as a record of the commands you have used. Double-clicking on a line in the Review window makes the command reappear in the Command window. Give it a try.

4 **Variables**. The Variables window lists the variables and variable labels in the dataset.

5 **Properties**. The Properties window provides more-detailed information about any selected variable (for example, is it numeric or string as well as detailed information about the loaded dataset (for example, number of variables, size).

2.3 Getting data in Stata

You have two options for opening Stata datasets: the **File** menu or the `use` command. To use the point-and-click method, choose **File ▷ Open**, and then navigate to the location of your data. That is it. Although you may choose the point-and-click method, internally Stata uses the `use` command. Indeed, after opening a data file, the Results window displays the specifics of the `use` command for the data.

I do not recommend using the point-and-click method because it does not lend itself to reproducible analyses (see section 2.8). Rather, I recommend the `use` command.

A challenge with the `use` command is specifying the file path. On a Mac or Linux, the file paths will typically start with /Users/*userid*/ (where *userid* is your username for the computer), followed by more file names (for example, Documents, Dropbox). On Windows, the file path will typically start with a letter, such as `C:\` or `D:\`, followed by more file names.

Suppose I had a Stata data file called `stress.dta` in my `Users` directory and that my user name was `scott`. The command `use "/Users/scott/stress"` loads the data. Note that the file extension (that is, .dta) is not included. If the data were stored within the `Documents` folder in my `Users` directory, I would type `use "/Users/scott/Documents/stress"`. If I were using Windows and `stress.dta` was saved on the `C:` drive within the `Documents` folder, I would load the data by typing `use "C:\Documents\stress"`.

There is a better way than typing the full file path: working directories. Working directories are a type of workspace or work area for your data analysis. Specifying a working directory lets Stata know where data and other files for your project are stored. Suppose you have a research project investigating the relationship between stress and athletic performance. Your `Documents` folder contains a folder named `Performance`, which contains your data. In this case, your working directory would be set to the `Performance` folder. Working directories can be set via a point-and-click menu or with a command.

To set the working directory via point-and-click, choose **File ▷ Change working directory...**. Then select the folder you want to be the working directory (in our example, `Performance`). To set the working directory with a command, use the `cd` command, which stands for "change directory". Thus, to make `Performance` your working directory via command, type `cd "/Users/scott/Documents/Performance"` (or on Windows, type `cd "C:\Documents\Performance"`). If you use Stata's Project Manager (see section 2.8.3) and launch Stata by clicking on the project file, Stata will automatically set the working directory to the folder where the Project Manager is saved.

Once the working directory has been changed, Stata looks for data within the working directory or within folders contained in the working directory. Figure 2.2 provides a visualization of a working directory within a `Documents` folder. In the `Documents` folder are three folders: 1) `Misc`, 2) `Performance`, and 3) `Teaching`. `Performance` is

2.3 Getting data in Stata

the working directory. After setting the working directory, typing `use data_orig` loads `data_orig.dta`. Stata does not require the full file path because by specifying the working directory as the `Performance` folder, Stata looks within `Performance` when loading data. Notice in figure 2.2 that the `Data` folder also contains three Stata datasets. Typing `use "/Data/time1"` loads `time1.dta`. Thus, rather than specifying the full file path, you specify the file path relative to the working directory.

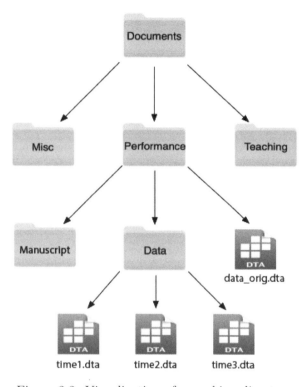

Figure 2.2. Visualization of a working directory

If you have to figure out the full file path to change the working directory, why bother with setting the working directory at all? That is, why not just load each data file using the full file path? In a word, portability. By setting the working directory, analyses become portable both within a computer (that is, moving your files to a new folder on your computer) and on a collaborator's computer.[3] This type of portability is important for reproducibility (see section 2.8).

---

[3]. This is how I work on my computer. It is nimble because if I change the location of my project folder (for example, because I moved to a new computer), then I only have to click on the project file to set the directory.

2.4 Viewing and describing data

If you are anything like me (let's hope not), it is important to see the loaded data. A common command to use after loading a dataset is the `describe` command. Load `auto.dta` and then type `describe`:

```
. sysuse auto
(1978 Automobile Data)
. describe
Contains data from /Applications/Stata/ado/base/a/auto.dta
  obs:            74                          1978 Automobile Data
  vars:           12                          13 Apr 2016 17:45
  size:        3,182                          (_dta has notes)

              storage   display    value
variable name   type    format     label      variable label

make            str18   %-18s                 Make and Model
price           int     %8.0gc                Price
mpg             int     %8.0g                 Mileage (mpg)
rep78           int     %8.0g                 Repair Record 1978
headroom        float   %6.1f                 Headroom (in.)
trunk           int     %8.0g                 Trunk space (cu. ft.)
weight          int     %8.0gc                Weight (lbs.)
length          int     %8.0g                 Length (in.)
turn            int     %8.0g                 Turn Circle (ft.)
displacement    int     %8.0g                 Displacement (cu. in.)
gear_ratio      float   %6.2f                 Gear Ratio
foreign         byte    %8.0g      origin     Car type

Sorted by: foreign
```

The `describe` command provides a high-level summary of the dataset, including variable names, observations, number of variables, and variable labels. Stata has three types of variables:

1. **Numeric**. Numeric variables are numbers stored as byte, integer, long, float, and double. Storage types are important but can be a source of confusion. Because computers represent all information as a series of 0s and 1s, numbers like 2.74 have to be translated into a sequence of 0s and 1s. The storage type refers to how Stata has encoded that variable. Stata is pretty smart about how variables get stored, so I would not change the storage type without a compelling reason. See [U] **12.2.2 Numeric storage types** for a detailed discussion of the storage types.

2. **Date and time**. Dates and times are stored in their own format so that time-based calculations are straightforward and correct. For details, see [U] **24 Working with dates and times**.

3. **String**. A string variable is composed of letters or other nonnumeric characters, perhaps in addition to numbers. String variables can be only letters, be a mix of numbers and letters, or be only numbers, which may be confusing to newer users. The key thing to remember is that with a string variable, the numbers will not be

treated like normal numbers (for example, you cannot add and subtract them). See [U] **12.4 Strings** for a detailed discussion of string variables.

Stata has a Data Editor, which is a spreadsheet view of your data. Once `auto.dta` is loaded, go to **Window ▷ Data Editor** to open the Data Editor. A window similar to figure 2.3 will appear. String variables are red while numeric and date/time variables are black. Numeric variables with labels are blue. See [GS] **6 Using the Data Editor** for a thorough discussion of the Data Editor. Although I only use it to view data, it has a lot of functionality. For example, you can change data in the Data Editor. (But do not do that! Use syntax if you change data so that you have a record of what you did.)

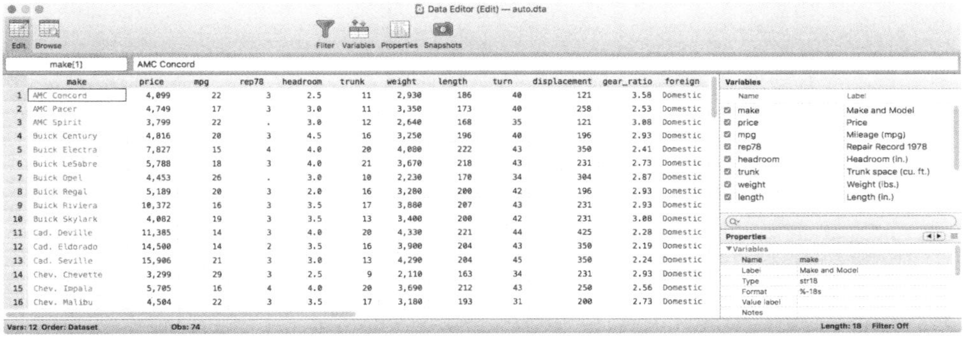

Figure 2.3. The Data Editor

2.4.1 list, in, and if

Sometimes it is useful to get a listing of data in the Results window, which is what the `list` command is for. To get a list of the `make` and `mpg` variables in the dataset, type

```
. list make mpg
```
(*output omitted*)

Suppose you want a list of `make` and `mpg` for just the first five observations. You can use the `in` qualifier.

```
. list make mpg in 1/5
```

	make	mpg
1.	AMC Concord	22
2.	AMC Pacer	17
3.	AMC Spirit	22
4.	Buick Century	20
5.	Buick Electra	15

Or you may want a list of `make` and `mpg` for observations 15 through 25.

```
. list make mpg in 15/25
```
(*output omitted*)

Now suppose you want a list of `make` and `mpg` only for foreign cars (that is, foreign relative to the United States). The data include the variable `foreign`, which is 1 for foreign cars and 0 for domestic cars. You can limit data to foreign cars with the `if` qualifier.

```
. list make mpg if foreign == 1
```

	make	mpg
53.	Audi 5000	17
54.	Audi Fox	23
55.	BMW 320i	25
56.	Datsun 200	23
57.	Datsun 210	35
58.	Datsun 510	24
59.	Datsun 810	21
60.	Fiat Strada	21
61.	Honda Accord	25
62.	Honda Civic	28
63.	Mazda GLC	30
64.	Peugeot 604	14
65.	Renault Le Car	26
66.	Subaru	35
67.	Toyota Celica	18
68.	Toyota Corolla	31
69.	Toyota Corona	18
70.	VW Dasher	23
71.	VW Diesel	41
72.	VW Rabbit	25
73.	VW Scirocco	25
74.	Volvo 260	17

Why the double equals (==) sign? In Stata, and in most computer languages, a single equals sign (=) is used for assignment (that is, assigning a variable a particular value, like x is equal to 2). In contrast, == is used for evaluation (that is, comparing one value to another, like does x equal 2?). In this example, when creating the list of `make` and `mpg`, Stata goes through each observation in the dataset and evaluates whether `foreign` is equal to 1.

2.5 Creating new variables

You can combine `in` and `if`. Suppose you want a list of `make` and `mpg` for the first 10 observations but only for the cars that get less than 20 mpg (miles per gallon).

```
. list make mpg in 1/10 if mpg < 20
```

	make	mpg
2.	AMC Pacer	17
5.	Buick Electra	15
6.	Buick LeSabre	18
9.	Buick Riviera	16
10.	Buick Skylark	19

2.5 Creating new variables

A major part of data analysis is creating new variables. The primary commands for creating new variables are `generate` and `replace`. For example, to generate a new variable that indicates whether a car gets above 20 mpg, create a new variable called `mpg20` that is equal to 1 if the car gets greater than 20 mpg and is equal to 0 otherwise. To create `mpg20`, use `generate` to create the new variable `mpg20` that is 0 for all rows in the dataset. Then replace the values of `mpg20` with 1 if `mpg` is greater than 20.

```
. generate mpg20 = 0
. replace mpg20 = 1 if mpg > 20
(36 real changes made)
```

It took two lines of code to create `mpg20` using `generate` and then `replace`. However, this variable can be created with just a single line of code:

```
. generate mpg20new = mpg > 20
```

In words, this code does the following: "Create a new variable called `mpg20new` that is equal to 1 if `mpg` is greater than 20 and is equal to 0 otherwise."

Now suppose you want to create a new variable with three levels, where 1 means the car gets less than 20 mpg, 2 means the car gets 20–30 mpg, and 3 means the car gets greater than 30 mpg.

```
. generate mpgnew = 1 if mpg < 20
(39 missing values generated)
. replace mpgnew = 2 if mpg >= 20 & mpg <= 30
(34 real changes made)
. replace mpgnew = 3 if mpg > 30
(5 real changes made)
```

Stata also has a command called `egen`, which stands for "extensions to `generate`". With `egen`, you can create new variables that are functions of existing variables. For example, we can create a new variable that is the mean value of `mpg` in the dataset.

```
. egen meanmpg = mean(mpg)
```

Some of the `egen` functions can be combined with `by`-processing to create new variables that are stratified by the levels of another variable (see section 2.6.1), such as the mean value of `mpg` for foreign and domestic cars.

```
. by foreign: egen groupmpg = mean(mpg)
```

See [D] **egen** for a description of the functions available to `egen`.

2.5.1 Missing data

When I first used Stata in earnest, I noticed unexpected results when I had to generate new variables in a dataset with missing data. When using `list` or the Data Editor, Stata presents missing data as a period (.). Internally, however, Stata stores missing data as very large, positive numbers (see [U] **12.2.1 Missing values**). This can be confusing if you forget.

To illustrate, change the first observation for `mpg` to missing.

```
. replace mpg = . in 1
(1 real change made, 1 to missing)
. list mpg in 1/5
```

	mpg
1.	.
2.	17
3.	22
4.	20
5.	15

Next, create a new variable, `mpg20a`, that is equal to 1 if `mpg > 20`. If you are like me, you expect the first value of `mpg20a` to be missing because `mpg` is missing. However, it is 1.

```
. generate mpg20a = 1 if mpg > 20
(38 missing values generated)
. list mpg mpg20a in 1/5
```

	mpg	mpg20a
1.	.	1
2.	17	.
3.	22	1
4.	20	.
5.	15	.

Because Stata stores missing values as large, positive numbers, a missing value is greater than 20, and `mpg20a` gets a 1 for the first observation. To exclude the missing values, change the `generate` code so that the new variable is equal to 1 if both `mpg > 20` and `mpg < .`; any number will be less than missing.

2.5.2 Labels

```
. generate mpg20b = 1 if mpg > 20 & mpg < .
(39 missing values generated)

. list mpg mpg20a mpg20b in 1/5
```

	mpg	mpg20a	mpg20b
1.	.	1	.
2.	17	.	.
3.	22	1	1
4.	20	.	.
5.	15	.	.

2.5.2 Labels

Stata supports labeling variables. Variable labels explain what a variable is. As demonstrated in section 2.4, `describe` lists the variable labels and so does the Variables window. All the variables in `auto.dta` have labels. For example, the label for `mpg` is `Mileage (mpg)`.

If `mpg` did not have a label or you wanted to change the label, you would use the `label variable` command.

```
. label variable mpg "Miles per gallon"
```

In section 2.5, we created the variable `mpg20`, which was 1 if the car got over 20 mpg and was 0 otherwise. For variables like `mpg20`, because 1 does not explicitly convey information about mpg (it is just the number 1 after all), it can be helpful to attach some labels to that variable. The `label define` and `label values` commands provide these labels.

First, you use `label define` to specify the value labels.

```
. label define mpgvalues 1 "Greater than 20" 0 "20 or less"
```

In words, this command says: "Create a label definition and call it `mpgvalues`. Associate the number 1 with the label 'Greater than 20' and the number 0 with the label '20 or less'." The name of the label definition is arbitrary and could be anything, but it is helpful to use a name that is associated with the labels themselves.

After creating a definition, you then apply the definition to a variable by using the `label values` command.

```
. label values mpg20 mpgvalues
```

In words, this command says: "Apply the label definition `mpgvalues` to the variable `mpg20`." Now the values of `mpg20` are labeled.

2.6 Summarizing data

The three most common methods for summarizing data are `summarize`, `table`, and `tabulate`.

2.6.1 summarize

To obtain basic descriptive statistics, such as the mean and standard deviation, use `summarize`:

```
. summarize mpg
    Variable |       Obs        Mean    Std. Dev.       Min        Max
-------------+--------------------------------------------------------
         mpg |        74     21.2973    5.785503         12         41
```

The `summarize` command can accommodate multiple variables at a time.

```
. summarize mpg price weight
    Variable |       Obs        Mean    Std. Dev.       Min        Max
-------------+--------------------------------------------------------
         mpg |        74     21.2973    5.785503         12         41
       price |        74    6165.257    2949.496       3291      15906
      weight |        74    3019.459    777.1936       1760       4840
```

The `detail` option to `summarize` provides additional statistics, such as percentiles, variance, skewness, and kurtosis.

```
. summarize mpg, detail
                        Mileage (mpg)
-------------------------------------------------------------
      Percentiles      Smallest
 1%           12             12
 5%           14             12
10%           14             14       Obs                  74
25%           18             14       Sum of Wgt.          74

50%           20                      Mean            21.2973
                        Largest       Std. Dev.      5.785503
75%           25             34
90%           29             35       Variance        33.47205
95%           34             35       Skewness        .9487176
99%           41             41       Kurtosis        3.975005
```

You can use the `if` or `in` qualifiers with `summarize`:

```
. summarize mpg if foreign == 1
    Variable |       Obs        Mean    Std. Dev.       Min        Max
-------------+--------------------------------------------------------
         mpg |        22    24.77273    6.611187         14         41
```

However, often when I use the `if` qualifier, what I want is summary statistics separately for groups. Thus, I write `summarize` several times, each with a different condition for if. An alternative is to use by-processing. For example, to compute summary statistics for mpg for both foreign and domestic cars, type the following:

```
. by foreign: summarize mpg

-> foreign = Domestic
    Variable |       Obs        Mean    Std. Dev.       Min        Max
-------------+--------------------------------------------------------
         mpg |        52    19.82692    4.743297         12         34

-> foreign = Foreign
    Variable |       Obs        Mean    Std. Dev.       Min        Max
-------------+--------------------------------------------------------
         mpg |        22    24.77273    6.611187         14         41
```

For by to work, the data must be sorted by the variable specified; in our case, the data need to be sorted by foreign. It turns out that auto.dta was already sorted by foreign. However, if it was not already sorted, you could type either by foreign, sort: summarize mpg or bysort foreign: summarize mpg. Both commands produce the same output as above.

2.6.2 table and tabulate

Another common data summary is frequencies. The most basic way to show frequencies is with table:

```
. table foreign
```

Car type	Freq.
Domestic	52
Foreign	22

Two-way tables (or even higher dimensions) can be created with table. Here is a cross-tabulation of the number of cars with 20 or more mpg (this variable was created in section 2.5) and whether the car is foreign or domestic:

```
. table foreign mpg20
```

Car type	mpg20 0	1
Domestic	33	19
Foreign	5	17

The `tabulate` command will automatically calculate the percentages to go along with the frequencies.

```
. tabulate foreign

    Car type |      Freq.     Percent        Cum.
-------------+-----------------------------------
    Domestic |         52       70.27       70.27
     Foreign |         22       29.73      100.00
-------------+-----------------------------------
       Total |         74      100.00
```

A two-way table using `tabulate` produces row and column frequencies.

```
. tabulate foreign mpg20

             |      mpg20
    Car type |         0          1 |     Total
-------------+----------------------+----------
    Domestic |        33         19 |        52
     Foreign |         5         17 |        22
-------------+----------------------+----------
       Total |        38         36 |        74
```

Suppose you want to combine the information from `table` and `summarize`. For example, suppose that in addition to the frequency of foreign and domestic cars in `auto.dta`, you want the mean and standard deviation of weight for foreign and domestic cars. To do this, use the `contents()` option for `table`.

```
. table foreign, contents(n weight mean weight sd weight)

----------------------------------------------------
    Car type |  N(weight)   mean(weight)   sd(weight)
-------------+--------------------------------------
    Domestic |         52       3,317.1      695.3638
     Foreign |         22       2,315.9      433.0034
----------------------------------------------------
```

The `contents()` option takes a list of statistics and the variable names that you want the statistics for. The list in this example is for the a) number of observations, b) mean, and c) standard deviation of `weight` for each value of `foreign`.

2.7 Graphing data

Stata's graphics are excellent, comprehensive, and flexible (Mitchell 2012b). Graphs can be created and tweaked using syntax, which makes reproducing graphs easy.[4] Stata ships with multiple graphical schemes, which control the general look of a graph (for example, black-and-white versus color).

---

[4]. The initial creation of the graph may be difficult and take time. However, once the commands are in place, it is easy to run the code.

2.7.1 Histograms 23

My goal for this section is to provide a high-level overview of graphing in Stata rather than providing comprehensive coverage of a graphing command. Throughout the book, we create graphs and learn more about the graphing system. Mitchell (2012b) and the Stata graphics documentation (StataCorp 2017c) are comprehensive resources for learning Stata graphics.

2.7.1 Histograms

Histograms are created using the `histogram` command, which takes a single variable as its argument (see figure 2.4).

```
. histogram mpg, title("Distribution of miles per gallon") normal scheme(lean2)
(bin=8, start=12, width=3.625)
```

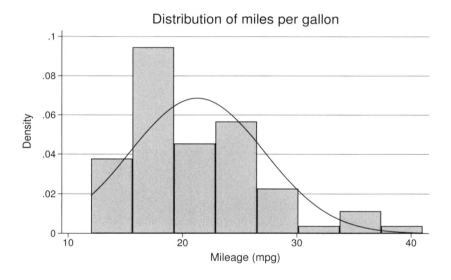

Figure 2.4. Example histogram with normal density plot

I added three options: a) a title, b) a normal curve overlay, and c) the graphics scheme `lean2`,[5] which is the black-and-white scheme I use in this text. Stata reports how many bins the histogram uses, what value on `mpg` the bins start, and the width of each of these bins. The `histogram` command provides options for controlling all three.

---

5. I am not a fan of Stata's default graphics schemes. In both the color and the black-and-white plots, the backgrounds are too dark. I prefer white backgrounds. The spacing of the titles in relation to the axes is too narrow, which makes the graph look cramped. Finally, the y-axis labels are rotated 90 degrees in the built-in schemes (that is, you read them from bottom to top). I prefer the labels to be in the typical direction for English words and numbers. If you do not have the `lean2` scheme installed, type `net install gr0002_3` (or type `findit lean2` in the Command window, click on gr0002_3, and tell Stata to install it).

2.7.2 Box plots

Box plots are created using the `graph box` command, which can take multiple variables (a *varlist* in Stata's terms) as arguments (see figure 2.5).

```
. graph box mpg, by(foreign,
> title("Miles per gallon for foreign and domestic cars"))
> scheme(lean2)
```

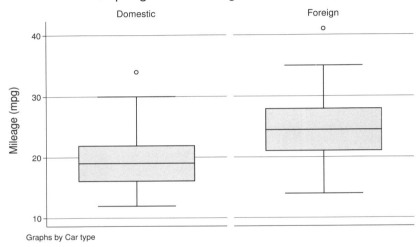

Figure 2.5. Example box plot

I added the `by()` option to create separate panels for the levels of `foreign`. The `by()` option includes a `title()` suboption to ensure that the graph title looks correct when the `by()` option is used.

2.7.3 Scatterplots 25

> **When your command extends over multiple lines**
>
> By default, Stata assumes that any text on a single line goes with one command. As soon as Stata sees a new line (that is, what happens when you press the *Return* key on a keyboard), Stata thinks that you want to start a new command. Sometimes, however, commands are lengthy, such as graphical commands. If you try to squeeze these lengthy commands onto a single line, the line would be long and hard to read.
>
> The `graph box` syntax above extends over three lines of type. Let's say we had typed this command in a do-file (see section 2.8.1) that we are now running. Stata needs to know when one command ends and another begins, and sometimes we need a way to organize long commands in do-files for easy modification later. You can use three slashes (///) to extend a command over multiple lines. The three slashes signal to Stata that the command continues onto the next line. The part of our do-file containing the `graph box` command above would have looked like this:
>
> ```
> graph box mpg, by(foreign, ///
> title("Miles per gallon for foreign and domestic cars")) ///
> scheme(lean2)
> ```
>
> If /// had not been included in the do-file, Stata would have assumed, upon running the do-file, that you were attempting to run three separate commands rather than `title()` and `scheme()` being options to `graph`. In that case, Stata would have produced an error.

2.7.3 Scatterplots

Scatterplots are created using the `twoway scatter` command. The keyword *twoway* refers to a family of graphs in Stata that are made for plotting two numeric variables on the x and y axes. See [G-2] **graph twoway** for a list of possible twoway graphs, and see individual manual entries in the *Stata Graphics Reference Manual* (StataCorp 2017c) for details about each graph.

```
. twoway scatter mpg weight, title("Relationship between MPG and weight")
> scheme(lean2)
```

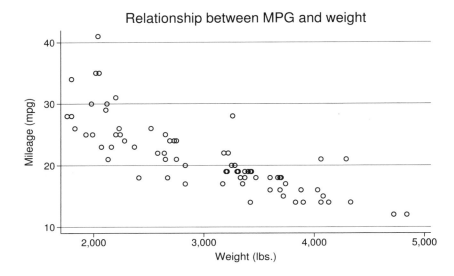

Figure 2.6. Example scatterplot

The variable for the y axis is listed first followed by the variable for the x axis. I have added the `title()` and `scheme()` options. Figure 2.6 is the resulting scatterplot.

2.8 Reproducible analysis

We may want to distinguish between data from foreign and domestic cars. This requires combining two plots (see figure 2.7).

```
. twoway (scatter mpg weight if foreign == 1)
> (scatter mpg weight if foreign == 0),
> legend(label(1 "Foreign") label(2 "Domestic"))
> title("Relationship between MPG and weight") scheme(lean2)
```

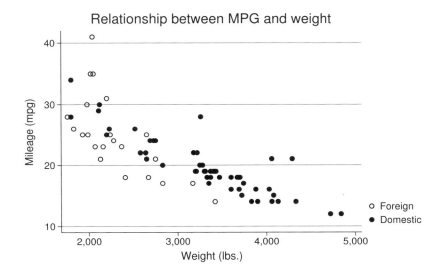

Figure 2.7. Example scatterplot with points that vary as a function of a variable in the dataset

Note that there are two scatterplots, each one surrounded by parentheses in the command. This is one method for combining two or more `twoway` plots. The first plot is limited to only foreign cars (`if foreign == 1`); the second plot is limited to only domestic cars (`if foreign == 0`). Because there are two kinds of points in the graph now, Stata produced a legend. However, the default legend labels will often not be useful. Thus, I added the `legend()` option to specify the labels. The order of the labels in the code corresponds to the order of the scatterplots.

2.8 Reproducible analysis

A hallmark of good data analysis is replicability. Can the analyses be reproduced by someone else or by you sometime in the future? I do a lot of consulting and collaboration, sometimes in my role as a teacher helping students with assignments or research projects and sometimes as a colleague. When I am consulting, students or colleagues typically show me output from analyses. Often, the output includes information I do not recognize, such as a variable they have not mentioned or output that comes from an option I am not familiar with. When I ask, "How did you create that variable?", I am often met with something like, "I think I combined these two variables. Maybe I took the average of these two or maybe I just added them together." This makes it challenging to help because steps of the data analysis are opaque. Opaque analyses are difficult to replicate.

The first step in a reproducible analysis is to create a file or set of files that does, at a minimum, the following six things:

1. Reads in a raw data file. This is the file that contains the data we need but needs some cleaning and organization before it is ready for analysis.
2. Performs any data cleaning, such as a) generation of new variables, b) recoding of variables, c) labeling of variables and values, d) identification of data-recording errors (for example, out-of-bound values), and e) merging of multiple datasets.
3. Saves any intermediate data files.
4. Performs analyses.
5. Creates and exports graphs.
6. Includes comments and explanations that increase transparency and understanding of the code (for example, why certain values were dropped or how missing data were handled).

2.8.1 Do-files

Stata commands can be collected into text files known as "do-files". Although do-files are plain text files that can be opened in many programs, Stata includes a Do-file Editor that is tightly integrated with the rest of the program (see figure 2.8). The Do-file Editor provides syntax highlighting (that is, commands are in one color, variables are in another color, etc.) and the ability to run commands. For example, to run line 11 from figure 2.8, you would highlight line 11 and click the **Do** button[6] in the top right corner.

---

6. You could use a key combination instead: *Command-Shift-D* on a Mac or *Control-D* on Windows.

2.8.1 Do-files

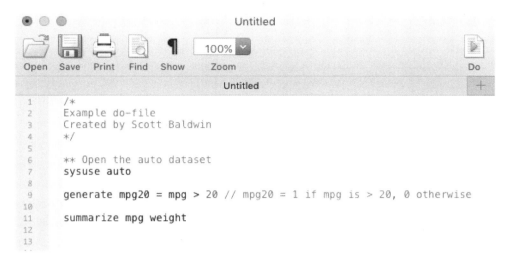

Figure 2.8. Screenshot of the Do-file Editor on a Mac

To my mind, do-files are a nonnegotiable, fundamental part of data analysis using Stata.[7] I rarely work only with the Command window. Furthermore, if I use the point-and-click interface, I always copy the command from the Results window and paste it into a do-file.

Figure 2.8 includes three comments. Comments are words, phrases, or lines from a do-file that you tell Stata to ignore. In other words, Stata will not try to run analyses or generate variables based on comments. This makes it possible to annotate your code. You can identify comments to Stata in three ways:

1. By using the /* and */ delimiters. This method is useful for long annotations, because Stata will ignore anything between these delimiters, even if that material crosses multiple lines in the do-file.

2. By using one or more * to begin a line. I usually include at least two to make it easier to spot when reading through a do-file.

3. By using // at the beginning of a line or after a command. This is useful for a brief comment following a command.

I strongly recommend annotating your code. Although it may seem like you will not forget what a particular aspect of your code does or why you chose to drop some observations, you likely will. Furthermore, if others need to use your code, they will appreciate your commentary.

---

7. Likewise, if using any other program, learn to use syntax. Point-and-click is too error-prone. For this reason, I do not believe Excel (or any other spreadsheet program) is a good option for scientific data analysis.

2.8.2 Log files

Stata can produce log files, which are copies of the Stata output you see in the Results window. All commands and results, except for graphs, are included in the log file. Log files are useful for keeping a record of your analysis and sharing output and results with colleagues. Having a record of an analysis is particularly nice if the analyses are time consuming. Rather than having to run the models again, you can just open the log file.

Log files can be created by choosing **File ▷ Log ▷ Begin...**. You then select where to save the log file and begin your analyses. When you are done with analyses, you choose **File ▷ Log ▷ Close**. There is also a button at the top of the Stata interface that does the same thing (it looks like a journal or log book).

As with all things in Stata, logs can also be started with commands. Given my preference for using do-files, I strongly recommend you begin and end log files in a do-file.

```
log using "auto_analysis"
sysuse auto
summarize mpg
log close
```

The first line opens the log file and names it `auto_analysis`. The second and third lines load the data and summarize `mpg`. The final line closes the log. Stata will save log files using SMCL, which stands for Stata Markup and Control Language ([P] **smcl**). If you have a copy of Stata available, you will be able to read the SMCL file. If you do not have Stata available, you will be able to open the log file but it could be challenging to read. Consequently, I prefer to save my log files as plain text files. To make the log plain text, use the option `text` when opening the log file.

```
log using "auto_analysis", text
```

2.8.3 Project Manager

Stata includes a Project Manager aimed to help users organize their Stata files. The help file for the Project Manager says the following:

> The Project Manager is a tool for organizing and navigating Stata files. It allows you to collect all the files associated with a given project into a single interface where you can have quick access to them without navigating through file dialogs. You can open do-files in the Do-file Editor, use Stata data, and draw saved Stata graphs by double-clicking on the files in the Project Manager. (StataCorp 2017h, 431)

To create a new project, you click on **File ▷ New ▷ Project**. This will open a file navigator window, and you choose the location of the project. This will create a Stata project file with a file extension of `.stpr`. My favorite aspect of the Project Manager is that if you launch Stata by double-clicking on a project file, it will automatically set the working directory to the location of the project file. Hence, I recommend that you place the project file in the root directory of your project (that is, where you store your data).

2.8.4 Workflow

As you become more experienced with data analysis and get involved in more complex projects, it is essential to establish a consistent, replicable workflow for your data analysis. Major projects require extensive data management and cleaning, which requires documentation and record keeping. A complete discussion of data analysis workflows is beyond the scope of this book. However, Long (2009) has written a book on best practices for data-analysis workflow using Stata.[8] His strategies and structure will help keep your analyses organized and replicable.

That being said, I have a few recommendations:

1. Where possible, avoid setting the working directory at the top of your do-file. Doing so makes your do-files "fragile"—if you change the structure of the folders on your computer (which is likely over time) or move the file to a different computer (or email to a colleague), you will get an error. To keep your files portable, a good solution is to use the Stata Project Manager and save the project file at the root folder of your desired working directory. When you open Stata by clicking on the project file, Stata sets the working directory to the folder where the project file is saved.

2. Name your files sensibly. If your project is examining bias in intelligence testing, consider naming your analysis files using numbers: bias_analysis1.do, bias_analysis2.do, and so on. This makes it easy to name new drafts by simply increasing the number. If you choose a name such as bias_analysis_draft.do or bias_analysis_final.do, you will end up having to change the whole name of the draft each time, such as bias_analysis_actual_final.do.

3. Use log files to document your work and save them as plain text files. Plain text log files can be easily read on any computer, even if you (or a colleague) do not have a copy of Stata. Do not count on the fact that you will have Stata forever, either to rerun your analysis or to read the log files in the default format (that is, SMCL).

4. Create your graphs using syntax because it makes reproducing, tweaking, and saving your graphs straightforward. The Graph Editor is great, but it is much more difficult to reproduce your graphs with it.[9]

---

[8]. Most of his suggestions can generalize to other software.
[9]. You can record changes made via the Graph Editor, so technically you can reproduce them. However, it can be slow, and the key changes you need to make nearly always can be done with syntax.

2.9 Getting help

One of Stata's best features is its thorough documentation. Stata ships with an extensive set of help documents as well as a complete set of documentation in PDF form. The help documents can be accessed by using the `help` command. For example, to obtain help for `summarize`, type `help summarize`.

2.9.1 Help documents

The help documents provide an overview of the command, syntax options for the command, and examples of how to use the command. The first part of a help file is the title, which provides a clickable link to the PDF documentation. This is followed by an overview of the syntax for the command. Stata's syntax is remarkably consistent across commands. This makes learning syntax relatively easy, because what you learn for one command will generalize to other commands.

The syntax for summarize as listed in the help document is

<u>su</u>mmarize [*varlist*] [*if*] [*in*] [*weight*] [, *options*]

Underlines indicate permissible abbreviations. For example, the underline under the `su` in `summarize` indicates that you can abbreviate `summarize` with just `su`. After the command name come arguments, which can be required or optional. Optional arguments are surrounded by brackets. Thus, all arguments for `summarize` are optional, including *varlist*, which is just a list of variables. Indeed, if you just type `summarize`, Stata will provide summary statistics for each variable in the dataset. The syntax will also show whether the command supports the `if` or `in` qualifiers as well as `weight`, which lets you weight some observations more than others. Finally, any argument that occurs after the comma is an option for the analysis. Options control how the estimation occurs, what gets printed, and even what gets estimated.

The next part of the help document tells you how to use `summarize` using the menus, followed by a brief description of what the command does. Next is a listing and description of the options for the command. For example, the help for `summarize` indicates that one option is `detail`, which "produces additional statistics, including skewness, kurtosis, the four smallest and four largest values, and various percentiles". We saw how `detail` works in section 2.6.1.

The help document then shows several examples of how to use the command. What is great about these examples is that they are accompanied by datasets that either come with Stata or can be downloaded from within Stata. This helps you to learn the commands and test options.

Finally, the help document concludes with stored results. Stata commands produce information, such as the mean or standard deviation of the variable. Stata saves these results in memory, and we can access them to use in a later analysis.

2.9.2 PDF documentation

The PDF documentation is most easily accessed by clicking on the title of an entry in a help document. For example, if you click on the word "summarize" at the top of the `summarize` help document, the PDF documentation for `summarize` will open.

The PDF documentation covers all aspects of Stata. Syntax for all commands is covered, as is a complete explanation of options. The PDF documentation also includes detailed examples with an explanation of how the command works and the output a command produces. You can also find technical details about how the command works, such as information about what estimation methods are used for a particular analysis. Finally, the PDF documentation provides details about things like the Stata interface, unique aspects of using Stata on specific platforms, and how to program your own analysis routines.

2.10 Extending Stata

2.10.1 Statistical Software Components

Occasionally, you will run into a data management or analysis problem that requires a routine not available in Stata. In this instance, there is a good chance that someone else has written a Stata program that will do what you want. For example, I regularly use meta-analysis, which is a method for combining results from multiple studies. Stata does not come with built-in meta-analysis routines. However, several researchers have written meta-analysis commands for Stata (Palmer and Sterne 2016). These commands can be installed and then used just like official Stata commands.

A common meta-analysis command is `metan`, which stands for meta-analysis. The files for `metan` are stored on the Statistical Software Components archive, which hosts hundreds of community-contributed Stata programs (http://www.repec.org). Stata includes a command called `ssc`, which allows programs stored on the Statistical Software Components website to be installed on your computer. To install `metan`, you would type `ssc install metan`. If you already have `metan` but an update is available, you would type `ssc install metan, replace`. See [U] **28.4 Downloading and managing additions by users** for more details about `ssc install` as well as other methods for installing community-contributed programs.

2.10.2 Writing your own programs

Finally, it may be that you need some functionality not available in Stata or through a community-contributed command. In that case, you will need to write your own program. My goal here is simply to raise your awareness that this can be done rather than teach you how to write your own program. We will write some simple programs to perform bootstrapping (see chapter 11) and Monte Carlo simulations (see chapter 3). The *Stata Programming Reference Manual* (StataCorp 2017h) provides details about writing your own programs, as does the book by Baum (2016).

Part II

Understanding relationships between variables

3 Regression with continuous predictors

Data analysis in psychology is fundamentally about the relationship between two or more variables. How is memory affected by brain injury? Is cognitive-behavioral therapy more effective than interpersonal therapy for eating disorders? The most commonly used tool for studying those relationships is regression.

In this chapter, you will learn:

1. How to use regression to explore the relationship between two or more variables.
2. How to interpret regression parameters.
3. How to generate predictions using regression.
4. How to partition the variance in the outcome into the part associated with the regression and the part unaccounted for by the model.
5. How to perform a null hypothesis significance test.
6. How to construct and interpret a confidence interval.
7. How to explore the implications of the regression model.

Stata commands featured in this chapter

- `regress`: for fitting regression models
- `histogram`: for plotting data and sampling distributions
- `program`: for writing a simulation program
- `simulate`: for executing a simulation program
- `tabulate`: for exploring categorical data
- `twoway scatter`: for producing scatterplots
- `summarize`: for computing descriptive data
- `predict`: for producing predicted values and residuals
- `margins`: to aid interpretation of regression models
- `marginsplot`: for visualizing regression results, especially models including interactions
- `lincom`: for creating linear combinations of parameter estimates

3.1 Data

The data for this chapter come from the General Social Survey (GSS):

. use http://www.stata-press.com/data/pspus/gss_attitude

The GSS is a major survey in the United States that covers a huge number of social topics, including emotional well-being, marriage and family, problems with alcohol, and political attitudes. Some questions are asked repeatedly for each wave of the GSS and some questions only appear once.

In this chapter, we investigate predictors of attitudes toward using medications to treat mental health problems. In 2002, survey respondents were asked to what degree they agreed with the following six statements about medications:

1. Giving children psychiatric medications when they are young only puts off dealing with their real problems.
2. Medications for behavior problems just prevent families from working out problems themselves.
3. These [psychotropic] medications often have unacceptable side effects.
4. Taking [psychotropic] medications is a sign of weakness.
5. These [psychotropic] medications are addictive.
6. These [psychotropic] medications help people control their symptoms.

Respondents answered as 1) strongly agree, 2) agree somewhat, 3) disagree somewhat, or 4) strongly disagree. Anyone who answered "Don't Know" or "Not Applicable" or who did not provide an answer was treated as missing.

I created a scale called `attitude` that is the average of the responses to the six items. Given that most items reflect negative attitudes toward psychotropic medications, higher values on `attitude` indicate more positive attitudes toward medications. I reverse coded the responses to "These medications help people control their symptoms" so that the valence of the responses to item 6 was identical to the other items. In addition to `attitude`, the `gss_attitude` dataset includes `educ`, which is the highest year of education.

3.2 Exploration

Suppose you are a university administrator and want to predict how a student will perform in an introductory statistics course. You do not know anything about this student, except that she will enroll in the statistics course next semester. What would be your best guess about her grade? You speak to the instructor and find out that students earn a range of grades but average a B−. Would you guess the new student would earn an A? How about a C? How about a B−? In the long run, your best guess

3.2 Exploration

for a new student would be a B−. Although this particular new student might earn an A (or F), every time you need to make a guess for a new student, you will be closest to the "truth" by guessing the average. Most students are going to be near the average, so predicting the average will get you closest—your errors in prediction will be minimized.

Now suppose you want to use the GSS data on medication attitudes to predict the attitude of a new respondent. As with the statistics course, you do not know anything about the new respondent, but you have the data for `attitude`. Figure 3.1 presents a histogram for `attitude`. The responses range from 1 to 4, with most responses in the 2.5 range.

```
. histogram attitude, scheme(lean2) bin(19)
(bin=19, start=1, width=.15789474)
```

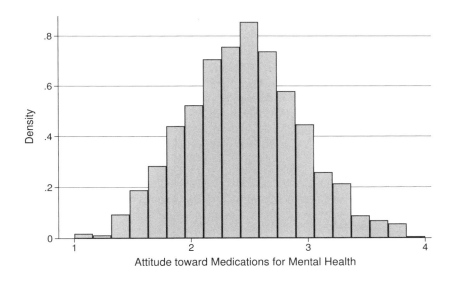

Figure 3.1. Histogram of `attitude`

The mean of `attitude` is 2.4.

```
. summarize attitude
```

Variable	Obs	Mean	Std. Dev.	Min	Max
attitude	1,006	2.444334	.5247805	1	4

Thus, your best guess of a new respondent's attitude toward medications for mental health would be 2.4 (that is, neither positive nor negative). Your prediction for this new respondent may not be close. However, over many new respondents, the mean will produce the best predictions.

3.2.1 Demonstration

An excellent way to explore the idea that the mean minimizes errors in prediction is to use simulation. We can simulate a new observation of `attitude`, which we will call `new_attitude`, and see how much the new observation differs from the mean of `attitude` (2.4). This difference is the error in prediction. We can also examine how much `new_attitude` differs from any other value of `attitude`, for example, the 75th percentile (3.33). We can simulate many values of `new_attitude` and compute these two differences. If the claim that the mean is the best guess is true, then over many values of `new_attitude` the average difference between `new_attitude` and the mean ought to be smaller than the difference between `new_attitude` and the 75th percentile. In this simulation, we compare the squared deviation `new_attitude` from the mean with the squared deviation of `new_attitude` from the 75th percentile. We use the squared deviation in this case to ensure that all deviation values are positive.

Simulation program

Simulations are conceptually simple. We generate data, perform analysis on the generated data, and collect the results. We then repeat that process many, many times, allowing us to examine how statistical analyses work in the long run under various scenarios. Stata includes a command called `simulate` for running simulations. To use `simulate`, we must define a program (that is, a Stata command) that a) generates the data we need, b) computes the quantities we need, and c) returns the information we want.

As noted in section 2.10.2, you can use the Stata language to write your own programs. The program for this simulation as well as the call to `simulate` is below.

```
. program define simdemo, rclass
  1.         version 15
  2.         drop _all
  3.         set obs 1
  4.         generate new_attitude = rnormal(2.4, .5)
  5.         generate diff_mean = (new_attitude - 2.4)^2
  6.         generate diff_p75 = (new_attitude - 3.33)^2
  7.         generate mean_small = diff_mean < diff_p75
  8.         return scalar mean_small = mean_small
  9. end
. simulate mean_smaller = r(mean_small), reps(10000) nodots seed(38499): simdemo
        command:  simdemo
  mean_smaller:  r(mean_small)
```

At this point, you do not need to understand how the program works.[1] You only need to understand the logic of the simulation. The program is defined using the `program define` command and is named `simdemo`. Following each numbered line, here is what the program does:

---

1. Baum (2016) provides additional detail about programming Stata.

3.2.1 Demonstration

1. Specifies that the program use Stata 15 commands.
2. Drops all existing observations.
3. Sets the number of observations to 1 (we are generating one new observation).
4. Generates `new_attitude`, which is created by randomly drawing a new observation from a normal distribution with a mean of 2.4 and standard deviation of 0.5.
5. Generates the squared deviation for the mean—$(\text{new\_attitude} - 2.4)^2$.
6. Generates the squared deviation for the 75th percentile—$(\text{new\_attitude} - 3.3)^2$.
7. Generates `mean_small`, which is a binary variable that is 1 if `diff_mean` is less than `diff_p75` and is 0 otherwise.
8. Returns `mean_small` (that is, makes it available for later analysis).
9. Ends the program.

The `simulate` command runs `simdemo`. The first argument for `simulate` indicates what output is saved from the simulation. In this case, we create a variable[2] called `mean_smaller` that is equal to the returned value from line 8 of the program above. The `reps(10000)` option indicates that we want to run `simdemo` 10,000 times (that is, Stata is going to execute lines 1–9 over and over 10,000 times), and the `nodots` option indicates that we do not want Stata to print a dot for every replication of the simulation. The `seed(34899)` option sets the seed for the random number generation (line 4 of `simdemo`). Because random number generation is random, if we do not use a seed, our results will vary (that is, you will not get the same results as presented in the book).[3] Finally, following the colon, we tell Stata to use the `simdemo` program for the simulation.

After running the `simulate` command, the active dataset will include 10,000 values of `mean_smaller`. The proportion of 1s indicates how often the squared deviation for the mean was smaller than the squared deviation for the 75th percentile for each of the 10,000 observations.

```
. tabulate mean_smaller
r(mean_smal
         l) |      Freq.     Percent        Cum.
------------+-----------------------------------
          0 |      1,754       17.54       17.54
          1 |      8,246       82.46      100.00
------------+-----------------------------------
      Total |     10,000      100.00
```

Thus, 82.5% of the time, the squared deviation for the mean was smaller than the squared deviation for the 75th percentile. Said another way, we have minimized the error in prediction by guessing the mean rather than the 75th percentile.[4]

---

2. The variable is created within `simulate` rather than with `generate`.
3. However, using a different seed is fine too; you will just get a different number than I do. Nevertheless, the outcome of the simulation should be about the same.
4. We used the 75th percentile, but any value would not perform as well as the mean.

3.3 Bivariate regression

The mean was our best guess regarding the value of `new_attitude`. Another term meaning best guess is *expected value*. That is, in the long run, what do we expect the value of `attitude` to be? For a single variable, such as `attitude`, the expected value is the mean. Regression, on the other hand, allows us to determine an expected value for `attitude` that incorporates information from other variables. Our best guess about `attitude` will be based on variables such as the respondent's level of education and mental health symptoms.

Figure 3.2 is a scatterplot of `educ` and `attitude`.

```
. use http://www.stata-press.com/data/pspus/gss_attitude, clear
. twoway scatter attitude educ, jitter(2) scheme(lean2)
```

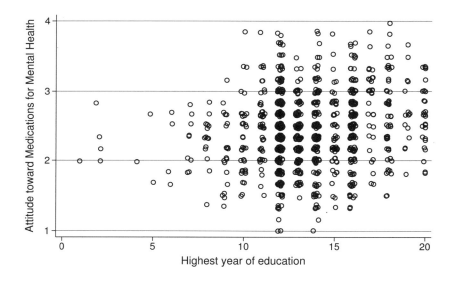

Figure 3.2. Relationship between `educ` and `attitude`

Based on figure 3.2, what do you think is the expected value for `attitude`? Does that expected value differ for someone who has 10 years of education compared with someone who has 20 years of education? That is, if we consider `educ`, does that affect what we expect `attitude` to be? If there is a relationship between education and attitude toward psychotropic medication, then expected values of `attitude` will vary as a function of `educ`. Regression provides information about the nature of that relationship in the form of a line, which is why regression is also called linear regression.

3.3.1 Lines

Let's go back in time and revisit junior high school algebra. You may recall the formula for a line:

$$y = mx + b$$

where x and y are variables, m is the slope, and b is the y intercept. We say that y is a function of two quantities: i) the product of m and x and ii) b. The slope describes changes in y with respect to changes in x (that is, "rise over run"), and the y intercept provides the value of y when $x = 0$. If we know m, x, and b, then we know y (see figure 3.3). Regression produces slope and intercept values that describe the relationship between an outcome and predictors.

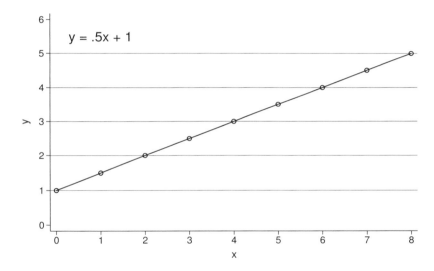

Figure 3.3. Illustration of $y = mx + b$

3.3.2 Regression equation

Compare figures 3.2 and 3.3. In figure 3.3, it is obvious what the line should be because the points all fall on a straight line. In figure 3.2, you could draw any number of lines through the data—a horizontal line, a vertical line, a line moving up from left to right, or a line moving down from left to right. All lines would provide some indication of the relationship between `attitude` and `educ`, and all lines will provide an intercept and slope. Which line is best? The best line is the line that provides the closest correspondence between the expected value of y and the actual value of y. In other words, the regression line will provide the best prediction of y given the x values across the entire dataset.

A regression equation for the relationship between `attitude` and `educ` is

$$\texttt{attitude}_i = \beta_0 + \beta_1 \texttt{educ}_i + \epsilon_i \tag{3.1}$$

This equation says that the observed value of `attitude` for person i is a function of an intercept (β_0), a slope (β_1) relating `educ` to `attitude`, and error in prediction (ϵ_i). Error is included in (3.1) because the observed value of `attitude` is not perfectly captured by β_0 and β_1 (that is, the data do fall on a straight line).

Commonly, the regression equation is written in terms of the expected value of y. For the example model, the equation is

$$E(\texttt{attitude}_i|\texttt{educ}_i) = \beta_0 + \beta_1 \texttt{educ}_i \tag{3.2}$$

which is read as, "The expected value of `attitude` given `educ` is equal to β_0 plus β_1 times `educ`." Alternatively, we can substitute $E(\texttt{attitude}_i|\texttt{educ}_i)$ with $\widehat{\texttt{attitude}_i}$:

$$\widehat{\texttt{attitude}_i} = \beta_0 + \beta_1 \texttt{educ}_i \tag{3.3}$$

The "hat" over $\texttt{attitude}_i$ denotes the predicted value of `attitude`. Equations (3.2) and (3.3) omit ϵ_i because these equations deal with the expected or predicted values of `attitude`—errors come into play when comparing the expected value to the actual value of y.

A general bivariate regression (one y and one x) equation is

$$y_i = \beta_0 + \beta_1 x_i + \epsilon_i \tag{3.4}$$

Further, the general form of (3.2) and (3.3) is

$$E(y_i|x_i) = \beta_0 + \beta_1 x_i \tag{3.5}$$

and

$$\widehat{y_i} = \beta_0 + \beta_1 x_i \tag{3.6}$$

Population parameters versus sample-based estimates

In research, we collect data on samples to learn about populations. In regression, we estimate slopes and intercepts based on samples to learn about what the slopes and intercepts are in the broader population. Slopes and intercepts in the population are called population parameters. Parameters are typically symbolized using Greek letters, such as β in the case of regression coefficients [see (3.4)]. In this book, sample-based estimates of those equations are symbolized using lowercase, Roman letters, such as b in the case of regression coefficients. Thus, the population slope is β_1 and the sample-based estimate of the slope is b_1. Other examples include σ and s for the population and sample-based standard deviation, and ρ and r for the population and sample-based correlation.

3.3.3 Estimation

The slope is computed as

$$b_1 = r_{yx}\frac{s_y}{s_x} \qquad (3.7)$$

where r_{yx} is the correlation between x and y, and s_y and s_x are the standard deviations for y and x, respectively. The intercept is computed as

$$b_0 = \overline{y} - b_1\overline{x} \qquad (3.8)$$

Of course, we rarely—if ever—compute these quantities by hand. Stata's **regress** command does this for us. The syntax for **regress** is straightforward. Following the keyword **regress**, you type the dependent variable followed by the independent variables. As discussed in section 2.4.1, you can use **if** and **in** with **regress**.

<u>regress</u> *depvar* [*indepvars*] [*if*] [*in*] [*weight*] [, *options*]

Estimating (3.3) with **regress** is done as follows:

```
. regress attitude educ
     Source |       SS           df       MS      Number of obs   =      1,006
------------+----------------------------------   F(1, 1004)      =      33.02
      Model |  8.81344558         1  8.81344558   Prob > F        =     0.0000
   Residual |  267.958141     1,004  .266890579   R-squared       =     0.0318
------------+----------------------------------   Adj R-squared   =     0.0309
      Total |  276.771587     1,005  .275394614   Root MSE        =     .51661

    attitude |      Coef.   Std. Err.      t    P>|t|     [95% Conf. Interval]
-------------+----------------------------------------------------------------
        educ |   .0335253   .005834     5.75   0.000     .0220771    .0449736
       _cons |   1.991708   .0804314   24.76   0.000     1.833876    2.149541
```

Focus on the bottom portion of the output for now. Stata calls the estimated intercept _cons (that is, constant). Thus, $b_0 = 1.99$. The slope for educ is $b_1 = 0.03$. We can now fill in the details of (3.3).

$$\widehat{\text{attitude}}_i = 1.99 + 0.03\text{educ}_i \qquad (3.9)$$

We can make predictions based on the coefficients. For example, if we want to know the predicted attitude for someone who has 15 years of education, we simply plug 15 into (3.9).

$$2.44 = 1.99 + 0.03 \times 15 \qquad (3.10)$$

The predicted attitude for someone who has 15 years of education is 2.44. Another way to say that is, "The expected value of attitude given that educ equals 15 is 2.44, or $E(\text{attitude}|\text{educ} = 15) = 2.44$."

3.3.4 Interpretation

Slope

The interpretation of a slope is

A one-unit difference[5] in x is associated with a b_1 difference in y.

In the context of the example, a one-unit difference in educ (that is, 1 year of education) is associated with a 0.03 difference in attitude. Why is this? Figure 3.4 plots (3.9). The inner triangle represents a one-unit difference in educ. The lower left corner of the inner triangle is $E(\text{attitude}|\text{educ}) = 10$, and the upper right corner of the inner triangle is $E(\text{attitude}|\text{educ}) = 11$. In other words, these corners represent the expected values of attitude when educ differs by one unit.

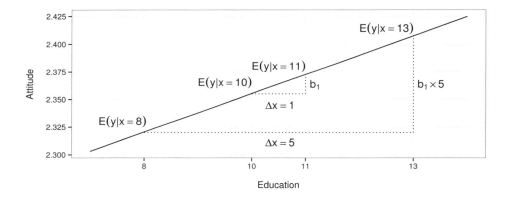

Figure 3.4. Interpretation of b_1; Δ = difference

Let's compute the two expected values for educ = 10 and educ = 11.

$$E(\text{attitude}|\text{educ} = 10) = 1.99 + 0.03 \times 10 = 2.29$$
$$E(\text{attitude}|\text{educ} = 11) = 1.99 + 0.03 \times 11 = 2.32$$

Further, $E(\text{attitude}|\text{educ} = 11) - E(\text{attitude}|\text{educ} = 10) = 2.32 - 2.29 = 0.03$, which is the same as b_1. This difference is illustrated by the vertical shift in attitude in figure 3.4.

---

5. I used to teach the definition by saying a one-unit *change* in x is associated with a b_1 change in y. I think *difference* is better. Change is often interpreted as meaning evidence that a variable was, in fact, changed in the sample. Within-person change is distinct from between-person differences. Thus, I use difference. It is correct and relevant to both longitudinal and cross-sectional designs.

3.3.4 Interpretation

We can formally demonstrate that for any one-unit difference in x, we expect a b_1 difference in y.

$$E(\texttt{attitude}_i|\texttt{educ}_i = x+1) - E(\texttt{attitude}_i|\texttt{educ}_i = x)$$
$$= \{b_0 + b_1(x+1)\} - \{b_0 + b_1(x)\}$$

After rearranging and simplifying, we have

$$E(\texttt{attitude}_i|\texttt{educ}_i = x+1) - E(\texttt{attitude}_i|\texttt{educ}_i = x) = b_0 - b_0 + b_1(x) - b_1(x) + b_1 = b_1$$

Figure 3.4 also shows the five-unit difference (that is, moving from 8 to 13) in x is associated with the $b_1 \times 5$ difference in y. We can formally demonstrate this:

$$E(\texttt{attitude}_i|\texttt{educ}_i = x+5) - E(\texttt{attitude}_i|\texttt{educ}_i = x) = b_0 + b_1(x+5) - \{b_0 + b_1(x)\}$$

After rearranging and simplifying, we have

$$E(\texttt{attitude}_i|\texttt{educ}_i = x+5) - E(\texttt{attitude}_i|\texttt{educ}_i = x)$$
$$= b_0 - b_0 + b_1(x) - b_1(x) + 5 \times b_1 = 5 \times b_1$$

Intercept

The interpretation of an intercept is

> The intercept is the expected value of y when x is 0.

In the context of our example, the expected `attitude` is 1.99 when `educ` $= 0$. Although it is possible for someone to have no formal education, sometimes it does not make sense for x to be equal to 0. For example, if the predictor was age, it would not make sense to predict `attitude` for someone who has not been born. You can center a predictor around its mean to increase interpretability.

To center the `educ` about its mean, you would subtract the mean `educ` from the observed `educ` for each observation.

$$\texttt{educ\_c}_i = \texttt{educ}_i - \overline{\texttt{educ}}$$

In Stata,[6]

```
. quietly summarize educ
. generate educ_c = educ - r(mean)
```

---

6. To center in this example, I used `summarize` and then centered `educ` using the returned mean from `summarize`. This is handy because I did not have to type the mean in by hand. Further, if for some reason my data change (for example, I have to drop outliers), this code will work just as it is supposed to. If I typed the mean as a number in my do-file, I would have to change it whenever my data change.

```
. regress attitude educ_c

      Source |       SS           df       MS      Number of obs   =     1,006
-------------+----------------------------------   F(1, 1004)      =     33.02
       Model |  8.81344563         1   8.81344563  Prob > F        =    0.0000
    Residual |  267.958141     1,004   .266890579  R-squared       =    0.0318
-------------+----------------------------------   Adj R-squared   =    0.0309
       Total |  276.771587     1,005   .275394614  Root MSE        =   .51661

    attitude |      Coef.   Std. Err.      t    P>|t|     [95% Conf. Interval]
-------------+----------------------------------------------------------------
      educ_c |   .0335254   .005834     5.75   0.000     .0220771    .0449736
       _cons |   2.444334   .016288   150.07   0.000     2.412372    2.476296
```

With education centered, b_0 is now 2.44, indicating that the expected `attitude` for someone with the average years of education is 2.44.

3.3.5 Residuals and predicted values

Recall that any number of lines can be drawn through a scatterplot to represent the relationship between x and y, but only one line is the regression line. Specifically, the regression line is the one that minimizes the errors in prediction across the entire sample. An error in prediction is the difference between actual and predicted values of y. Data are rarely, if ever, perfectly captured by the regression line. That is, the expected or predicted value of y is rarely equal to the actual value of y. Thus, errors are to be expected.

Figure 3.5 is a scatterplot of `educ` and `attitude`, and the dashed line is the regression line.[7] Because predicted values of `attitude` fall directly on the regression line, the solid lines represent the distance between the actual value of `attitude` and the predicted value of `attitude`. The distance between the predicted and actual values is referred to as a residual.

$$e_i = y_i - \widehat{y_i} \quad \text{or} \quad e_i = y_i - E(y_i|x_i)$$

---

7. This scatterplot includes 5% of the total dataset so that the figure can be readable.

3.3.5 Residuals and predicted values

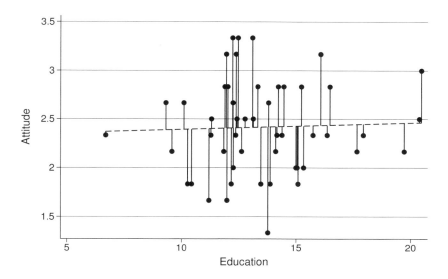

Figure 3.5. Illustration of errors in prediction

Regression minimizes the size of the residuals. More precisely, regression minimizes the sum of the squared residuals.

$$\sum_{i=1}^{N} e_i^2$$

This is why (3.7) and (3.8) are called least-squares estimators of β_1 and β_0. See Rencher and Schaalje (2008, 128) for a formal proof that the estimator that underlies regression is the least-squares estimator.

In Stata, residuals are obtained using the postestimation command `predict` with the `residual` option.

```
. quietly regress attitude educ
. predict res_att, residual
```

This will create a new variable called `res_att` that contains the residuals themselves. The name of the new variable is determined by the user.

We have already defined predicted values. Namely, they are expected values of y given a particular value of x [see (3.5) and (3.6)]. We can obtain two kinds of predicted values in Stata. First, we can obtain the predicted values of y given the specific values of x observed in the dataset. For example, we can obtain the predicted `attitude` given the observed `educ` in the data. These types of predicted values are obtained using the `predict` command.

```
. quietly regress attitude educ
. predict pred_att
(option xb assumed; fitted values)
```

This will create a new variable called `pred_att` that contains the predicted values.

The second kind of predicted value does not produce a predicted value for all observations but rather produces predicted values for specific values of x. We already did this with (3.10), where we computed the predicted `attitude` for someone with 15 years of education. Although we manually computed that quantity above, we can also use the `margins` command.

The `margins` command is an incredibly flexible, powerful command. The help documentation defines margins as "statistics calculated from predictions of a previously fit model at fixed values of some covariates and averaging or otherwise integrating over the remaining covariates". To see what this means for our analysis, use `margins` after `regress`:

```
. quietly regress attitude educ
. margins, at(educ = 15)
Adjusted predictions                           Number of obs    =      1,006
Model VCE    : OLS

Expression   : Linear prediction, predict()
at           : educ            =           15
```

	Margin	Delta-method Std. Err.	t	P>|t|	[95% Conf. Interval]	
_cons	2.494589	.0184872	134.94	0.000	2.458311	2.530867

The call to `margins` computes the predicted value of `attitude` when `educ` is fixed at 15.[8]

What is great about `margins` is that you can compute multiple predicted values at once. For example, the command to obtain the predicted `attitude` when `educ` is 10 through 20 increasing by 2s is

---

8. The value differs slightly from what we did by hand above because of rounding differences.

3.3.6 Partitioning variance

```
. quietly regress attitude educ
. margins, at(educ = (10(2)20))
Adjusted predictions                    Number of obs    =      1,006
Model VCE    : OLS
Expression   : Linear prediction, predict()
1._at        : educ            =    10
2._at        : educ            =    12
3._at        : educ            =    14
4._at        : educ            =    16
5._at        : educ            =    18
6._at        : educ            =    20
```

	Margin	Delta-method Std. Err.	t	P>|t|	[95% Conf. Interval]	
_at						
1	2.326962	.0261242	89.07	0.000	2.275698	2.378226
2	2.394013	.0184927	129.46	0.000	2.357724	2.430301
3	2.461063	.0165461	148.74	0.000	2.428594	2.493532
4	2.528114	.0218598	115.65	0.000	2.485218	2.57101
5	2.595165	.0308904	84.01	0.000	2.534548	2.655782
6	2.662215	.0412658	64.51	0.000	2.581238	2.743193

The output provides the six predicted values, one for each of the six values of educ we requested. The top part of the margins output provides the key for what the six values refer to. For example, the first predicted value is for when educ is equal to 10, the second predicted value is for when educ is equal to 12, and so on.

3.3.6 Partitioning variance

People clearly vary in their attitudes toward psychotropic medication. How much of the variability is associated with education levels? Are 10% of the differences among people associated with education? 5%? 50%? The larger this percentage, the more powerful education is as a predictor.

To determine this percentage, we can partition the total variability in attitude into two parts: i) variability associated with educ and ii) variability independent of educ. Consider the observation on the far right of figure 3.6. This person has 20 years of education and averaged 3 ("somewhat disagree") on attitude. The difference between the observed value of attitude and the mean of attitude is $y_i - \bar{y} = 3 - 2.4 = 0.6$, which represents the total variability for this person. This difference can be divided into the two parts—the part that is associated with educ and the part that is not.

First, consider the part of the variability associated with educ. This is measured by the difference between the predicted value $\widehat{y}_i$ and the mean $\bar{y}$: $\widehat{y}_i - \bar{y} = 2.5 - 2.4 = 0.1$. This difference is also referred to as the variability associated with the model or the regression. Second, consider the part of the variability independent of educ, which is measured by the difference between the observed value of y and the predicted value: $y_i - \widehat{y}_i = 3 - 2.5 = 0.5$. This is the residual discussed in section 3.3.5.

Thus, the difference between the observation and the mean is 0.6, which is divided into the part associated with educ (0.1) and the part independent of educ (0.5). If we square these differences and sum them over all observations, we obtain three sums of squares.

The total sums of squares (TSS) is

$$\text{TSS} = \sum_{i=1}^{N} (y_i - \bar{y})^2$$

The model sums of squares (MSS) is

$$\text{MSS} = \sum_{i=1}^{N} (\widehat{y}_i - \bar{y})^2$$

Finally, the residual sums of squares (RSS) is

$$\text{RSS} = \sum_{i=1}^{N} (y_i - \widehat{y}_i)^2$$

Recall, we were interested in determining what proportion of the total variability in attitude is associated with educ. TSS provides an estimate of total variability in attitude, and MSS provides an estimate of the variability in attitude associated with educ. Thus, the proportion of variability in attitude associated with educ is the ratio of MSS to TSS. This quantity is known as R^2:

$$R^2 = \frac{\text{MSS}}{\text{TSS}}$$

The sums of squares and R^2 are reported in Stata's regression output.

```
. regress attitude educ

      Source |       SS       df       MS              Number of obs =    1,006
-------------+------------------------------           F(1, 1004)    =    33.02
       Model |  8.81344558        1  8.81344558        Prob > F      =   0.0000
    Residual |  267.958141    1,004  .266890579        R-squared     =   0.0318
-------------+------------------------------           Adj R-squared =   0.0309
       Total |  276.771587    1,005  .275394614        Root MSE      =   .51661

------------------------------------------------------------------------------
    attitude |      Coef.   Std. Err.      t    P>|t|     [95% Conf. Interval]
-------------+----------------------------------------------------------------
        educ |   .0335253    .005834     5.75   0.000     .0220771    .0449736
       _cons |   1.991708   .0804314    24.76   0.000     1.833876    2.149541
------------------------------------------------------------------------------
```

3.3.6 Partitioning variance

$$R^2 = \frac{8.81}{276.77} = 0.03$$

Multiplying R^2 by 100 indicates that 3% of the variability in `attitude` is associated with `educ`.

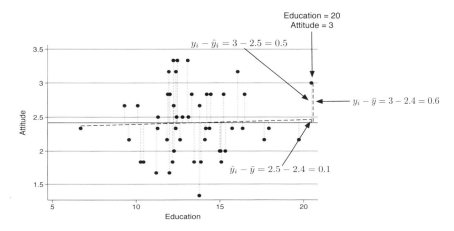

Figure 3.6. Partitioning the variability of a single score

The RSS can also be used to compute the conditional variance of y. The conditional variance is variability in the outcome after accounting for the predictor(s) in the model. Or, said another way, the conditional variance is the variability of the outcome around the regression line. This is in contrast to the unconditional variance (that is, just the regular variance), which is the variability of the outcome around the mean. The formula for the conditional variance is

$$s^2_{y|x} = \frac{\text{RSS}}{N - k - 1}$$

where $y|x$ means y given x, and k is the number of predictors in the regression model, which in this case is 1. The conditional variance is also called the error or residual variance because it is the variance of the residuals.

The square root of the conditional variance,

$$s_{y|x} = \sqrt{\frac{\text{RSS}}{N - k - 1}}$$

produces what is called the root mean squared error (MSE). The root MSE is the average error in prediction (that is, the difference between y and $\hat{y}$) for your model. It is the standard deviation of the residuals. Ideally, if your model is doing a good job reproducing the data, then the root MSE will be small. Stata reports the root MSE in the top portion of the output. In this model, the root MSE is 0.5, which indicates that the observed `attitude` and predicted `attitude` differed by 0.5 on average.

3.3.7 Confidence intervals

In section 3.3.3, the intercept (b_0) was estimated as 2.44 and the slope for the mean-centered education (b_1) as 0.03. These estimates are based on $N = 1006$ people who were sampled from a broader population. In other words, the sample was used to learn something about the population. The population coefficient for the education–attitude relationship (β_1) is not likely to be exactly 0.03 as was estimated in the model. In fact, if we were to gather data on a new sample of $N = 1006$ people, we can expect the estimates of b_0 and b_1 to change. We can use simulation methods to explore this variability.

Assume we know that the population value for the intercept is $\beta_0 = 2.5$ and the slope for the mean-centered education is $\beta_1 = 0.04$. Thus, the population regression equation is
$$\texttt{attitude}_i = 2.5 + 0.04 \texttt{educ\_c}_i + \epsilon_i \tag{3.11}$$
The estimates from section 3.3.3 are one realization from this population model.

Because we do not have access to 1,006 new people, we can simulate 1,006 new observations.

```
. drop _all
. set seed 49833
. set obs 1006
number of observations (_N) was 0, now 1,006
. generate educ = floor((20 - 1 + 1)*runiform() + 1)
. quietly summarize educ
. generate educ_c = educ - r(mean)
. generate attitude = 2.5 + 0.04*educ_c + rnormal(0, 0.5)
```

The only thing new in this code is the method for generating the education variable. In the GSS data, education ranged from 1 to 20 and was recorded in whole integers. To generate whole integers between minimum and maximum values, you can use the following code:

```
generate newvar = floor((max - min + 1)*runiform() + min)
```

The key at this point is to know where to put the minimum and maximum values. I leave it to you to use `summarize` and `table` to examine the specifics of the variables we created.

3.3.7 Confidence intervals

You can now predict the simulated `attitude` from the simulated `educ_c`.

```
. regress attitude educ_c
```

Source	SS	df	MS
Model	57.9613019	1	57.9613019
Residual	249.302288	1,004	.248309052
Total	307.26359	1,005	.305734915

Number of obs	= 1,006
F(1, 1004)	= 233.42
Prob > F	= 0.0000
R-squared	= 0.1886
Adj R-squared	= 0.1878
Root MSE	= .49831

attitude	Coef.	Std. Err.	t	P>|t|	[95% Conf. Interval]	
educ_c	.0415086	.0027168	15.28	0.000	.0361773	.04684
_cons	2.497187	.0157108	158.95	0.000	2.466357	2.528017

The estimates are slightly different from what we saw in our actual sample. You can change the seed and run the code a few times to see how the estimates bounce around from sample to sample.

Rather than run the code over and over again, let's create a program to run a simulation.

```
. program define simsample, rclass
  1.        version 15.1
  2.        syntax [, obs(integer 1)]
  3.        drop _all
  4.        set obs `obs'
  5.        generate educ = floor((20 - 1 + 1)*runiform() + 1)
  6.        quietly summarize educ
  7.        generate educ_c = educ - r(mean)
  8.        generate exp_att = 2.5 + 0.04*educ_c
  9.        generate att = rnormal(exp_att, 0.5)
 10.        regress att educ_c
 11.        return scalar b0 = _b[_cons]
 12.        return scalar b1 = _b[educ_c]
 13.        return scalar se_b0 = _se[_cons]
 14.        return scalar se_b1 = _se[educ_c]
 15.        return scalar ul_b1 = _b[educ_c]
> + _se[educ_c]*invttail(e(df_r), 0.025)
 16.        return scalar ll_b1 = _b[educ_c]
> - _se[educ_c]*invttail(e(df_r), 0.025)
 17. end
```

Line 2 adds to `simsample` an `obs()` argument, which allows you to specify the number of observations in a given dataset. Lines 11 and 12 save the intercept and slope from each replication of the simulation. Lines 13–16 save some other stuff we will need shortly.

Next we run the simulation. We save all the estimates from lines 11–16 of the simulation program. We also set the number of observations for each dataset in the simulation to 1,006.

```
. simulate b0 = r(b0) b1 = r(b1) se_b0 = r(se_b0) se_b1 = r(se_b1)
>               ul_b1 = r(ul_b1) ll_b1 = r(ll_b1),
>               reps(2000) seed(39744) nodots: simsample, obs(1006)

     command:  simsample, obs(1006)
          b0:  r(b0)
          b1:  r(b1)
       se_b0:  r(se_b0)
       se_b1:  r(se_b1)
       ul_b1:  r(ul_b1)
       ll_b1:  r(ll_b1)
```

Now we have a dataset of estimates from 2,000 replications of the simulation. Let's look at a histogram of the slope values.

```
. histogram b1, scheme(lean2) xtitle("b1")
(bin=33, start=.0303988, width=.00055159)
```

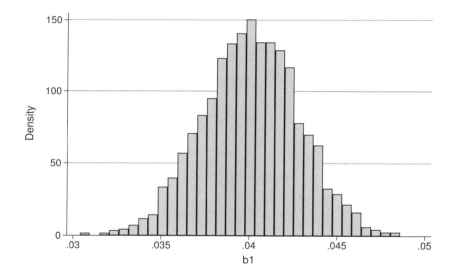

Figure 3.7. Sampling distribution of the slope for mean-centered education when $N = 1006$

Figure 3.7 represents the sampling distribution of the slope b_1. That is, it represents the variability in b_1 across repeated samples from the population. Sampling distributions are important because they describe the uncertainty in our estimates. If a sampling distribution is wide—if the slope estimates vary a lot from one sample to another—then our estimates contain a great deal of uncertainty.

3.3.7 Confidence intervals

I want to draw your attention to three characteristics of sampling distributions. First, the distribution is normally shaped. Second, the mean of the distribution is equal to the population value—in this case, the mean of the distribution is $\beta_1 = 0.04$ [see (3.11)]. Third, the standard deviation of the sampling distribution, called the standard error (SE), indicates how much estimates will vary from sample to sample. In other words, the standard error indicates the level of uncertainty in an estimate due to sampling error. The SE of the sampling distribution from the simulation is 0.003:

```
. summarize b1
    Variable |        Obs        Mean    Std. Dev.       Min        Max
          b1 |      2,000    .0400804    .0027378   .0303988   .0486013
```

This is very close to the estimate Stata provided when we analyzed the GSS data (see above).

A formula for the SE of a slope (SE_{b_1}) is (Cohen et al. 2003, 42)

$$\text{SE}_{b_1} = \frac{s_y}{s_x}\sqrt{\frac{1-r_{xy}^2}{n-2}} \quad (3.12)$$

which indicates that SEs get larger—there is more uncertainty in our estimates—when there is a) relatively more variability in y, b) relatively less variability in x, c) a small relationship between x and y, and d) small sample sizes.

Rerun the simulation, but this time reduce the sample size for each replication from 1,006 to 30. This illustrates the impact that sample size has on the uncertainty of an estimate.

```
. simulate b0 = r(b0) b1 = r(b1) se_b0 = r(se_b0) se_b1 = r(se_b1)
>               ul_b1 = r(ul_b1) ll_b1 = r(ll_b1),
>               reps(2000) seed(39744) nodots: simsample, obs(30)

      command:  simsample, obs(30)
           b0:  r(b0)
           b1:  r(b1)
        se_b0:  r(se_b0)
        se_b1:  r(se_b1)
        ul_b1:  r(ul_b1)
        ll_b1:  r(ll_b1)
```

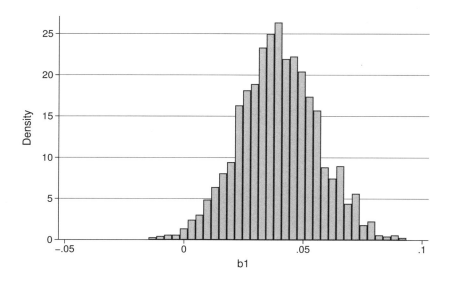

Figure 3.8. Sampling distribution of the slope for mean-centered education when $N = 30$

Comparing figures 3.7 and 3.8, you can see that $N = 30$ produces a much wider sampling distribution and thus more uncertainty than $N = 1006$. The range of values with $N = 30$ is $-0.01 - 0.09$, whereas the range of values when $N = 1006$ is $0.03 - 0.05$. The SE when $N = 30$ is approximately 0.02, which is about 10 times larger than the SE when $N = 1006$.

```
. summarize b1

    Variable |       Obs        Mean    Std. Dev.        Min        Max
-------------+--------------------------------------------------------
          b1 |     2,000     .0396184    .0166277   -.0148426   .0933865
```

A common method for quantifying uncertainty is to use confidence intervals. In short, confidence intervals combine the information from the slope estimate and the SE to provide an index of the uncertainty in the estimate. Stata automatically reports the 95% confidence interval for estimates. The Stata output from the GSS data indicated that $b_1 = 0.04$ and the 95% confidence interval is $[0.036, 0.047]$. This confidence interval is fairly narrow (that is, the limits do not differ much from the estimate), suggesting small uncertainty in the estimate. Figure 3.7 supports this notion, because estimates do not vary much from sample to sample.

The formulas for the lower and upper limits of a confidence interval are

$$\text{LL}_{b_1} = b_1 - (t_{\text{crit}} \times \text{SE}_{b_1}) \tag{3.13}$$
$$\text{UL}_{b_1} = b_1 + (t_{\text{crit}} \times \text{SE}_{b_1}) \tag{3.14}$$

3.3.7 Confidence intervals

where t_{crit} is the critical value from a t distribution with $N-2$ degrees of freedom that bounds the middle 95% of the distribution for a 95% confidence interval. If you compute a 90% confidence interval, then t_{crit} corresponds to the middle 90%.

A confusing aspect of confidence intervals is how to interpret them. The 95% confidence interval for the `educ_c` coefficient in the GSS data was $[0.036, 0.047]$. It seems intuitive to say that we are 95% confident that the population value β_1 falls within the range of $[0.036, 0.047]$. This is wrong and this mistaken interpretation is common. Indeed, even a popular regression textbook in psychology gets it wrong (Cohen et al. 2003, 43).

The correct interpretation is a "long-run" interpretation. If we compute the 95% confidence interval over many samples using (3.13) and (3.14), then 95% of the time, the confidence interval will contain the population value (Morey et al. 2016). The 95% confidence is in the method over the long run, not in any particular interval. It is not a particularly satisfying interpretation, but it is the correct one.

We can use the simulation to learn more about what this interpretation means. We will calculate how often, out of the 2,000 simulation replications, the 95% confidence interval contains the population slope, $\beta_1 = 0.04$. We expect 95% of the confidence intervals (1,900 of them) to contain 0.04. This is a called the coverage rate. A 95% interval should have 95% coverage. Note that lines 15 and 16 of the `simsample` program calculate and return the upper and lower limits of the confidence interval. We generate a new variable called `cover` that is 1 if an interval includes 0.04 and 0 if it does not.

```
. generate cover = ul_b1 >= 0.04 & ll_b1 <= 0.04
. tabulate cover

      cover |      Freq.     Percent        Cum.
------------+-----------------------------------
          0 |        105        5.25        5.25
          1 |      1,895       94.75      100.00
------------+-----------------------------------
      Total |      2,000      100.00
```

Thus, within rounding error, 95% of the intervals contained the population value.

Figure 3.9 displays the point estimate and confidence interval for the first 100 replications of the simulation (Carsey and Harden 2014). To create this graph, I first generated a new variable, `simrep`, that indexes the replication within the simulation. This is the x axis of the graph. The solid horizontal line at 0.04 represents the population value. The black points and intervals are the replications where the confidence interval included 0.04, and the gray points and intervals are the replications where the confidence interval did not include 0.04. Figure 3.9 underscores the long-run interpretation of a confidence interval: over many samples, a confidence interval will contain the population value 95% of the time.

```
.  generate simrep = _n
.  twoway rcap ul_b1 ll_b1 simrep in 1/100 if cover == 0, lcolor(gs8) ||
>          rcap ul_b1 ll_b1 simrep in 1/100 if cover == 1, lcolor(black) ||
>          scatter b1 simrep in 1/100 if cover == 0, msymbol(diamond) mcolor(gs8) ||
>          scatter b1 simrep in 1/100 if cover == 1, msymbol(circle) mcolor(black)
>          scheme(lean2) legend(off) yline(0.04)
```

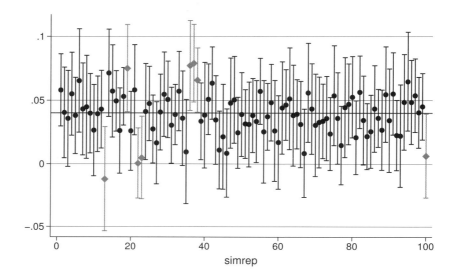

Figure 3.9. Illustration of 95% confidence interval coverage in the first 100 simulation replications

Given this awkward interpretation of confidence intervals, I recommend you focus more on the width of the confidence interval rather than the specific numbers. In other words, use the confidence interval to learn about uncertainty rather than making probability statements about the population value.

3.3.8 Null hypothesis significance testing

The SE is also used to perform null hypothesis significance testing (NHST). You have likely already been exposed to NHST, so my coverage will be brief. NHST is by far the most common method psychologists use to perform statistical inference. The general idea behind NHST is to examine whether the results observed in a specific study are due to chance. It sounds simple enough, but when it comes down to it, the conclusions we can draw from NHST are pretty limited.

NHST begins with specifying a null hypothesis and an alternative hypothesis. These hypotheses are mutually exclusive, meaning that together they have to accommodate

3.3.8 Null hypothesis significance testing

all possible outcomes. The null hypothesis can, in principle, be anything, but it is most common to set the null hypothesis equivalent to "no relationship". In regression, the null hypothesis for a slope is that the population value for the slope is 0:

$$H_0: \beta_1 = 0$$

The alternative hypothesis is that the population slope is not equal to 0:

$$H_a: \beta_1 \neq 0$$

In other words, when evaluating these hypotheses,[9] we aim to answer the following question:

> Was our sample drawn from a population where the predictor and outcome have no relationship (the null hypothesis) or from a population where the predictor and outcome do have a relationship (the alternative hypothesis)?

The problem with NHST is that we do not actually compute the probability that the null is true or the probability that the alternative is true. To see why, we need to specify the steps in NHST.

First, understand that NHST requires that we start by assuming the null hypothesis is true. Our question is this: is our sample drawn from a population consistent with the null hypothesis? Thus, step 1 is to specify a sampling distribution of b_1. The sampling distribution has a mean equal to the null hypothesis, $\beta_1 = 0$, and a standard deviation equal to the SE estimated from the sample [see (3.12)]. Because we do not have a probability distribution for the slope relating `educ` to `attitude`, we must approximate it with a known probability distribution. We can approximate it with a t distribution with $N - k - 1$ degrees of freedom (Cohen et al. 2003).

Step 2 is to translate the slope from the metric of the outcome variable to a t-value, so we can use the t distribution as an approximate of the sampling distribution. We standardize the slope value with respect to the SE to produce t_{obt}, which is the obtained t-value for the estimated slope:

$$t_{\text{obt}} = \frac{b_1 - 0}{\text{SE}_{b_1}}$$

In the GSS example, $t_{\text{obt}} = 5.8$, reported by Stata in the column with the heading `t`.

Step 3 is to determine the probability that t_{obt} was sampled from a population where the null hypothesis, $\beta_1 = 0$, is true. To do this, we assume the null hypothesis is true and then determine the probability of observing t_{obt} or something more extreme.[10] We do this by comparing t_{obt} to the sampling distribution from step 1. It is always possible to observe $t_{\text{obt}} = 5.8$ in the null hypothesis sampling distribution, but is it probable? If it is not probable, then we reject the null hypothesis.

---

9. Note that both hypotheses reference the population value of slope β_1, not the sample estimate b_1.
10. This performs a two-tailed test. One-tailed tests are also possible, and interested readers can review any introductory statistics text for a discussion of the difference (Cohen et al. 2003).

Step 4 is to apply the α level, which is the probability threshold for rejecting the null hypothesis.[11] Most commonly, α is set at 0.05, but it could be set to anything between 0 and 1. An α of 0.05 means that we are going to divide the sampling distribution of the slope under the null hypothesis into two parts: the inner 95% and the extreme 5% (see figure 3.10). If α were 0.01, the distribution would be divided into the inner 99% and the extreme 1%. If t_{obt} falls within the extreme portion of the sampling distribution, then we reject the null hypothesis. That is, if t_{obt} is in the extreme tails of the sampling distribution, then we conclude that it is unlikely that our sample results were drawn from a population where the null hypothesis is true.

Figure 3.10 represents the sampling distribution of b_1 assuming the null is true. It is a t distribution with 1,004 degrees of freedom. The biggest portion of this distribution is between t-values of -1.96 to 1.96. Indeed, 95% of the distribution is between these values, and 5% of the distribution exceeds these values (with 2.5% below -1.96 and 2.5% above 1.96). In other words, t-values more extreme than ± 1.96 would be fairly rare if the null hypothesis were true. The t-value for the slope comparing educ to attitude was $t_{\text{obt}} = 5.8$, which is more extreme than ± 1.96. This indicates that, if the null hypothesis is true, it is unlikely that we would observe a slope for educ as big as we did by chance alone.

---

11. I say to apply the α level rather than determine it because the α level should be established before the analysis is run and before you have seen the data. Fudging with the α level after you have run your analysis is generally considered poor scientific practice (that is, misconduct) and can lead to serious problems with the research literature (Simmons, Nelson, and Simonsohn 2011).

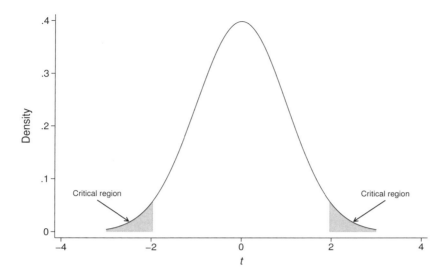

Figure 3.10. Sampling distribution of the slope under the null hypothesis that $\beta_1 = 0$. The sampling distribution is a t distribution with 1,004 degrees of freedom. The shaded areas are the critical areas of rejection for a two-tailed test and $\alpha = 0.05$.

Step 5 involves computing the probability of our sample results, if the null hypothesis is true. This probability is called a p-value. The probability of observing a result as extreme or more extreme than t_{obt}, if the null hypothesis is true, is tiny:

```
. display 2*ttail(1004, 5.8)
8.880e-09
```

Stata reports the p-value in the regression results in the column with the heading P>|t|. Given that the p-value is smaller than the α level, we conclude that it is unlikely that the observed slope was sampled from a population where the null hypothesis is true. We reject the null hypothesis.

3.3.9 Additional methods for understanding models

Without question, null hypothesis significance testing is the most common way psychologists and other social scientists evaluate their results. For example, psychologists may use regression to evaluate several predictors of delinquency, such as socioeconomic status and education. They will scan the results of the regression looking for coefficients that have p-values less than 0.05 and judge those to be worthy of attention. One might say something like, "socioeconomic status has a negative relationship with delinquency ($p < 0.05$)" or "higher levels of education are associated with lower levels of delinquency ($p = 0.01$)".

The problems with null hypothesis testing are legion and have been well documented for many years. Some problems include that a) the null hypothesis is often uninteresting and may be implausible (Meehl 1978)[12]; b) the p-value gives the probability of observing results as extreme or more extreme as those in the study if the null hypothesis is true, when what is often desired is the probability that the null is true or a probable range of plausible values (Cohen 1994); c) null hypothesis testing leads to binary decisions—an effect is important if $p < 0.05$ and uninteresting otherwise (Gelman and Carlin 2014); and d) p-values do not provide assurance that an effect is real or important (Gelman and Carlin 2014; Ioannidis 2005; Pashler and Harris 2012).

Fortunately, there are many alternatives to hunting for p-values in regression output. In this section, I focus on using predicted scores to unpack the inferences we make from our models. I use Stata's `margins` and `predict` postestimation commands to create the predicted scores and estimates of the uncertainty in prediction. These tools help us make more precise statements about our results. For example, after mastering this material, instead of saying that "education has a negative relationship with delinquency, $p < 0.05$", we can say, "Those with 12 years of education are predicted to have 20% lower probability of delinquency, with a 95% margin of error of ±5%, than those with 5 years or less of education." In short, this material aims to move you beyond sterile, difficult to understand statistical reasoning to powerful, transparent reasoning. It is well worth your time.

Using predicted scores to understand model implications

Using margins

I introduced predicted scores in section 3.3.5 and showed how to use `predict` and `margins` to compute them—`predict` for predicted scores based on the values of x in the observed data and `margins` for arbitrary values of x. Regardless of which method we use, predicted scores help us understand what a model implies about the observed data and about future data. What does a slope of $b = 0.04$ mean and actually look like? Create predicted scores and see for yourself. Rather than just saying there is a relationship between two variables ($p < 0.05$), be specific regarding what the relationship is and what the relationship means with respect to specific values of the predictor(s) and outcome. The Stata commands also help us quantify uncertainty in our predictions, which helps us maintain perspective on the precision of our inferences.

Let's return to the regression of `attitude` on `educ`. We will use the uncentered version of `educ` to make things easier on us. Our research question is to compare the attitude toward psychotropic medication for people with no high school education (less

---

12. Meehl (1978) was not one to mince words about the role of null hypothesis testing in psychology: "I suggest to you that Sir Ronald [Ronald Fisher developed NHST] has befuddled us, mesmerized us, and led us down the primrose path. I believe that the almost universal reliance on merely refuting the null hypothesis as the standard method for corroborating substantive theories in the soft areas is a terrible mistake, is basically unsound, poor scientific strategy, and one of the worst things that ever happened in the history of psychology." (p. 817)

3.3.9 Additional methods for understanding models

than 9 years of education), people with at least some high school education (9–12 years of education), and those with at least some college education (13 or more years of education).

```
. regress attitude educ

      Source |       SS           df       MS      Number of obs   =      1,006
-------------+----------------------------------   F(1, 1004)      =      33.02
       Model |  8.81344558         1  8.81344558   Prob > F        =     0.0000
    Residual |  267.958141     1,004  .266890579   R-squared       =     0.0318
-------------+----------------------------------   Adj R-squared   =     0.0309
       Total |  276.771587     1,005  .275394614   Root MSE        =     .51661

------------------------------------------------------------------------------
    attitude |      Coef.   Std. Err.      t    P>|t|     [95% Conf. Interval]
-------------+----------------------------------------------------------------
        educ |   .0335253   .005834     5.75   0.000     .0220771    .0449736
       _cons |   1.991708   .0804314   24.76   0.000     1.833876    2.149541
------------------------------------------------------------------------------
```

The expected `attitude` for someone with 8, 12, and 16 years of education is

```
. display _b[_cons] + _b[educ]*8
2.2599113
. display _b[_cons] + _b[educ]*12
2.3940126
. display _b[_cons] + _b[educ]*16
2.528114
```

All three values of education are expected to have attitudes between "Agree Somewhat" and "Disagree Somewhat" with the negative statements about psychotropic medications. As expected, given the slope value, those with higher amounts of education have slightly more positive attitudes.

A problem with these estimates is that they do not account for inferential uncertainty (Gelman and Hill 2007), which refers to uncertainty in the regression coefficients (for example, b_0 and b_1). In other words, when we computed the predicted values above, we treated b_0 and b_1 as known (that is, fixed to the estimated values); however, as we saw when computing the confidence interval for the coefficients, the estimated values for b_0 and b_1 are just that, estimated values. These estimates are uncertain because they would vary from sample to sample. Our exploration of predicted scores should reflect that uncertainty.

We can obtain confidence intervals for the predicted values by using `margins`:

```
. margins, at(educ = (8 12 16))
Adjusted predictions                            Number of obs   =      1,006
Model VCE    : OLS
Expression   : Linear prediction, predict()
1._at        : educ            =        8
2._at        : educ            =       12
3._at        : educ            =       16
```

	Margin	Delta-method Std. Err.	t	P>|t|	[95% Conf. Interval]	
_at						
1	2.259911	.0359896	62.79	0.000	2.189288	2.330535
2	2.394013	.0184927	129.46	0.000	2.357724	2.430301
3	2.528114	.0218598	115.65	0.000	2.485218	2.57101

The `at(educ = (8 12 16))` option means that we compute the predicted value of `attitude` when `educ` is 8, 12, and 16. The predicted values are identical to what we computed before, and `margins` automatically produces confidence intervals for each predicted value, which reflects the inferential uncertainty in the slope.

Plotting the predicted values and confidence intervals helps solidify the model-implied relationship between `attitude` and `educ`. The `marginsplot` command automates plotting these relationships. `marginsplot` is a postestimation command and must follow `margins`.

```
. marginsplot, scheme(lean2)
  Variables that uniquely identify margins: educ
```

Other than specifying the color scheme for the plot, we added no options to the command `marginsplot`. Figure 3.11 is the resulting plot and provides the predicted values of `attitude` and their confidence intervals. I do not love the default plot because the y axis has a narrow range, which makes the relationship between `educ` and `attitude` look stronger than it is when we consider the entire range of `attitude` (1–4). Likewise, the confidence intervals appear large, even though the numerical range is quite small.

3.3.9 *Additional methods for understanding models*   67

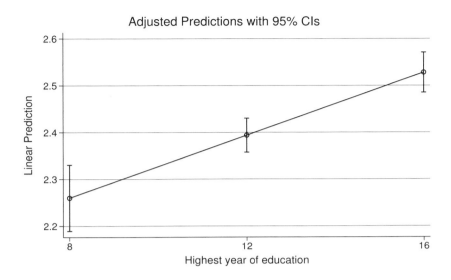

Figure 3.11. `marginsplot` showing the expected relationship (including 95% confidence intervals) between `attitude` and `educ` accounting for uncertainty in model parameters

We can adjust the range of the y axis and change the labels of the axes with options (see figure 3.12).

```
. marginsplot, ylabel(1(1)4 1 "SA" 2 "AS" 3 "DS" 4 "SD")
>         ytitle("Attitude") xtitle("Education") title("") scheme(lean2)
    Variables that uniquely identify margins: educ
```

The `ylabel()` option changes the range of the y axis and changes the marker labels from numeric to string. `ytitle()` changes the y-axis title, `xtitle()` changes the x-axis title, and `title()` changes the overall title (removed in this case). The main problem with figure 3.12 is that changing the range of the y axis has made the confidence intervals so narrow they are hard to see. Although I believe this is more accurate than figure 3.11, it is still not optimal.

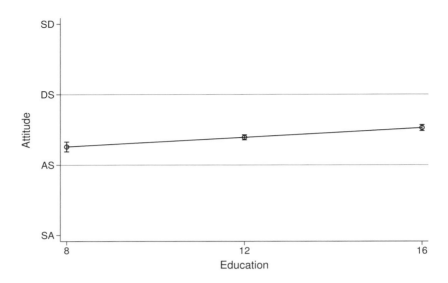

Figure 3.12. `marginsplot` showing the expected relationship (including 95% confidence intervals) between `attitude` and `educ` accounting for uncertainty in model parameters but with adjusted axis titles and marker labels

Fortunately, the plot can be changed to use a line plot rather than a scatterplot for the predicted values and a shaded ranged plot rather than a ranged cap plot for the confidence intervals. In `marginsplot`, these changes are called "recasting" because you are recasting the plot from the defaults to a custom form. The syntax is

```
. marginsplot, ylabel(1(1)4 1 "SA" 2 "AS" 3 "DS" 4 "SD")
>       ytitle("Attitude") xtitle("Education") title("") scheme(lean2)
>       recast(line) recastci(rarea) ciopts(color(gs8) fcolor(gs8%50))
  Variables that uniquely identify margins: educ
```

`recast(line)` changes the plot for the predicted values to a line plot, `recastci(rarea)` changes the confidence interval plot to a ranged area plot, and `ciopts()` controls options for the area plot (for example, the color of the outline of the area [`color()`] and the shading of the area [`fcolor()`]). The option `fcolor(gs8%50)` changes the shading of the area plot to a light gray and increases the transparency of the shading to 50%. Leaving the shading opaque makes it difficult, sometimes impossible, to see the predicted values. Figure 3.13 is the resulting plot, which I believe clearly depicts the nature of the implied relationship between `attitude` and `educ` as well as the inferential uncertainty in that relationship.

3.3.9 Additional methods for understanding models

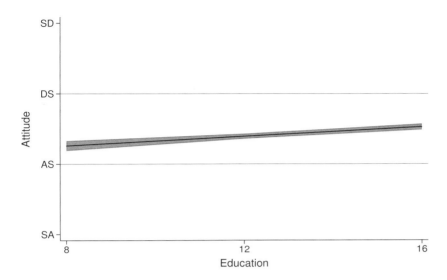

Figure 3.13. `marginsplot` showing the expected relationship (including 95% confidence intervals) between `attitude` and `educ` accounting for uncertainty in model parameters but with adjusted axis titles, marker labels, and plottypes

Finally, to create a plot across the entire observed range of `educ` and that adds the raw data to the plot, we can do the following:

```
. quietly margins, at(educ = (1(1)20))
. marginsplot, ylabel(1(1)4 1 "SA" 2 "AS" 3 "DS" 4 "SD")
>       ytitle("Attitude") xtitle("Education") title("") scheme(lean2)
>       recast(line) recastci(rarea) ciopts(color(gs8) fcolor(gs8%50))
>       addplot(scatter attitude educ, below color(gs4%50)
>       msymbol(circle_hollow) jitter(4))
>       legend(off)
   Variables that uniquely identify margins: educ
```

The `margins` command uses a number list to produce predicted values across the range of `educ`. The only change to `marginsplot` is the `addplot()` option, which allows you to add twoway plots (for example, scatterplots, line plots, histograms) to a `marginsplot`. We have added a scatterplot of the raw data for `attitude` and `educ`. The `below` option ensures the data points are plotted behind the line and area plots. The `color()` and `msymbol()` options change the color and symbol type for the points. The `jitter(4)` option adds a small amount of noise to the data to prevent overplotting.

Figure 3.14 is the final plot. Adding the raw data keeps us honest. Based on the range of the data, I would not feel comfortable drawing inferences about the `attitude` and `educ` relationship for people with less than 8 or 9 years of education. We do not know much about them.

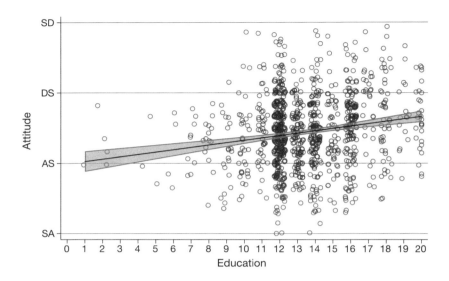

Figure 3.14. `marginsplot` showing the expected relationship (including 95% confidence intervals) between `attitude` and `educ` accounting for uncertainty in model parameters and across the entire range of `educ`; the circles are the raw data

Using predict with the stdp option

We can produce this same kind of plot by using `predict`. As we saw previously, using `predict` after `regress` produces predicted values based on the values of the covariates (that is, the x's) in the dataset. This is just like `margins` except we use the observed values of the covariates rather than specifying the values of the covariates as we did with `margins`. Why would we not want to just specify them with `margins`? In many ways, this is a distinction without a practical difference. When we manually specify the covariate values with `margins`, we typically will use the range of the observed data given that it is not good practice to make inferences beyond the scope of the data. Thus, `margins` and `predict` will give the same answers.

We can replicate the `margins` results with `predict` (or vice versa) by computing the predicted scores and their SE. To obtain the SE that accounts for inferential uncertainty in the predicted scores, use the `stdp` option for `predict` (the option name stands for the SE of the prediction). Then construct the upper and lower limits of the confidence interval, and summarize and plot the predictions as you wish.

```
. predict e_att
(option xb assumed; fitted values)
. predict e_att_se, stdp
. generate ell = e_att - 1.96*e_att_se
. generate eul = e_att + 1.96*e_att_se
```

3.3.9 Additional methods for understanding models

To create a plot similar to figure 3.14 but that uses the output from `predict` rather than `margins`, the code is

```
. twoway scatter attitude educ, jitter(4) mcolor(gs4) ||
>        rarea ell eul educ, sort color(black) fcolor(gs8%50) ||
>        line e_att educ, lcolor(black) scheme(lean2)
>        legend(off)
>        ylabel(1(1)4 1 "SA" 2 "AS" 3 "DS" 4 "SD")
>        ytitle("Attitude") xtitle("Education")
```

Figure 3.15 is the resulting plot.

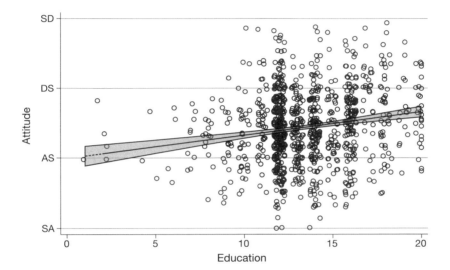

Figure 3.15. Plot based on the output of `predict` rather than `margins` showing the expected relationship (including 95% confidence intervals) between `attitude` and `educ` accounting for uncertainty in model parameters and across the entire range of `educ`; the circles are the raw data

Using predict with the stdf option

The predictions we have produced are expected values, which are the average value of the outcome variable for people with a specific level of education. The expected `attitude` for people with 16 years of education is 2.5. However, people with 16 years of education will certainly vary with respect to their attitudes. The confidence interval for the expected value only provides an index of the uncertainty in the expected value (that is, an average across many people); it does not tell us how much variability there will be among people with 16 years of education. To estimate this variability, we need to generate a point prediction. Point prediction gets at the following question: What is the predicted `attitude` of a single person (not the average) with 16 years of education? The point prediction will take into account the expected value (average) as well as variability around that average (error in prediction). Point predictions involve a new data point or new observation, whereas the expected value predictions involve the average of those new observations.

Like expected value predictions, point predictions account for inferential uncertainty, but point predictions also account for predictive uncertainty (Gelman and Hill 2007). Predictive uncertainty refers to uncertainty due to any variables or random noise not accounted for by predictor variables (Gelman and Hill 2007). Thus, the width of interval estimates for expected value predictions is influenced by a single source of uncertainty, and the width for point predictions is influenced by two sources of uncertainty. Consequently, interval estimates for point predictions will be wider than those for expected value predictions. The mean of the point predictions is the same as that of the expected values because the average point prediction is the expected value in regression (Gelman and Hill 2007). However, the interval estimates for the point predictions are wider than the interval estimates for the expected values because both inferential and prediction uncertainty are involved.

Stata refers to point predictions as forecast, presumably because Stata is providing a forecast of what a new value would be based on a particular value of the covariate. To obtain the SE of the forecast, use the `stdf` option for `predict` (the option name stands for the SE of the forecast). Then create the upper and lower limits of the point prediction interval (or forecast interval) and add the prediction interval to figure 3.15.

```
. predict pp_att_se, stdf
. generate ppll = e_att - 1.96*pp_att_se
. generate ppul = e_att + 1.96*pp_att_se
. twoway scatter attitude educ, jitter(4) mcolor(gs4) ||
>        rarea ppll ppul educ, sort fcolor(gs12%70) ||
>        rarea ell eul educ, sort fcolor(gs8%50) ||
>        line e_att educ, lcolor(black) scheme(lean2)
>        legend(off)
>        ylabel(1(1)4 1 "SA" 2 "AS" 3 "DS" 4 "SD")
>        ytitle("Attitude") xtitle("Education")
```

The resulting plot is shown in figure 3.16. As you can see, the prediction interval is much wider than the standard confidence interval. This is because predicting a single data point is much more difficult than predicting the average. This is important to remember because we make decisions, such as in medical or clinical settings, based on research. The recommendations are typically based on the average patient. That may be the best prediction for a given patient, but we also need to recognize that there is some uncertainty in that prediction based on both inferential and predictive uncertainty. It is fairly humbling.

3.3.9 Additional methods for understanding models

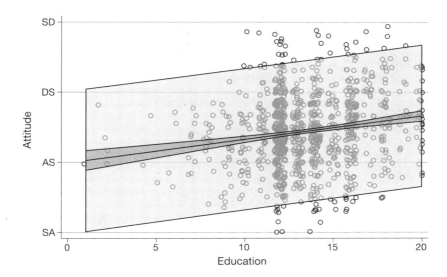

Figure 3.16. Plot of the expected relationship (including 95% confidence interval and prediction interval) between `attitude` and `educ` across the entire range of `educ`; the circles are the raw data

Composite contrasts

Once the relationship between `attitude` and `educ` has been established using regression, we may also want to compare specific levels or combinations of levels of `educ` with respect to `attitude`. For example, we may want to examine whether there is a difference between the expected attitude for people with only a high school education and those with at least 4 years of college. The null hypothesis for this contrast is

$$H_0: E\,(\texttt{attitude}|\texttt{educ} \geq 16) - E\,(\texttt{attitude}|\texttt{educ} = 12) = 0$$

To test this hypothesis, we want to see if the average expected `attitude` for participants with 16–20 years of education is different from the expected `attitude` for participants with 12 years of education.

$$\begin{aligned}\{E\,(\texttt{attitude}|\texttt{educ} = 16) + E\,(\texttt{attitude}|\texttt{educ} = 17)\\ + E\,(\texttt{attitude}|\texttt{educ} = 18) + E\,(\texttt{attitude}|\texttt{educ} = 19)\\ + E\,(\texttt{attitude}|\texttt{educ} = 20)\}/5 - E\,(\texttt{attitude}|\texttt{educ} = 12) = 0\end{aligned} \quad (3.15)$$

We can test this after `regress` by using the `lincom` postestimation command (the command name stands for linear combination). `lincom` allows us to create a linear combination of parameters, which is what (3.15) is, and perform inferences for the combination of parameters. For example, we can compute the difference between the average expected `attitude` for participants with at least 4 years of college and those with only a high school degree. `lincom` produces not only the difference but also the SE of that difference, which allows us to perform a null hypothesis test and compute a confidence interval. The `lincom` command for this test is

```
. lincom (_b[educ]*16 + _b[educ]*17 + _b[educ]*18 + _b[educ]*19 + _b[educ]*20)/5
>    - _b[educ]*12
 ( 1)  6*educ = 0
```

attitude	Coef.	Std. Err.	t	P>|t|	[95% Conf. Interval]
(1)	.2011521	.0350041	5.75	0.000	.1324626 .2698416

The difference in `attitude` for participants with at least 4 years of college and those with only a high school degree is 0.20, which indicates that those with more years of education have a more positive view of medications than those with fewer years. This difference was statistically significant ($p < 0.001$), and the confidence interval ranged from $[0.13, 0.27]$.

Composite contrasts using `lincom` (or `nlcom` for nonlinear combinations of parameters) provide users with considerable flexibility in examining the implications of their regression models as well as other types of models (see chapter 11 for a use of `nlcom`). Consequently, users can unpack the models and make much more precise statements about the results than just a binary significant or nonsignificant relationship.

3.4 Conclusions

This chapter has introduced regression and discussed the fundamentals of regression using models with just a single predictor. Chapter 4 expands the regression model to include multiple predictors, categorical predictors, and interactions. This chapter did not cover regression diagnostics, which is a set of analyses and exploratory statistics that help the analyst evaluate the assumptions of regression. Stata has extensive features for exploring the assumptions. Readers wanting to understand the regression diagnostics can consult Cohen et al. (2003), and researchers wanting to understand how to use Stata's diagnostic features can consult [R] **diagnostic plots** (StataCorp 2017a).

4 Regression with categorical and continuous predictors

In chapter 3, we covered regression with a single predictor, also known as bivariate regression. It is more common to include several predictors in a regression equation, which is called multiple regression. This chapter discusses how to interpret models that include several predictors. Additionally, this chapter discusses how to include both continuous and categorical predictors as well has how to examine and probe interactions among predictors.

Specifically, you will learn the following:

1. Why categorical predictors require special treatment in regression.
2. Coding strategies and tools for dealing with categorical predictors.
3. How to interpret regression models with multiple predictors.
4. How to examine interactions between predictors.

Stata commands featured in this chapter

- `regress`: for fitting regression models
- `tabulate`: for producing dummy-coded variables
- `tabstat`: for creating tables of summary statistics
- factor variables: for automatically creating indicator variables and interactions
- `egen` combined with the `std` option: for creating standardized variables
- `margins`: to aid interpretation of regression models
- `marginsplot`: for visualizing regression results, especially models including interactions
- `regress` option `coeflegend`: for obtaining coefficient labels for use with postestimation commands
- `lincom`: for creating linear combinations of parameter estimates
- `regress` option `noconstant`: for suppressing the intercept in separate intercept, separate slope models
- `test`: for evaluating a joint null hypothesis

4.1 Data for this chapter

The data for this chapter describe the performance and winnings of men playing in the Professional Golf Association in 2004.[1] These data will allow us to examine which performance variables predict a golfer's position on the money list. The variables in the dataset are

- `name`: the golfer's name
- `avg_drive`: average drive in yards (that is, how far the golfer hits the ball on his first hit per hole)
- `regulation`: the percentage of greens the golfer reaches in regulation (if the hole is a par 5, then regulation is reaching the green in 3 shots; if the hole is a par 4, then regulation is reaching the green in 2 shots; and if the hole is a par 3, then regulation is reaching the green in 1 shot)
- `avg_putts`: average number of putts per hole
- `money_rank`: rank on the official money list (lower values are better)

---

1. The data are a modified version of the dataset posted at http://users.stat.ufl.edu/~winner/data/pga2004.dat. I dropped some observations to aid the pedagogical aims of the chapter.

- `reg_cat`: a categorical version of `regulation`, with three levels representing the frequency in hitting the green regulation (1 = low, 2 = mid, and 3 = high)

```
. use http://www.stata-press.com/data/pspus/golf_data
. describe
Contains data from golf_data.dta
  obs:             145
 vars:               6                          28 Nov 2018 12:46
 size:           6,090

              storage   display    value
variable name   type    format     label      variable label
name           str22    %22s                  Golfer's name
avg_drive      float    %9.0g                 Average drive in yards
regulation     float    %9.0g                 The percentage of greens reached
                                                in regulation
avg_putts      float    %9.0g                 Average number of putts
money_rank     float    %9.0g                 Rank on the money list
reg_cat        float    %9.0g      regul      Greens in regulation - Category

Sorted by:
```

4.2 Why categorical predictors need special care

In golf, hitting a green in regulation is important to success. Given that the goal of golf is to get the ball in the hole in as few strokes as possible and given that it is easier to get the ball in the hole by putting the ball while on the green, it follows that getting the ball onto the green in a small number of strokes will lead to a better score. Consequently, we should expect the `reg_cat` variable, which categorizes golfers by the frequency with which they reach the green in regulation, to predict `money_rank`.

Suppose we start by predicting `money_rank` from `reg_cat`:

$$\texttt{money\_rank}_i = b_0 + b_1 \texttt{reg\_cat}_i$$

In Stata, we type

```
. table reg_cat
```

Greens in regulation - Category	Freq.
Low	63
Mid	74
High	8

```
. regress money_rank reg_cat

      Source |       SS       df       MS              Number of obs =     145
-------------+------------------------------           F(1, 143)     =   17.29
       Model |  28677.8702        1  28677.8702        Prob > F      =  0.0001
    Residual |  237231.757      143  1658.96334        R-squared     =  0.1078
-------------+------------------------------           Adj R-squared =  0.1016
       Total |  265909.628      144  1846.59464        Root MSE      =   40.73

  money_rank |      Coef.   Std. Err.      t    P>|t|     [95% Conf. Interval]
-------------+----------------------------------------------------------------
     reg_cat |  -23.91609   5.752216    -4.16   0.000    -35.28645   -12.54573
       _cons |   113.395    9.917217    11.43   0.000     93.79176    132.9983
```

The slope for `money_rank` is $b_1 = -23.9$, indicating that for every one-unit difference in `reg_cat`, a golfer's rank on the money list goes down by about 24 spots. This seems simple enough, but there is a problem. The one-unit difference does not make sense for a categorical variable like `reg_cat` because there is not a well-defined one-unit difference. The numbers attached to `reg_cat` are as arbitrary as you can get—we could have coded them as 10, 50, and 150 or just used the words "low", "mid", and "high". We cannot assume that the distance between low and mid is the same as the distance between mid and high or that the distance between low and high is twice that of low to mid or mid to high. Consequently, it is incorrect to use `reg_cat` as a predictor without modification (for example, see section 4.3.1).

4.3 Dummy coding

Categorical variables, such as `reg_cat`, must be transformed into a set of contrast variables. These contrast variables will contain the same information as the categorical variables (that is, the contrast variables will tell us which observations belong in which category) while still allowing us to use the regression methods we know and love. The most common method for creating contrast variables is called dummy coding. Dummy coding is possible for categorical variables with two or more categories and can easily be generated in Stata. Furthermore, Stata has built-in features for creating dummy codes automatically.

Dummy codes are named as such because we create new variables that are placeholders for the levels of the categorical variable. They are just a transformation of the categorical variable and do not provide any new information. The rules for creating dummy codes are

1. For a variable with j levels, create $j-1$ contrasts (that is, dummy variables).
2. Choose one of the j levels as a reference or base level. Which one you choose is arbitrary, but keep track of which level it is for correct interpretation.
3. The contrast variables have only 0s and 1s. Give the comparison level a 1, the reference level a 0, and all unused levels a 0.

4.3 Dummy coding

4. When using dummy codes in a regression model, include all $j - 1$ contrasts in the model.

For example, the `reg_cat` variable has $j = 3$ levels (low, mid, high), and thus we need to create two contrasts. We select the low category to be the reference category. Thus, we create two new variables: `mid_dummy` and `high_dummy`. Any golfer who is in the mid category on `reg_cat` is coded to have a 1 on `mid_dummy`; all other golfers are given a 0 on `mid_dummy`. Any golfer who is in the high category on `reg_cat` is coded to have a 1 on `high_dummy`; all other golfers are given a 0 on `high_dummy`.

This can be done in many ways in Stata. We could use `generate` and `replace`.

```
. generate mid_dummy = 1 if reg_cat == 2
(71 missing values generated)
. replace mid_dummy = 0 if reg_cat != 2
(71 real changes made)
. generate high_dummy = 1 if reg_cat == 3
(137 missing values generated)
. replace high_dummy = 0 if reg_cat !=3
(137 real changes made)
. list name reg_cat mid_dummy high_dummy in 20/40, clean

              name      reg_cat    mid_du~y    high_d~y
 20.    Brett Quigley      Mid          1           0
 21.    Brian Bateman      Mid          1           0
 22.       Brian Gay       Low          0           0
 23.     Briny Baird      High          0           1
 24.   Cameron Beckman     Mid          1           0
 25.   Carl Pettersson     Low          0           0
 26.    Carlos Franco      Mid          1           0
 27.    Chad Campbell      Mid          1           0
 28.  Charles Howell II    Mid          1           0
 29.     Chris DiMarco     Mid          1           0
 30.      Chris Riley      Low          0           0
 31.      Chris Smith     High          0           1
 32.      Corey Pavin      Low          0           0
 33.     Craig Barlow      Mid          1           0
 34.     Craig Bowden      Low          0           0
 35.      Craig Parry      Low          0           0
 36.    Daniel Chopra      Low          0           0
 37.      Danny Ellis      Low          0           0
 38.    Darren Clarke      Low          0           0
 39.    David Peoples      Low          0           0
 40.       David Toms      Mid          1           0
```

Golfers in the low category on `reg_cat` (for example, Brian Gay, Corey Pavin) are coded as 0 on `mid_dummy` and 0 on `high_dummy`; those in the mid category on `reg_cat` (for example, Brett Quigly, Brian Bateman) are coded as 1 on `mid_dummy` and 0 on `high_dummy`; and those in the high category on `reg_cat` (for example, Briny Baird, Chris Smith) are coded as 0 on `mid_dummy` and 1 on `high_dummy`. The two contrast variables, `mid_dummy` and `high_dummy`, contain all the information from `reg_cat`: 1s on `reg_cat` are represented in the contrast variables by a 0 and 0, 2s are represented by a 1 and 0, and 3s are represented by a 0 and 1.

A second method for creating dummy variables only requires **generate**.

```
. generate mid_dummy = reg_cat == 2
. generate high_dummy = reg_cat == 3
. table mid_dummy
```

mid_dummy	Freq.
0	71
1	74

```
. table high_dummy
```

high_dumm y	Freq.
0	137
1	8

This particular syntax for **generate** was made for dummy codes. The line **generate mid_dummy = reg_cat == 2** can be read as saying, "Generate a new variable called **mid_dummy**; set it equal to 1 if **reg_cat** is equal to 2 and equal to 0 otherwise."

A third method for creating dummy variables is to use **tabulate** with the option **generate()**. The **generate()** option will create dummy variables (Stata calls them indicator variables) for all levels of the categorical variables. We must supply a *stubname* for the new variables.

```
. tabulate reg_cat, generate(dummy)
```

Greens in regulation - Category	Freq.	Percent	Cum.
Low	63	43.45	43.45
Mid	74	51.03	94.48
High	8	5.52	100.00
Total	145	100.00	

4.3 Dummy coding

```
. list name reg_cat dummy1 dummy2 dummy3 in 20/40, clean
                 name    reg_cat   dummy1   dummy2   dummy3
 20.     Brett Quigley       Mid        0        1        0
 21.     Brian Bateman       Mid        0        1        0
 22.         Brian Gay       Low        1        0        0
 23.       Briny Baird      High        0        0        1
 24.    Cameron Beckman      Mid        0        1        0
 25.    Carl Pettersson      Low        1        0        0
 26.     Carlos Franco       Mid        0        1        0
 27.     Chad Campbell       Mid        0        1        0
 28.   Charles Howell II     Mid        0        1        0
 29.     Chris DiMarco       Mid        0        1        0
 30.        Chris Riley      Low        1        0        0
 31.        Chris Smith     High        0        0        1
 32.        Corey Pavin      Low        1        0        0
 33.       Craig Barlow      Mid        0        1        0
 34.       Craig Bowden      Low        1        0        0
 35.        Craig Parry      Low        1        0        0
 36.      Daniel Chopra      Low        1        0        0
 37.        Danny Ellis      Low        1        0        0
 38.      Darren Clarke      Low        1        0        0
 39.      David Peoples      Low        1        0        0
 40.         David Toms      Mid        0        1        0
```

Typing `tabulate reg_cat, generate(dummy)` tells Stata to use `dummy` as the *stubname*, and Stata will create three new variables: `dummy1`, `dummy2`, and `dummy3`. These are dummy variables (that is, they only take on the values of 0 and 1) for the three levels of `reg_cat`. Stata will produce one more dummy variable than we need for regression, so when we perform the regression, we need to select two. To match the `mid_dummy` and `high_dummy` variables from above, we select `dummy2` and `dummy3`.

Now that we have created dummy variables, we are set to construct the regression model where rank on the money list is predicted from greens in regulation category. The regression model is

$$\texttt{money\_rank}_i = b_0 + b_1 \texttt{mid\_dummy}_i + b_2 \texttt{high\_dummy}_i$$

We can compare the expected values of `money_rank` to help us understand the interpretation of the coefficients. For example, the expected value of `money_rank` for golfers in the low category is

$$E\left(\texttt{money\_rank}|\texttt{reg\_cat} = \text{low}\right) = b_0 + b_1 \times 0 + b_2 \times 0 = b_0$$

Because golfers in the low category are coded as 0 on both `mid_dummy` and `high_dummy`, the expected value is simply the b_0 (that is, b_1 and b_2 are multiplied by 0). Therefore, the intercept is interpreted as the expected value of the reference category. The expected value for golfers in the mid category is

$$E\left(\texttt{money\_rank}|\texttt{reg\_cat} = \text{mid}\right) = b_0 + b_1 \times 1 + b_2 \times 0 = b_0 + b_1$$

Finally, the expected value for golfers in the high category is

$$E\left(\texttt{money\_rank}|\texttt{reg\_cat} = \text{high}\right) = b_0 + b_1 \times 0 + b_2 \times 1 = b_0 + b_2$$

The slopes, b_1 and b_2, represent the difference between the expected values of the mid category and the low category and the difference between the expected values of the high category and the low category, respectively.

You can prove this to yourself by computing the difference between the expected values. The difference between the expected values of the mid and low categories is

$$E(\texttt{money\_rank}|\texttt{reg\_cat} = \text{mid}) - E(\texttt{money\_rank}|\texttt{reg\_cat} = \text{low}) =$$
$$b_0 + b_1 - b_0 = b_1$$

The difference between the expected values of the high and low categories is

$$E(\texttt{money\_rank}|\texttt{reg\_cat} = \text{high}) - E(\texttt{money\_rank}|\texttt{reg\_cat} = \text{low}) =$$
$$b_0 + b_2 - b_0 = b_2$$

In summary, the coefficients from a regression using dummy coding are interpreted as follows:

- Intercept = the expected value of the reference category.
- Slope = the difference between the category of interest and the reference category.

The category of interest is the group coded with a 1 on the dummy variable.

The means for each reg_cat group as well as the differences mentioned above are

```
. tabstat money_rank, by(reg_cat)
Summary for variables: money_rank
     by categories of: reg_cat (Greens in regulation - Category)

 reg_cat |      mean
---------+----------
     Low |  91.90476
     Mid |  61.43243
    High |     60.75
---------+----------
   Total |  74.63448
------------------

. display "The difference between Mid and Low is " round(61.43243 - 91.90476, .01)
The difference between Mid and Low is -30.47
. display "The difference between High and Low is " round(60.74 - 91.90476, .01)
The difference between High and Low is -31.16
```

4.3 Dummy coding

The regression model shows that the parameter estimates are equal to the values displayed above:

```
. regress money_rank mid_dummy high_dummy
```

Source	SS	df	MS			
Model	33230.5369	2	16615.2684	Number of obs	=	145
Residual	232679.091	142	1638.58515	F(2, 142)	=	10.14
				Prob > F	=	0.0001
				R-squared	=	0.1250
				Adj R-squared	=	0.1126
Total	265909.628	144	1846.59464	Root MSE	=	40.479

money_rank	Coef.	Std. Err.	t	P>|t|	[95% Conf. Interval]	
mid_dummy	-30.47233	6.939188	-4.39	0.000	-44.18979	-16.75487
high_dummy	-31.15476	15.19317	-2.05	0.042	-61.18879	-1.120735
_cons	91.90476	5.09993	18.02	0.000	81.82316	101.9864

The null hypotheses for the significance tests in the output are as follows:

- Intercept: $\beta_0 = 0$ (or the expected value of the reference category is 0 in the population).
- Slope: β_1 or $\beta_2 = 0$ (or there is no difference between the category of interest and the reference category in the population).

The significance tests reported by a simple regression with no additional covariates will be equivalent to a two-sample t test comparing those two groups. Regression often includes additional variables as control variables (which we discuss shortly), which means that the significance tests for the slopes will no longer be equivalent to two-sample t tests.

Stata can build dummy variables on the fly without the need to create new variables as we did above. To create dummy variables on the fly, we can use what Stata calls factor variables. Stata automatically creates the correct number of indicator variables and sets a reference category, which is called the base category. Factor variables can be included in most estimation commands. For example, we can rerun our analysis using factor variables by replacing mid_dummy and high_dummy with i.reg_cat. The i. before reg_cat signals to Stata that we want to create indicator variables, which is Stata's term for dummy variables. The regress command is

```
. regress money_rank i.reg_cat
      Source |       SS           df       MS      Number of obs   =       145
-------------+------------------------------        F(2, 142)       =     10.14
       Model |  33230.5369         2  16615.2684    Prob > F        =    0.0001
    Residual |  232679.091       142  1638.58515    R-squared       =    0.1250
-------------+------------------------------        Adj R-squared   =    0.1126
       Total |  265909.628       144  1846.59464    Root MSE        =    40.479

  money_rank |      Coef.   Std. Err.      t    P>|t|     [95% Conf. Interval]
-------------+----------------------------------------------------------------
     reg_cat |
         Mid |  -30.47233   6.939188    -4.39   0.000    -44.18979   -16.75487
        High |  -31.15476   15.19317    -2.05   0.042    -61.18879   -1.120735
             |
       _cons |   91.90476   5.09993    18.02   0.000     81.82316    101.9864
```

This estimates the same regression model as before. Stata chose the low category from `reg_cat` to be the base category. Unless we tell it otherwise, Stata will choose the category with the lowest numerical value—the low category had a 1 on `reg_cat` and thus was chosen. To select a different reference category, we can change `i.reg_cat` to `ib#.reg_cat`, where we replace # with the value on `reg_cat` that we want to be the base category. Thus, if we want to use the high category as the base, we would type `ib3.reg_cat`.

```
. regress money_rank ib3.reg_cat
      Source |       SS           df       MS      Number of obs   =       145
-------------+------------------------------        F(2, 142)       =     10.14
       Model |  33230.5369         2  16615.2684    Prob > F        =    0.0001
    Residual |  232679.091       142  1638.58515    R-squared       =    0.1250
-------------+------------------------------        Adj R-squared   =    0.1126
       Total |  265909.628       144  1846.59464    Root MSE        =    40.479

  money_rank |      Coef.   Std. Err.      t    P>|t|     [95% Conf. Interval]
-------------+----------------------------------------------------------------
     reg_cat |
         Low |   31.15476   15.19317     2.05   0.042     1.120735    61.18879
         Mid |   .6824324   15.0654      0.05   0.964    -29.09901    30.46387
             |
       _cons |      60.75   14.31164     4.24   0.000     32.45859    89.04141
```

As you can see, altering the base level changes the coefficients. Given the interpretations listed above, this makes sense.

I strongly recommend that you use factor variables when possible. I think it is important to learn to create dummy variables manually to aid understanding of what Stata is doing. However, factor variables make it simpler to create interaction variables and to write readable code (see section 4.5.1). Further, factor variables are needed for some postestimation commands, such as `margins`.

4.3.1 Example: Incorrect use of categorical variable

Stata and other software typically assume that variables are numeric unless we indicate that they are categorical. As demonstrated above, using `reg_cat` as a predictor in `regress` causes Stata to treat `reg_cat` as a continuous variable. This mistake does not just show up in books teaching about regression. I see this mistake regularly in homework assignments or in consultations with other researchers. This mistake also occasionally shows up in the published literature.

Decety et al. (2015) examined the relationship between altruism and religious participation among 1,170 children from six countries. Altruism was assessed via a task during which the participants could share stickers with other children who the participants believed did not receive stickers. A key outcome was the number of stickers each participant shared. One of their key predictors was religious identification dummy coded as 1 for religious and 0 for nonreligious. Additional predictors were participant's age, birth sex, socioeconomic status, and country of origin. Their results suggested a negative relationship between religiousness and altruism—kids from homes that identify as religious had lower rates of sharing. Decety et al. (2015) conclude: "Our findings robustly demonstrate that children from households identifying as either of the two major world religions (Christianity and Islam) were less altruistic than children from nonreligious households" (p. 2952).

Their analysis had a major problem, however. Shariff et al. (2016) noted that Decety et al. (2015) include country of origin as a continuous variable rather than a categorical variable:

> But when they included their categorically coded country (1 = US, 2 = Canada, and so on) in their models, it was entered not as fixed effects, with dummy variables for all of the countries except one, but as a continuous measure. This treats the variable as a measure of "country-ness" (for example, Canada is twice as much a country as the US) instead of providing the fixed effects they explicitly intended. (p. R699)

This proved to be a critical mistake. Shariff et al. (2016) reran the analysis but treated country as a categorical variable rather than a continuous variable. In the reanalysis, religious identification was not a statistically significant predictor. They also ran several variants of the model to see if alternative ways of examining the data produced a relationship. The other models were consistent with no relationship. The incorrect use of country in the original analysis masked critical differences among countries. Specifically, Turkey and South Africa had low levels of sharing as compared to other countries (Shariff et al. 2016).

The moral of this story is to be careful with statistical models and carefully consider the meaning of the variables in models. A lot of statistical software has features to keep us from making mistakes or that alert us when something is wrong. Nevertheless, software cannot divine our intentions for a variable and sometimes will give us answers that are not useful. I often ask my students to explicitly interpret every estimate in their output, even when the interpretations will not make it in their articles. This can help us identify problems in our models and avoid including variables that represent the "country-ness" of countries.

4.4 Multiple predictors

A golfer's rank on the money list is not dictated solely by how often that golfer hit a green in regulation. Two additional variables that could predict winnings are the average number of putts a golfer takes per hole (avg_putts) and how far, on average, the golfer hits his drives (avg_drive). We can create a more complicated regression that considers these variables. An initial model could look like

$$\text{money\_rank}_i = b_0 + b_1 \text{mid\_dummy}_i + b_2 \text{high\_dummy}_i$$
$$+ b_3 \text{avg\_putts}_i + b_4 \text{avg\_drive}_i$$

The Stata code and output for this model is

```
. regress money_rank i.reg_cat avg_putts avg_drive
```

Source	SS	df	MS		Number of obs	=	145
					F(4, 140)	=	10.85
Model	62934.7833	4	15733.6958		Prob > F	=	0.0000
Residual	202974.844	140	1449.82032		R-squared	=	0.2367
					Adj R-squared	=	0.2149
Total	265909.628	144	1846.59464		Root MSE	=	38.077

money_rank	Coef.	Std. Err.	t	P>|t|	[95% Conf.	Interval]
reg_cat						
Mid	-33.44758	6.617995	-5.05	0.000	-46.53171	-20.36345
High	-45.42503	14.95489	-3.04	0.003	-74.99166	-15.85841
avg_putts	718.8169	159.0554	4.52	0.000	404.3559	1033.278
avg_drive	.0407023	.3920221	0.10	0.917	-.7343465	.8157511
_cons	-1190.963	299.3962	-3.98	0.000	-1782.885	-599.0402

4.4.1 Interpretation

Model fit

The top portion of the regress output is the same as it was with bivariate regression. The F test is used to examine the overall model. However, the difference is that the F test examines the null hypothesis that there is no relationship between the set of predictors and the outcome. In this example, the null hypothesis is that a golfer's category for greens in regulation, average number of putts, and average driving distance are not related to rank on the money list. For this model, $F(4, 140) = 10.6$, $p < 0.001$. Thus, we reject the null hypothesis.

4.4.1 Interpretation

R^2 is also used with multiple regression and has the same interpretation as before, except now we consider all predictors. Specifically, R^2 is interpreted as the proportion of variance in the outcome accounted for by all the predictors. In this example, $R^2 = 0.24$, which means that a golfer's category for greens in regulation, average number of putts, and average driving distance account for 24% of the variability in rank on the money list. R^2 should increase as we add predictors to the model—unless we add predictors that are unrelated to the outcome or that are redundant (see section 4.4.2).

Finally, the root mean squared error is the same as before and represents the standard deviation of the errors in prediction. The root mean squared error should get smaller as we add predictors, if the predictors have a relationship and are not redundant.

Intercept

The intercept is interpreted as before. Namely, it is the expected value of `money_rank` when all other predictors are 0. In other words, a golfer in the low greens in regulation category (that is, the reference group is 0 on both dummy variables), who averages no putts per hole, and averages 0 yards per drive is expected to be ranked $-1{,}191$ on the money list. We can verify that this is the correct interpretation by producing the expected value for this group:

$$E(\texttt{money\_rank}|\texttt{reg\_cat} = \text{low}, \texttt{avg\_putts} = 0, \texttt{avg\_drive} = 0) =$$
$$b_0 + b_1 \times 0 + b_2 \times 0 + b_3 \times 0 + b_4 \times 0 = b_0$$

This is weird, but it is the correct interpretation of the estimate. The problem is that this prediction is beyond the observed data—no one averages 0 putts and 0 yards per drive.[2] Consequently, we get a funny expected value.

A method for obtaining a more interpretable intercept is to use mean centering, which rescales variables so that the mean is equal to 0. This is done by taking the observed score on a variable and subtracting the mean. Thus, the mean-centered putts and average drive are

$$\texttt{c\_putts}_i = \texttt{avg\_putts}_i - \overline{\texttt{avg\_putts}}$$
$$\texttt{c\_drive}_i = \texttt{avg\_drive}_i - \overline{\texttt{avg\_drive}}$$

If we fit the regression with the centered variables rather than the original variables, the intercept will be the expected value for a golfer in the low greens in regulation category, who has the average number of putts, and who has the average number of yards per drive.

Note that creating a z score is also centering, except that we also divide by the standard deviation. Nevertheless, a z score has a mean equal to 0. Consequently, it is common to use z scores as a method of centering, which is what we do in this chapter.

---

[2] I have to admit, I have had a few rounds of golf where I might as well have not been hitting it anywhere on my drives.

The `std()` function from `egen` can be used to create z scores for `avg_putts` and `avg_drive`.

```
. egen z_putts = std(avg_putts)
. egen z_drive = std(avg_drive)
```

Reestimate the regression model with the new variable.

```
. regress money_rank i.reg_cat z_putts z_drive
```

Source	SS	df	MS			
Model	62934.7835	4	15733.6959	Number of obs	=	145
Residual	202974.844	140	1449.82031	F(4, 140)	=	10.85
				Prob > F	=	0.0000
				R-squared	=	0.2367
				Adj R-squared	=	0.2149
Total	265909.628	144	1846.59464	Root MSE	=	38.077

money_rank	Coef.	Std. Err.	t	P>|t|	[95% Conf. Interval]	
reg_cat						
Mid	-33.44758	6.617995	-5.05	0.000	-46.53171	-20.36345
High	-45.42504	14.95489	-3.04	0.003	-74.99166	-15.85841
z_putts	14.69217	3.250992	4.52	0.000	8.264781	21.11955
z_drive	.3392129	3.267109	0.10	0.917	-6.120037	6.798463
_cons	94.21049	4.863184	19.37	0.000	84.59571	103.8253

A golfer who a) is in the low greens in regulation category, b) has the average number of putts, and c) has the average number of yards per drive is expected to be $b_0 = 94.2$ (that is, 94th) on the money list.

Slopes

The interpretation of the slopes in a model with multiple predictors is the same as before with one important exception: the slopes are now referred to as partial slopes. This means that the coefficient for `high_dummy` is the difference between the high and low categories on `reg_cat` while holding the other variables in the model constant. Look back at the model with putts and driving distance in their original metric. The slope for `avg_putts` ($b_4 = 719$) can be interpreted as follows: a one-unit difference in average drive length is associated with being 719 spots higher on the money list, holding constant the greens in regulation category and average number of putts.

Holding constant means what it sounds like. Consider again the slope for `avg_putts` ($b_4 = 719$). Holding constant the other variables means that regardless of whether a golfer is in the low, mid, or high category for greens in regulation and regardless of how many putts the golfer averages, a one-unit difference between golfers in the number of putts per hole is associated with being 719 spots higher on the money ranking. The interpretation of `avg_drive` ($b_3 = 0.04$) is that regardless of whether a golfer is in the low, mid, or high category for greens in regulation and regardless of how many putts the golfer averages, a one-unit difference in drive length is associated with being 0.04 spots higher on the money ranking.

4.4.2 Unique variance

An increase of 719 spots is implausible given that there are only 145 observations in the dataset. The problem arises because of the scaling of `avg_putts`.

```
. summarize avg_putts, detail
                    Average number of putts

         Percentiles      Smallest
  1%        1.724          1.723
  5%        1.74           1.724
 10%        1.746          1.733      Obs                 145
 25%        1.758          1.736      Sum of Wgt.         145

 50%        1.769                     Mean           1.771607
                          Largest     Std. Dev.       .0204394
 75%        1.787          1.811
 90%        1.799          1.811      Variance        .0004178
 95%        1.807          1.819      Skewness        .199207
 99%        1.819          1.829      Kurtosis       2.728189
```

The average number of putts per hole ranges from 1.72 to 1.83, with a mean of 1.77. However, the slope for `avg_putts` ($b_4 = 719$) is scaled assuming a difference between 1 to 2 putts (or 2 to 3, or any other one-unit difference). That is why the slope is so large—it is assuming a change in `avg_putts` that exceeds what is observed in the data.

Fortunately, the solution to this problem is the same as the solution to dealing with the intercept. Rescaling `avg_putts` and using the z scores means that a one-unit difference in z score for putts is equal to a one-standard-deviation difference in putts, which is 0.02 putts per hole. The regression model using z scores for putts and drive length produces a more understandable slope for putts: a one-unit difference in `z_putts` is associated with being 14.7 spots higher on the money list, holding the other variables constant.

Note that using the z scores does not affect the model fit at all. Look at the analysis of variance table, F test, and R^2 values for both regression models—they are identical. Therefore, we can conclude that rescaling does not affect the predictive relationship among the variables; it simply produces more interpretable results.

4.4.2 Unique variance

The ideal kinds of predictors in a regression model are predictors that have a strong relationship with the outcome and are not strongly correlated with one another. If we use predictors that are also strongly related to one another, we do not gain much benefit from adding predictors. Strongly correlated predictors are redundant with one another, and redundant information does not help with prediction. Essentially, the redundant information gets discarded in regression. Therefore, if we include two predictors that are strongly related, the model will be only slightly better than a model with just one of the predictors. In contrast, if we include two predictors that are unrelated to one another—predictors that have no redundancy—then the model with two predictors will be better than the model with just one.

If we compare R^2 from two models, model 1 and model 2, where model 2 has more predictors than model 1, then the increase in R^2 from model 1 to model 2 represents the unique variance in the outcome that the additional predictors account for. Any redundant information in the additional predictors does not get counted. The R^2 from the model with just `mid_dummy` and `high_dummy` was 0.13, and the R^2 from the model that added `z_putts` and `z_drive` was 0.24. Thus, the unique information contained in a golfer's putting and driving distance improves the model.

Figure 4.1 shows how R^2 is affected by the correlation between predictors. The regression models underlying this graph have the form

$$y_i = \beta_0 + \beta_1 x_1 + \beta_2 x_2 \tag{4.1}$$

The x axis represents the correlation between the predictors, x_1 and x_2. The three lines present the correlation between y and each predictor. The correlation between y and x_1 was fixed to be the same as the correlation between y and x_2; that is, $r_{yx_1} = r_{yx_2}$. When the predictors are uncorrelated, $R^2 = r_{yx_1}^2 + r_{yx_2}^2$. For example, if $r_{yx_1} = r_{yx_2} = 0.6$, then R^2 from (4.1) is $0.36 + 0.36 = 0.72$. As the predictors are more correlated, moving left to right on the graph, R^2 goes down because the predictors do not provide as much unique information.

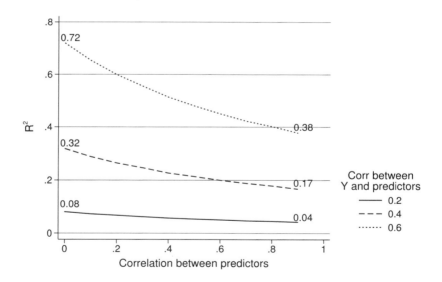

Figure 4.1. The impact of correlation among two predictors on R^2

4.5 Interactions

The models considered thus far have assumed that the relationships between greens in regulation, average number of putts, and money rank are additive; that is, we can take the relationship between greens in regulation and money rank and add to it the relationship between average number of putts and money rank. Regardless of which greens in regulation category the golfer is in, the additional effect of number of putts will be the same: a 1-standard-deviation difference in average putts is associated with a $b_3 = 14.7$ point difference in rank on the money list. This is what it means to hold a variable constant.

It is also possible for two or more predictors to have a multiplicative relationship with the outcome. That is, it is possible for the relationship between a predictor and the outcome to be different for specific values of another predictor. When this happens, we say that there is an interaction between predictors or that one predictor moderates the relationship of another predictor with the outcome.

For example, it is possible that the relationship between the average number of putts and money rank is particularly strong for golfers who often reach the green in regulation. However, if a golfer does not hit the green in regulation as often, the number of putts may be less important for money rank. Using only one putt when a golfer hits the green in regulation means the golfer made birdie; using one putt when a golfer takes one additional shot to hit the green in regulation means the golfer made par. All things being equal, golfers who make more birdies than pars are going to earn more money. Therefore, it is the combination of hitting a green in regulation and using few putts that earns golfers money and improves their rank on the money list.

4.5.1 Categorical by continuous interactions

Dichotomous by continuous interactions

Regression can be adapted to include interactions by adding an additional term to the model. We will start by simplifying the model in two ways. These simplifications are not required for interactions but will help facilitate the explanation. First, we are going to remove driving distance. Second, we are going to reduce reg_cat to two levels by combining the mid and high levels into a single level. Thus, reg_cat will be represented by a single dummy variable, which we will call reg_cat2, that is 1 for the mid and high levels and 0 for the low level.

```
. recode reg_cat (1 = 0) (2 = 1) (3 = 1), gen(reg_cat2)
(145 differences between reg_cat and reg_cat2)
. label define regul2 0 "Low" 1 "High"
. label values reg_cat2 regul2
. label variable reg_cat2 "Greens in Regulation -- Recoded"
```

The regression model without the interaction is

$$\texttt{money\_rank}_i = b_0 + b_1 \texttt{reg\_cat2} + b_2 \texttt{z\_putts} \tag{4.2}$$

The **regress** output for this model is

```
. regress money_rank i.reg_cat2 c.z_putts
```

Source	SS	df	MS		Number of obs	=	145
					F(2, 142)	=	21.57
Model	61952.4176	2	30976.2088		Prob > F	=	0.0000
Residual	203957.21	142	1436.31838		R-squared	=	0.2330
					Adj R-squared	=	0.2222
Total	265909.628	144	1846.59464		Root MSE	=	37.899

money_rank	Coef.	Std. Err.	t	P>|t|	[95% Conf. Interval]	
reg_cat2						
High	-34.37959	6.407211	-5.37	0.000	-47.04543	-21.71374
z_putts	14.25238	3.186992	4.47	0.000	7.952302	20.55246
_cons	94.07673	4.799436	19.60	0.000	84.58915	103.5643

Thus, those in the high greens in regulation category were expected to be about 34 spots less than those in the low category, holding constant average number of putts. Likewise, a 1-standard-deviation increase in average number of putts is associated with being 14 spots higher on the money list, holding greens in regulation constant (this last bit is important because it means that regardless of which greens in regulation category a golfer is in, the relationship between the number of putts and money is the same). We can use **margins** and **marginsplot** to create predicted scores and visualize the relationship.

```
. margins, at(z_putts = (-2(1)2) reg_cat2 = (0 1))
Adjusted predictions                              Number of obs    =    145
Model VCE     : OLS

Expression    : Linear prediction, predict()
1._at         : reg_cat2         =           0
                z_putts          =          -2

2._at         : reg_cat2         =           0
                z_putts          =          -1

3._at         : reg_cat2         =           0
                z_putts          =           0

4._at         : reg_cat2         =           0
                z_putts          =           1

5._at         : reg_cat2         =           0
                z_putts          =           2

6._at         : reg_cat2         =           1
                z_putts          =          -2

7._at         : reg_cat2         =           1
                z_putts          =          -1

8._at         : reg_cat2         =           1
                z_putts          =           0

9._at         : reg_cat2         =           1
                z_putts          =           1

10._at        : reg_cat2         =           1
                z_putts          =           2
```

4.5.1 Categorical by continuous interactions

	Margin	Delta-method Std. Err.	t	P>|t|	[95% Conf. Interval]	
_at						
1	65.57197	7.580954	8.65	0.000	50.58585	80.55808
2	79.82435	5.485965	14.55	0.000	68.97964	90.66907
3	94.07673	4.799436	19.60	0.000	84.58915	103.5643
4	108.3291	6.023886	17.98	0.000	96.42103	120.2372
5	122.5815	8.35785	14.67	0.000	106.0596	139.1034
6	31.19238	7.93976	3.93	0.000	15.49697	46.88778
7	45.44476	5.4946	8.27	0.000	34.58298	56.30655
8	59.69714	4.201823	14.21	0.000	51.39094	68.00335
9	73.94953	5.043196	14.66	0.000	63.98008	83.91897
10	88.20191	7.316159	12.06	0.000	73.73925	102.6646

```
. marginsplot, noci scheme(lean2)
>         legend(order(1 "Low" 2 "High" 3 "y") pos(6) row(1))
>         legend(subtitle("Greens in regulation category"))
>         title("") ytitle("Predicted money rank")
>         xtitle("Average number of putts - {it:z} score")
>         addplot((pcarrowi 62 0 92 0) (pcarrowi 92 0 62 0))
>         text(89 -0.1 "b{subscript:0}")
>         text(77 -0.1 "b{subscript:1}")
>         text(63 -0.2 "b{subscript:0} + b{subscript:1}")
  Variables that uniquely identify margins: z_putts reg_cat2
. gr_edit .legend.plotregion1.key[3].draw_view.setstyle, style(no)
. gr_edit .legend.plotregion1.key[4].draw_view.setstyle, style(no)
. gr_edit .legend.plotregion1.label[3].draw_view.setstyle, style(no)
. gr_edit .legend.plotregion1.label[3].fill_if_undrawn.setstyle, style(no)
. gr_edit .legend.plotregion1.label[4].draw_view.setstyle, style(no)
. gr_edit .legend.plotregion1.label[4].fill_if_undrawn.setstyle, style(no)
. gr_edit .legend.plotregion1.DragBy 0 10
```

Figure 4.2[3] shows that the greens in regulation categories have different intercepts, where the intercept for the low category is $b_0 = 94$, the intercept for the high category is $b_0 + b_1 = 60$, and the difference is $b_1 = -34$ [see (4.2)]. The slope for putts is the same for both groups: $b_1 = 14$.

---

3. Because of some challenges with formatting the legend of this graph, I had to use the Graph Editor and record the changes made. For more details on how to do this in your own analyses, type **help graph editor**.

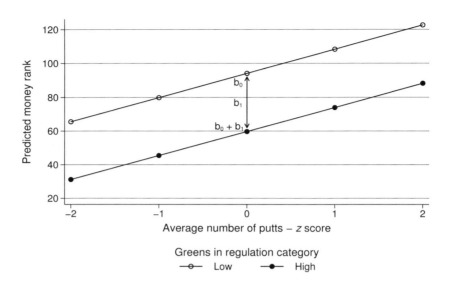

Figure 4.2. Visualization of no interaction between average number of putts and greens in regulation category

If we believe that the relationship between putts and money rank is different for the high and low groups, then we can use an interaction. The regression model becomes

$$\texttt{money\_rank}_i = b_0 + b_1 \texttt{reg\_cat2} + b_2 \texttt{z\_putts} \qquad (4.3)$$
$$+ b_3 \texttt{reg\_cat2} \times \texttt{z\_putts}$$

The interaction is just the product of `reg_cat2` and `z_putts`. This can be created using `generate` (for example, `generate high_putts = reg_cat2*z_putts`). My preference is to use factor-variable notation because Stata will create the interaction on the fly.

4.5.1 Categorical by continuous interactions

```
. regress money_rank i.reg_cat2 c.z_putts i.reg_cat2#c.z_putts

      Source |       SS           df       MS      Number of obs   =       145
-------------+----------------------------------   F(3, 141)       =     17.46
       Model |  72020.3009         3   24006.767   Prob > F        =    0.0000
    Residual |  193889.327       141  1375.10161   R-squared       =    0.2708
-------------+----------------------------------   Adj R-squared   =    0.2553
       Total |  265909.628       144  1846.59464   Root MSE        =    37.082

    money_rank |      Coef.   Std. Err.      t    P>|t|     [95% Conf. Interval]

      reg_cat2 |
         High  | -33.82358   6.272551    -5.39   0.000    -46.22399   -21.42318
       z_putts |  4.810639   4.679735     1.03   0.306    -4.440878    14.06215

     reg_cat2#|
      c.z_putts|
         High  |  16.98219   6.276132     2.71   0.008     4.57471    29.38968

         _cons |  92.63787   4.726056    19.60   0.000    83.29478    101.981
```

The additional term i.reg_cat2#c.z_putts creates the interaction term.[4]

We can return to expected values to unpack what the coefficients mean. The expected value for a golfer in the low greens in regulation category is

$$E(\texttt{money\_rank}|\texttt{reg\_cat2} = \text{low}) =$$
$$b_0 + b_1 \times 0 + b_2 \texttt{z\_putts} + b_3 0 \times \texttt{z\_putts} = b_0 + b_2 \texttt{z\_putts} \quad (4.4)$$

Thus, b_0 is the expected rank on the money list for a golfer in the low category for greens in regulation and who has a 0 on z_putts (that is, who is at the average). Compare (4.4) with the expected value for the high category:

$$E(\texttt{money\_rank}|\texttt{reg\_cat2} = \text{high}) =$$
$$b_0 + b_1 \times 1 + b_2 \texttt{z\_putts} + b_3 1 \times \texttt{z\_putts} = b_0 + b_1 + b_2 \texttt{z\_putts} + b_3 \texttt{z\_putts} \quad (4.5)$$

Simplifying (4.5), we get

$$E(\texttt{money\_rank}|\texttt{reg\_cat2} = \text{high}) = b_0 + b_1 + (b_2 + b_3)\texttt{z\_putts} \quad (4.6)$$

Thus, $b_0 + b_1$ is the expected rank on the money list for a golfer in the high category for greens in regulation and who has a 0 on z_putts, which means that b_1 is the difference between intercepts between the high and low categories:

$$E(\texttt{money\_rank}|\texttt{reg\_cat2} = \text{high}, \texttt{z\_putts} = 0)-$$
$$E(\texttt{money\_rank}|\texttt{reg\_cat2} = \text{low}, \texttt{z\_putts} = 0) =$$
$$(b_0 + b_1) - b_0 = b_1$$

---

[4]. There is another option. The syntax i.reg_cat2##c.z_putts is identical to i.reg_cat2 c.z_putts and i.reg_cat2#c.z_putts. Consequently, the syntax regress money_rank i.reg_cat2##c.z_putts can also be used to fit the same model as above. Although the syntax I use in the main text requires more typing, I prefer it for learning as well as day-to-day use because it forces me to be explicit regarding the model I'm fitting. Also, when reviewing syntax at a later date, I think it is easier to quickly see the model without having to expand the syntax in my head.

Equation (4.4) indicates that b_2 is the relationship between putts and money rank for golfers in the low category. Equation (4.6) indicates that $b_2 + b_3$ is the relationship between putts and money rank for golfers in the high category. Consequently, b_3 is the difference in the slope for z_putts between the high and the low category. The null hypothesis for b_3 is that there is no difference between the slopes in each category in the population; that is, the relationship between putts and money rank does differ across the greens in regulation category. Given that b_3 is statistically significant, we reject that null hypothesis and conclude that there is a difference in the relationship between money rank and number of putts between the high and low categories.

This discussion underscores that regression coefficients have a unique interpretation in the presence of an interaction. In (4.2), b_1 represents the difference between the high and low categories, holding constant average number of putts. In (4.3), b_1 represents the difference between the high and low categories when the average number of putts is 0. Because we standardized number of putts, the 0 represents the mean. This is another motivation for standardizing continuous variables when interactions are included.

Coefficients such as b_1 and b_2 from (4.3) are called first-order coefficients because they involve variables that are part of an interaction. Interactions are called higher-order coefficients: second order if it is a two-way interaction, third order if a three-way interaction, and so on (Aiken and West 1991). First-order coefficients should not be interpreted the same way as coefficients from a regression that does not include an interaction. Doing so leads to a misinterpretation and is a common mistake. The motivation for including an interaction is to test whether the relationship between variables is influenced by the specific values of another variable. The interpretation of first-order variables includes specific levels of another variable. Thus, the seemingly peculiar interpretation of first-order variables is consistent with the substantive aims that lead to using interactions in the first place.

Including an interaction only affects coefficients involving variables that are part of an interaction. If the regression model includes additional covariates, those coefficients are interpreted normally. For example, if we extended (4.3) to include driving distance,

$$\text{money\_rank}_i = b_0 + b_1 \text{high\_dummy2} + b_2 \text{z\_putts} \\ + b_3 \text{high\_dummy2} \times \text{z\_putts} + b_4 \text{z\_drive}$$

then b_4 is interpreted as the expected difference in money rank given a one-unit difference in (standardized) driving distance, holding constant greens in regulation and average number of putts.

There are multiple methods for interpreting and probing interactions. Most commonly, researchers create simple regression equations (Aiken and West 1991), which are regression equations describing the specific relationships represented by the interaction model. If we have a regression model involving y predicted by x, z, and their interaction, then we can form several simple regression equations. We could form an equation where y is regressed on x at specific levels of z. Another way to say this is y regressed on x conditional on the value of z. We could also form the compliment: y regressed

4.5.1 Categorical by continuous interactions

on z conditional on the value of x. In the golf example, we can form equations for `money_rank` regressed on `z_putts` for either the high or the low group of `reg_cat2`. For the low group, the simple regression equation is (4.4); for the high group, the simple regression equation is (4.6).

Visualizing the simple regression equations, and thus the interaction, is straightforward with `margins` and `marginsplot`. With `margins`, we create the expected values of `money_rank` for the high and low categories of `reg_cat2` as well as for a range of `z_putts` values. We then use `marginsplot` to visualize the lines.

```
. quietly margins, at(z_putts = (-2(1)2) reg_cat = (0 1))
. marginsplot, noci scheme(lean2)
>         legend(order(1 "Low" 2 "High") pos(6) row(1))
>         legend(subtitle("Greens in regulation category"))
>         title("") ytitle("Predicted money rank")
>         xtitle("Average number of putts - {it:z} score")
>         text(20 0 "y{superscript:´} = b{subscript:0}
>         + b{subscript:1}high_dummy2 + b{subscript:2}z_putts
>         + b{subscript:3}high_dummy2 x z_putts")
>         text(80 -1.5 "y{superscript:´} = b{subscript:0}
>         + b{subscript:2}z_putts")
  Variables that uniquely identify margins: z_putts reg_cat2
```

Figure 4.3 shows the results of `marginsplot`. I added the simple regression equation for each line as reference. This figure underscores what the coefficient and significance test for b_3 indicated, which is that the relationship between money rank and average number of putts is different for the high and low categories. Specifically, when a golfer is in the low category for greens in regulation, being below average on number of putts is only slightly associated with money rank. However, when a golfer is in the high category for greens in regulation, being below average on number of putts is fairly strongly associated with an improvement in money rank. This figure, coupled with the significant interaction, suggests it is important to model this interaction when considering the relationship between putting and money rank.

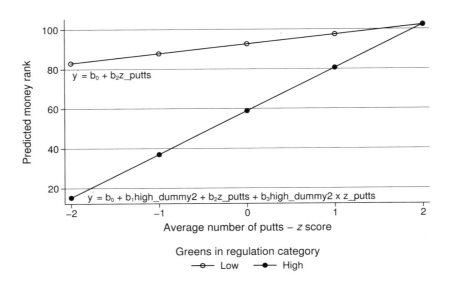

Figure 4.3. Plot of the simple regression equations describing the relationship between `money_rank` and `z_putts` conditional on `reg_cat2`

What if we want to know the specific intercept and slope values for the high and low categories? For the low category, the `regress` output produces them; they are just b_0 and b_1. For the high category, we add b_0 to b_1 for the intercept and add b_2 to b_3 for the slope. However, that will not provide us with a significance test or confidence intervals for either coefficient because we do not have a standard error for either. We have two options to produce the coefficients and standard errors for the high category: use `lincom` to construct them or reparameterize the regression equation to estimate them directly.

4.5.1 Categorical by continuous interactions

Using `lincom` requires that we know the coefficient name for each parameter. Adding the `coeflegend` option to `regress` will print the coefficient names.

```
. regress money_rank i.reg_cat2 c.z_putts i.reg_cat2#c.z_putts, coeflegend

      Source |       SS           df       MS      Number of obs   =       145
-------------+----------------------------------   F(3, 141)       =     17.46
       Model |  72020.3009         3   24006.767   Prob > F        =    0.0000
    Residual |  193889.327       141  1375.10161   R-squared       =    0.2708
-------------+----------------------------------   Adj R-squared   =    0.2553
       Total |  265909.628       144  1846.59464   Root MSE        =    37.082

  money_rank |      Coef.   Legend
-------------+----------------------------------
    reg_cat2 |
       High  | -33.82358    _b[1.reg_cat2]
     z_putts |  4.810639    _b[z_putts]
             |
    reg_cat2#|
   c.z_putts |
       High  |  16.98219    _b[1.reg_cat2#c.z_putts]
             |
       _cons |  92.63787    _b[_cons]
```

lincom, which stands for linear combination, means that we want to add two or more parameters together. The intercept for the high category is $b_0^* = b_0 + b_1$, which are labeled _b[_cons] and _b[2.reg_cat2], respectively. The slope for the high category is $b_2^* = b_2 + b_3$, which are labeled _b[z_putts] and _b[2.reg_cat2#c.z_putts], respectively. The lincom syntax is

```
. lincom _b[_cons] + _b[1.reg_cat2]   //intercept -- b0 + b1
 ( 1)  1.reg_cat2 + _cons = 0

  money_rank |      Coef.   Std. Err.      t    P>|t|     [95% Conf. Interval]
-------------+----------------------------------------------------------------
         (1) |   58.81429   4.124232    14.26   0.000     50.66096    66.96761

. lincom _b[z_putts] + _b[1.reg_cat2#c.z_putts]   //slope -- b2 + b3
 ( 1)  z_putts + 1.reg_cat2#c.z_putts = 0

  money_rank |      Coef.   Std. Err.      t    P>|t|     [95% Conf. Interval]
-------------+----------------------------------------------------------------
         (1) |   21.79283   4.182094     5.21   0.000     13.52512    30.06054
```

The output provides the coefficient, standard error, significance test, and confidence interval for each linear combination (see Aiken and West [1991] for an explanation of how to generate the standard errors for these coefficients). Thus, the simple regression equation for the high category is

$$E(\texttt{money\_rank}|\texttt{reg\_cat2} = \text{high}) = 58.8 + 21.8 \texttt{z\_putts}$$

An alternative method is to reformulate the regression equation to estimate the parameters directly. This can be done by suppressing the intercept for the model. In (4.3), we estimate an intercept, b_0, and a difference from the intercept, b_1. We also estimate a slope, b_2, and a difference from the slope, b_3. We want to reformulate the model so that we estimate two distinct intercepts and two distinct slopes rather than an intercept, a slope, and two differences. This is called a separate intercepts, separate slopes model.[5] The model is

$$\texttt{money\_rank} = b_0 \texttt{reg\_cat2} = 0 + b_0^* \texttt{reg\_cat2} = 1$$
$$+ b_2 \texttt{z\_putts} \times \texttt{reg\_cat2} = 0 + b_2^* \texttt{z\_putts} \times \texttt{reg\_cat2} = 1$$

where b_0 and b_2 are the intercept and the slope for the low category, and b_0^* and b_2^* are the intercept and the slope for the high category. This model appears odd because it includes both dummy variables for `reg_cat2`, which is contrary to the rules about dummy variables. Likewise, the model also includes two interactions. Note that all coefficients are multiplied by at least one predictor; that is, there is not an overall intercept. Nor is there a coefficient for `z_putts` only. Rather, we estimate two distinct intercepts and two distinct slopes. There are still four parameter estimates, just as there were in (4.3), so we have reexpressed the model.

We can fit this model with `regress` by suppressing the overall intercept via the `noconstant` option and by suppressing the reference level in the factor variables (that is, by using the `ibn.` prefix).

```
. regress money_rank ibn.reg_cat2 ibn.reg_cat2#c.z_putts, nocons
```

Source	SS	df	MS		Number of obs	=	145
					F(4, 141)	=	159.94
Model	879714.673	4	219928.668		Prob > F	=	0.0000
Residual	193889.327	141	1375.10161		R-squared	=	0.8194
					Adj R-squared	=	0.8143
Total	1073604	145	7404.16552		Root MSE	=	37.082

money_rank	Coef.	Std. Err.	t	P>|t|	[95% Conf. Interval]	
reg_cat2						
Low	92.63787	4.726056	19.60	0.000	83.29478	101.981
High	58.81429	4.124232	14.26	0.000	50.66096	66.96761
reg_cat2# c.z_putts						
Low	4.810639	4.679735	1.03	0.306	-4.440878	14.06215
High	21.79283	4.182094	5.21	0.000	13.52512	30.06054

Using the `ibn.` prefix for `reg_cat2` signals to Stata that we want to treat `reg_cat2` as a categorical variable but we do not want a reference category. This combined with the `noconstant` option will ensure that we obtain separate intercepts for each group.

---

5. Readers familiar with structural equation modeling may recognize this model as a multiple-group model. In fact, it would be identical to a fully unconstrained multiple-group model if we estimate separate residual errors for the low and high categories.

4.5.1 Categorical by continuous interactions

If we forget `noconstant`, Stata will estimate an intercept and fix one of the regression coefficients to 0. To estimate the separate slopes, we include `ibn.reg_cat2#z_putts`. As with the intercept, using the `ibn.` prefix ensures that no base category is used. Note also that we do not include `z_putts` without the interaction. Together, these two aspects of the syntax mean that we get both slopes. Compare the results with what we saw from `lincom`.

We can obtain b_1, b_3, standard errors, significance tests, and confidence intervals from (4.3) via `lincom`:

```
. regress money_rank ibn.reg_cat2 ibn.reg_cat2#c.z_putts, nocons coeflegend
```

Source	SS	df	MS		Number of obs	=	145
					F(4, 141)	=	159.94
Model	879714.673	4	219928.668		Prob > F	=	0.0000
Residual	193889.327	141	1375.10161		R-squared	=	0.8194
					Adj R-squared	=	0.8143
Total	1073604	145	7404.16552		Root MSE	=	37.082

money_rank	Coef.	Legend
reg_cat2		
Low	92.63787	_b[0bn.reg_cat2]
High	58.81429	_b[1.reg_cat2]
reg_cat2# c.z_putts		
Low	4.810639	_b[0bn.reg_cat2#c.z_putts]
High	21.79283	_b[1.reg_cat2#c.z_putts]

```
. lincom _b[1.reg_cat2] - _b[0bn.reg_cat2] // b1
( 1)  - 0bn.reg_cat2 + 1.reg_cat2 = 0
```

money_rank	Coef.	Std. Err.	t	P>|t|	[95% Conf. Interval]
(1)	-33.82358	6.272551	-5.39	0.000	-46.22399	-21.42318

```
. lincom _b[1.reg_cat2#c.z_putts] - _b[0bn.reg_cat2#c.z_putts] // b3
( 1)  - 0bn.reg_cat2#c.z_putts + 1.reg_cat2#c.z_putts = 0
```

money_rank	Coef.	Std. Err.	t	P>|t|	[95% Conf. Interval]
(1)	16.98219	6.276132	2.71	0.008	4.57471	29.38968

Polytomous by continuous interactions

Interactions can also be between categorical variables with three or more levels (that is, polytomous) and continuous variables.[6] For example, we can also fit a model with

---

[6]. In fact, interactions can be between any kind of variable and any number of variables. I focus only on two-way interactions, but we can also have interactions between any number of variables. Interpretation of multiway interactions is challenging. Consequently, tread carefully with interactions that are not clearly predicted by theory or previous data.

the original greens in regulation categorical variable, reg_cat, which has three levels. The model is similar in form to the models we have used up to this point, but it now includes multiple interaction terms:

$$\texttt{money\_rank}_i = b_0 + b_1 \texttt{mid\_dummy}_i + b_2 \texttt{high\_dummy}_i + b_3 \texttt{z\_putts}_i$$
$$+ b_4 \texttt{mid\_dummy}_i \times \texttt{z\_putts}_i + b_5 \texttt{high\_dummy}_i \times \texttt{z\_putts}_i \quad (4.7)$$

It is critical that we include both of the first-order dummy variables and both of the interaction terms. Excluding any of them changes the meaning of the model and the coefficients. Consequently, be careful not to drop, for example, one of the interaction terms because it is not significant. Using factor variables in Stata ensures that the necessary terms are included.

The Stata code for (4.7) is as follows:

```
. regress money_rank i.reg_cat c.z_putts i.reg_cat#c.z_putts
      Source |       SS           df       MS      Number of obs   =       145
-------------+----------------------------------   F(5, 139)       =     10.91
       Model |  74928.4395         5  14985.6879   Prob > F        =    0.0000
    Residual |  190981.188       139  1373.96538   R-squared       =    0.2818
-------------+----------------------------------   Adj R-squared   =    0.2559
       Total |  265909.628       144  1846.59464   Root MSE        =    37.067

   money_rank |      Coef.   Std. Err.      t    P>|t|     [95% Conf. Interval]
      reg_cat |
          Mid |  -32.17852   6.397006    -5.03   0.000    -44.82654   -19.5305
         High |  -59.79689   19.71897    -3.03   0.003    -98.7848   -20.80897

      z_putts |   4.810639   4.677802     1.03   0.306    -4.438207   14.05948

     reg_cat#
    c.z_putts |
          Mid |   17.45918   6.441323     2.71   0.008     4.723542   30.19482
         High |    30.2545   18.14791     1.67   0.098    -5.627142   66.13615

        _cons |   92.63787   4.724103    19.61   0.000     83.29748   101.9783
```

The coefficients are interpreted as follows:

- $b_0 = 92.6$: The expected money rank for a golfer in the low greens in regulation category and at the mean number of putts (that is, z_putts = 0).
- $b_1 = -32.2$: The expected difference in money rank for a golfer in the mid versus the low category and who is at the mean number of putts.
- $b_2 = -59.8$: The expected difference in money rank for a golfer in the high versus the low category and who is at the mean number of putts.
- $b_3 = 4.8$: The slope describing the relationship between number of putts and money rank for a golfer in the low category.
- $b_4 = 17.5$: The expected difference in the slope for z_putts between golfers in the mid category and the low category.

4.5.1 Categorical by continuous interactions

- $b_5 = 30.3$: The expected difference in the slope for z_putts between golfers in the high category and the low category.

As before, the first-order coefficients represent relationships or differences when the other variable is 0, and the interaction terms represent differences in slopes.[7]

The margins and marginsplot commands can be used to visualize the interaction.

```
. quietly margins, at(z_putts = (-2(1)2) reg_cat = (1 2 3))
. marginsplot, noci scheme(lean2)
>         legend(pos(6) row(1))
>         legend(subtitle("Greens in regulation category"))
>         title("") ytitle("Predicted money rank")
>         xtitle("Average number of putts - {it:z} score")
>         addplot((scatter money_rank z_putts if reg_cat == 3,
>                 legend(order(1 "Low" 2 "Mid" 3 "High" 4 "High data"))))
   Variables that uniquely identify margins: z_putts reg_cat
```

Figure 4.4 shows one peculiarity about the relationship implied by the model. Namely, the model predicts that the golfers in the high category of greens in regulation and 1-standard-deviation or more below the mean in number of putts are predicted to have negative values on the money ranking list.

---

7. Just as in the dichotomous by continuous interaction model, we can estimate the simple regression coefficient either by using lincom and summing the appropriate coefficients (for example, the intercept for the high category is $b_0 + b_2$) or by using regress in combination with the noconstant option (for example, regress money_rank ibn.reg_cat c.z_putts ibn.reg_cat#c.z_putts, noconstant).

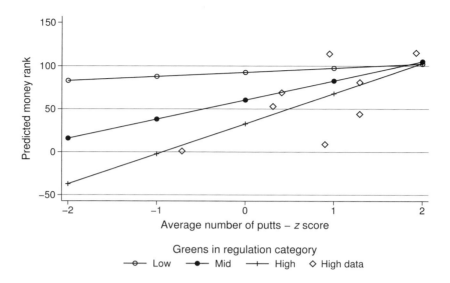

Figure 4.4. Visualization of a polytomous by continuous interaction using golf data

This is an artifact of two things: predicting outside the range of data and a small sample size in the high category. Figure 4.4 includes the data for the high category as diamond symbols. There are 8 observations, and 7 golfers are above the mean in number of putts. No golfer is more than 1 standard deviation below the mean. Nevertheless, the `margins` command produces expected values for golfers 2 standard deviations below the mean, and `marginsplot` plots those values.[8] Given that the expected values are negative, we know that these predictions are dubious. For outcomes where the scale is not as well defined as rank, it can be more difficult to know whether the predictions are meaningful. Consequently, it is best to avoid making predictions beyond the range of observed data.

Furthermore, the sample size in the high category is tiny, which means the uncertainty around the regression line is high. Therefore, the simple regression equation for the high group is estimated with low precision and should not be trusted. Try rerunning the `marginsplot` command without the `noci` option so that the confidence intervals will be printed.[9] The width of the confidence intervals demonstrates how uncertain the expected values in the high category are.

---

8. Because we asked it to do so. This underscores the need to understand the raw data and not just run analyses and report the results. We may end up reporting nonsense if we are not careful.
9. I suppressed the confidence intervals in figure 4.4 because including them makes the figure difficult to read.

4.5.1 Categorical by continuous interactions

Joint test for interactions with polytomous variables

As noted above, we need to include both `mid_dummy x z_putts` and `high_dummy x z_putts` when testing the interaction between the greens in regulation category and average number of putts. In doing so, we obtain significance tests for each of the interactions—for b_4 and b_5—but we do not get a test for whether the coefficients are jointly significant. That is, the null hypothesis for the significance test for b_4 is that there is no difference in the slope (in the population) for `z_putts` between the mid and low categories. The null hypothesis for the joint test is that there is no difference in the slope for `z_putts` for both the mid versus low and the high versus low categories. Specifically,

$$H_0 : \beta_4 = 0$$
$$\beta_5 = 0$$

This joint test is a test of the interaction between `reg_cat` and `z_putts`, and the tests for the individual slopes indicate which levels of `reg_cat` interact with `z_putts`.

We can test this null hypothesis in one of two ways. In the first method, we fit two regression models, one without the interaction and one with it. Then we use the following formula to compute an F test:

$$F_0 = \frac{N - k - 1}{q} \times \frac{R_1^2 - R_0^2}{1 - R_1^2}$$

where N is the sample size, k is the number of predictors in the model including the interaction, q is the number of coefficients in the null hypothesis (two in this case), R_1^2 is the R^2 from the model with the interaction, and R_0^2 is the R^2 from the model without the interaction. This F has q numerator degrees of freedom and $N - k - 1$ denominator degrees of freedom (Fox 2008, 109–110). To compute the F in Stata, we type

```
. regress money_rank i.reg_cat c.z_putts

      Source |       SS           df       MS      Number of obs   =       145
-------------+----------------------------------   F(3, 141)       =     14.57
       Model |  62919.1545         3  20973.0515   Prob > F        =    0.0000
    Residual |  202990.473       141  1439.64875   R-squared       =    0.2366
-------------+----------------------------------   Adj R-squared   =    0.2204
       Total |  265909.628       144  1846.59464   Root MSE        =    37.943

  money_rank |      Coef.   Std. Err.      t    P>|t|     [95% Conf. Interval]
-------------+----------------------------------------------------------------
     reg_cat |
        Mid  |  -33.35549   6.535241    -5.10   0.000    -46.27521   -20.43576
       High  |  -45.09812    14.5683    -3.10   0.002    -73.89864    -16.2976
             |
     z_putts |   14.70334   3.237794     4.54   0.000     8.30244    21.10423
       _cons |   94.14546   4.805729    19.59   0.000    84.64486    103.6461
```

```
. regress money_rank i.reg_cat c.z_putts i.reg_cat#c.z_putts
```

Source	SS	df	MS		Number of obs	=	145
					F(5, 139)	=	10.91
Model	74928.4395	5	14985.6879		Prob > F	=	0.0000
Residual	190981.188	139	1373.96538		R-squared	=	0.2818
					Adj R-squared	=	0.2559
Total	265909.628	144	1846.59464		Root MSE	=	37.067

money_rank	Coef.	Std. Err.	t	P>|t|	[95% Conf.	Interval]
reg_cat						
Mid	-32.17852	6.397006	-5.03	0.000	-44.82654	-19.5305
High	-59.79689	19.71897	-3.03	0.003	-98.7848	-20.80897
z_putts	4.810639	4.677802	1.03	0.306	-4.438207	14.05948
reg_cat#						
c.z_putts						
Mid	17.45918	6.441323	2.71	0.008	4.723542	30.19482
High	30.2545	18.14791	1.67	0.098	-5.627142	66.13615
_cons	92.63787	4.724103	19.61	0.000	83.29748	101.9783

```
. display "F_0 = " ((145-5-1)/2)*((0.2818-0.2366)/(1-.2818))
F_0 = 4.3739905
. display "p-value = " 1 - F(2, 139, 4.37)
p-value = .01443481
```

The second method (which I recommend) for testing the joint null hypothesis is to use the `test` command. Use the `coeflegend` option for `regress` to obtain the names for b_4 and b_5, and then set each coefficient equal to 0 with `test`.

```
. regress money_rank i.reg_cat c.z_putts i.reg_cat#c.z_putts, coeflegend
```

Source	SS	df	MS		Number of obs	=	145
					F(5, 139)	=	10.91
Model	74928.4395	5	14985.6879		Prob > F	=	0.0000
Residual	190981.188	139	1373.96538		R-squared	=	0.2818
					Adj R-squared	=	0.2559
Total	265909.628	144	1846.59464		Root MSE	=	37.067

money_rank	Coef.	Legend
reg_cat		
Mid	-32.17852	_b[2.reg_cat]
High	-59.79689	_b[3.reg_cat]
z_putts	4.810639	_b[z_putts]
reg_cat#		
c.z_putts		
Mid	17.45918	_b[2.reg_cat#c.z_putts]
High	30.2545	_b[3.reg_cat#c.z_putts]
_cons	92.63787	_b[_cons]

4.5.2 Continuous by continuous interactions

```
. test (_b[2.reg_cat#c.z_putts] = 0) (_b[3.reg_cat#c.z_putts] = 0)

 ( 1)  2.reg_cat#c.z_putts = 0
 ( 2)  3.reg_cat#c.z_putts = 0

       F(  2,   139) =    4.37
            Prob > F =    0.0144
```

The result is $F(2, 139) = 4.37$, $p = 0.01$, indicating that we should reject the null hypothesis that both coefficients are 0. The numerator degrees of freedom for the F test is equal to the number of coefficients constrained to 0 (in this case, 2), and the denominator degrees of freedom is equal to the residual degrees of freedom from the model (in this case, 139).

4.5.2 Continuous by continuous interactions

A final way to consider the greens in regulation and putting interaction is to use the continuous version of the greens in regulation percentage rather than the category. Generally speaking, it is best not to categorize continuous variables because doing so means we lose information. For example, when we combine anybody above 70% and make them the high category, we lose any distinctions among those golfers. Discarding information usually is a bad thing and can unduly affect studies and analyses.[10] MacCallum et al. (2002) note that the problems with dichotomizing continuous variables include the following:

> These [problems] include loss of information about individual differences, loss of effect size and power, the occurrence of spurious significant main effects or interactions, risks of overlooking nonlinear effects, and problems in comparing and aggregating findings across studies. To our knowledge, there have been no findings of positive consequences of dichotomization. (p. 29)

I categorized greens in regulation to teach the concepts. Fortunately, we can fit the model with the continuous version of greens in regulation, **regulation**. As with the other continuous variables we have used, we will start by standardizing **regulation**.[11]

```
. egen z_reg = std(regulation)
```

---

10. Most researchers know this. However, there is something appealing about separating people into groups, even when the grouping is arbitrary, so many researchers ignore the methodological problems with categorizing continuous variables. Keep an eye out to see if researchers justify their use of categories in their articles.
11. A disadvantage of standardizing **regulation** in this instance is that interpreting a one-unit change is fairly straightforward given that the metric of **regulation** is the percentage of greens hit in regulation. Unfortunately, there is still the problem of needing a meaningful 0 value for **regulation** (no professional golfer misses all greens in regulation). Consequently, we will standardize so that the mean is equal to 0.

The model is the same as before except now includes the continuous (and standardized) version of greens in regulation:

$$\texttt{money\_rank}'_i = b_0 + b_1 \texttt{z\_reg} + b_2 \texttt{z\_putts} + b_3 \texttt{z\_reg} \times \texttt{z\_putts} \tag{4.8}$$

The Stata code for (4.8) is

```
. regress money_rank c.z_reg c.z_putts c.z_reg#c.z_putts
```

Source	SS	df	MS			
Model	79397.0214	3	26465.6738	Number of obs	=	145
Residual	186512.606	141	1322.78444	F(3, 141)	=	20.01
				Prob > F	=	0.0000
				R-squared	=	0.2986
				Adj R-squared	=	0.2837
Total	265909.628	144	1846.59464	Root MSE	=	36.37

money_rank	Coef.	Std. Err.	t	P>|t|	[95% Conf. Interval]	
z_reg	-19.50542	3.09963	-6.29	0.000	-25.63318	-13.37767
z_putts	15.86881	3.091257	5.13	0.000	9.757609	21.98002
c.z_reg#c.z_putts	8.489794	3.154147	2.69	0.008	2.254262	14.72533
_cons	72.99279	3.081341	23.69	0.000	66.90119	79.08439

The interpretation of the coefficients is as follows:

- $b_0 = 73$: The expected value of money_rank for golfers with a 0 on z_reg (that is, at the mean) and for golfers with a 0 on z_putts (that is, at the mean).
- $b_1 = -19.5$: The slope for z_reg for golfers at the mean of z_putts (that is, where z_putts $= 0$).
- $b_2 = 15.9$: The slope for z_putts for golfers at the mean of z_reg (that is, where z_reg $= 0$).
- $b_3 = 8.5$: The difference in the slopes for z_reg and z_putts.

As before, b_1 and b_2 are first-order coefficients.

The margins and marginsplot commands can be used to visualize the relationship. As before, we will create simple regression equations to aid in the visualization. When we treated greens in regulation as a categorical variable, we created as many simple regression equations as we had categories; for example, we created three equations when the categories were low, mid, and high. Now that greens in regulation is treated as continuous, we need to select specific values of z_reg and z_putts at which we will create the simple equations. A common choice is to use 3–5 values for each variable. If we choose to use three values, we could do 1 standard deviation below the mean, the mean, and 1 standard deviation above the mean, that is, $z = -1, 0$, and 1. Five values could range from 2 standard deviations below the mean to 2 standard deviations above the mean.

4.6 Summary

```
. quietly margins, at(z_putts = (-2(1)2) z_reg = (-1(1)1))
. marginsplot, noci scheme(lean2)
>        legend(pos(6) row(1))
>        legend(subtitle("Greens in regulation category"))
>        title("") ytitle("Predicted money rank")
>        xtitle("Average number of putts - {it:z} score")
>        addplot((scatter money_rank z_putts,
>            legend(order(1 "-1 SD" 2 "Mean" 3 "1 SD" 4 "Data"))))
   Variables that uniquely identify margins: z_putts z_reg
```

Figure 4.5 shows the results of `marginsplot`. I chose three values for `z_reg` and five values for `z_putts`. The relationship between putting and money rank is stronger for those above the mean in greens in regulation than those below the mean. When the number of putts is above average, the expected difference in money rank among golfers is not heavily influenced by how often they reach the green in regulation. However, when the number of putts is below average, the expected difference in money rank is influenced by how often they reach the green in regulation.

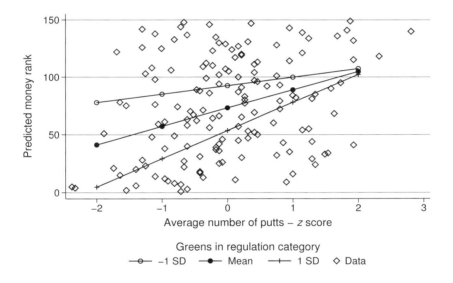

Figure 4.5. Visualization of a continuous by continuous interaction using the golf data

4.6 Summary

This chapter has shown how to generalize bivariate regression to accommodate a) categorical predictors, b) two or more predictors, and c) interactions among predictors. Here are some key takeaway points:

1. Treating categorical variables as numeric/continuous is a major problem and is misleading.
2. Including multiple predictors in a model changes the interpretation of a slope coefficient—the slopes are partial coefficients.
3. Interpreting coefficients from models with interactions requires care. Graphing the relationships can assist with interpretation. Adding the raw data is also valuable because it establishes where the data are thin.

I have not covered nonlinear relationships, such as quadratic or cubic relationships. Readers interested in these topics can consult Cohen et al. (2003), Fox (2008), or Mitchell (2012a).

5 t tests and one-way ANOVA

Chapter 4 covered using regression to explore how categorical variables are related to a continuous outcome. Regression is advantageous because it allows both categorical and continuous predictors in the same model. Often, we want to compare the levels of a categorical variable without any additional covariates. In that case, you can use either a t test or a one-way analysis of variance (ANOVA). This chapter teaches the fundamentals of between-groups t tests and one-way ANOVA.

Specifically, you will learn the following:

1. How to use a between-groups t test to compare two means.
2. How to use a one-way ANOVA to compare three or more means.
3. Why using ANOVA is preferable to running many t tests.
4. How to use planned comparisons and post hoc tests following ANOVA to compare groups.

Stata commands featured in this chapter

- `tabstat`: for producing a table of descriptive statistics
- `vioplot`: for producing a violin plot
- `esize`: for computing effect sizes
- `ttest`: for estimating a between-groups t test
- `anova`: for estimating a one-way ANOVA
- `margins`: for obtaining condition-specific means
- `marginsplot`: for producing plots of the condition-specific means
- `test`: for testing linear combinations
- `lincom`: for testing linear combinations
- `pwcompare`: for performing post hoc comparisons following an ANOVA.

5.1 Data

Dimidjian et al. (2006) report the results of a randomized trial comparing four treatments for depression. The treatments were cognitive behavioral therapy (CBT), behavioral activation (BA), antidepressant medication (ADM), and pill placebo. Patients were randomly assigned to these treatments and assessed at baseline, 8 weeks, and 16 weeks. Patients were divided into high and low depression severity groups. Treatment conditions were compared on the Beck depression inventory (BDI) and Hamilton rating scale for depression. Dimidjian et al. (2006) found that BA and ADM had comparable effects and were superior to CBT among patients in the high-severity condition. The dataset bdi_week8.dta provides simulated BDI data for the 8-week timepoint for the high-severity group.[1] Specifically, I created data for four conditions based on the means and standard deviations reported in table 2 of Dimidjian et al. (2006), except that I set the sample sizes for each condition to be equal to one another. Higher values on the BDI represent higher levels of depression symptoms.

```
. use http://www.stata-press.com/data/pspus/bdi_week8
. describe
Contains data from bdi_week8.dta
  obs:           152
 vars:             2                               7 Nov 2018 18:32
 size:           760
─────────────────────────────────────────────────────────────────────────
              storage   display    value
variable name   type    format     label       variable label
─────────────────────────────────────────────────────────────────────────
tx            byte      %9.0g      tx_cond     Treatment Condition
bdi           float     %9.0g                  Beck Depression Inventory
─────────────────────────────────────────────────────────────────────────
Sorted by: tx
```

5.2 Comparing two means

Many patients seeking treatment for depression wonder whether they should take medication or try psychotherapy. Thus, a primary question is whether CBT and BA are better than ADM. A t test is a statistical method for investigating these differences.

---

[1]. Readers familiar with the BDI may notice some differences between my simulated data and typical BDI data, such as negative values and some large values (> 70). I did not worry about these differences because they are not critical for understanding the statistical material.

5.2 Comparing two means

A good place to start examining differences is with descriptive statistics. The means, standard deviations, and sample size for each condition are

```
. tabstat bdi, by(tx) stat(mean sd n)
Summary for variables: bdi
     by categories of: tx (Treatment Condition)

     tx  |     mean         sd          N
---------+------------------------------------
    CBT  |  21.7232    17.51851         38
     BA  | 16.84499    8.881432         38
    ADM  | 13.50326    10.77835         38
Placebo  | 22.26988    14.46674         38
---------+------------------------------------
  Total  | 18.58533    13.69061        152
```

Patients taking medication had lower BDI scores than either CBT or BA. Figure 5.1 is a violin plot of the data across conditions. Violin plots are just like box plots (see section 2.7.2), but in addition to displaying the median and interquartile range, violin plots also include the probability density for each variable. Thus, violin plots allow you to see how the data for each treatment condition are distributed (for example, where is the center, what is the skew). Violin plots can be created using the community-contributed command `vioplot` (Winter and Nichols 2008), which can be installed by typing `ssc install vioplot`.

```
. vioplot bdi, over(tx) scheme(lean2) ytitle("Beck depression inventory")
```

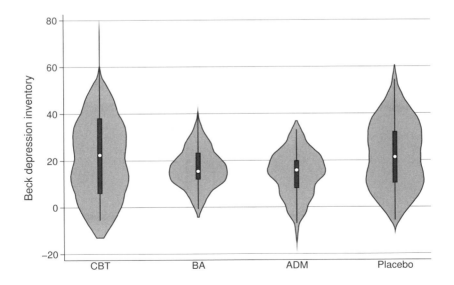

Figure 5.1. Violin plot of the BDI scores across treatment condition

Based on the descriptive statistics and the violin plots, CBT does not seem to produce as large an effect as ADM. BA also does not seem to produce as big an effect as ADM, though the differences between BA and ADM appear smaller than the differences between CBT and ADM. CBT is more variable than either BA or ADM, as indicated by the relatively large standard deviation for CBT and the longer violin plot for the CBT condition.

5.2.1 t test

The most common statistical method for comparing two groups is the independent-groups t test, hereafter just referred to as a t test. The null hypothesis for a t test is that the means of the CBT and ADM groups do not differ in the population.

$$H_0 : \mu_{\text{CBT}} = \mu_{\text{ADM}}$$

Said another way, the null hypothesis is that the difference between the means of the CBT and ADM groups is 0 in the population:

$$H_0 : \mu_{\text{CBT}} - \mu_{\text{ADM}} = 0 \tag{5.1}$$

Finally, a third way of stating the null hypothesis is that the CBT and ADM groups are sampled from the same population and as a consequence will be equal.

The formula for a t test describes the difference between the means of two groups [see (5.2)]. Consequently, we will use the null hypothesis in (5.1). The two-tailed alternative hypothesis is that the difference between the means of CBT and ADM is not 0 in the population:

$$H_a : \mu_{\text{CBT}} - \mu_{\text{ADM}} \neq 0$$

To test the null hypothesis, the difference between the means of CBT and ADM is transformed into a t statistic,

$$t_{\text{obt}} = \frac{\overline{y}_1 - \overline{y}_2 - (\mu_1 - \mu_2)}{s_p} \tag{5.2}$$

where s_p is the pooled standard error:

$$s_p = \sqrt{\frac{s_1^2}{n_1} + \frac{s_2^2}{n_2}} \tag{5.3}$$

The degrees of freedom (DF) is DF $= n_1 + n_2 - 2$. We lose 2 DF because we have to estimate the variance in both groups [see (5.3)].

To implement a t test in Stata, you can use the `ttest` command, which can perform one-sample t tests, paired t tests, and independent-groups (two-sample) t tests. It can even perform the independent-groups t test using two different syntaxes. The help documents call these two methods the a) two-sample t test using groups and the b) two-sample t test using variables. The first method is appropriate when your dataset is stacked or in long format—where there are as many rows in the dataset as there are observations and where there is a categorical variable that indicates which observations belong to particular conditions. This is how the dataset in this chapter is configured. The second method is appropriate when there is a unique outcome variable for every group. For example, we could use the second form if the outcome for the CBT condition was `bdi1` and the outcome for the ADM condition was `bdi2`. If you want to try this approach with the depression example, use `reshape`.

Given the structure of the dataset, I use the group method. The code to compute t_{obt} from (5.2) and perform the significance test is (note the use of `if` to limit the analysis to the CBT and ADM conditions):

```
. ttest bdi if tx == 1 | tx == 3, by(tx)
Two-sample t test with equal variances
-----------------------------------------------------------------------------
   Group |     Obs        Mean    Std. Err.   Std. Dev.   [95% Conf. Interval]
---------+-------------------------------------------------------------------
     CBT |      38     21.7232    2.841878    17.51851    15.96501    27.48139
     ADM |      38    13.50326    1.748479    10.77835    9.960508    17.04602
---------+-------------------------------------------------------------------
combined |      76    17.61323    1.723797    15.02771    14.17925    21.04721
---------+-------------------------------------------------------------------
    diff |             8.219935    3.336683                1.57145    14.86842
-----------------------------------------------------------------------------
    diff = mean(CBT) - mean(ADM)                              t =   2.4635
Ho: diff = 0                                   degrees of freedom =       74

    Ha: diff < 0                 Ha: diff != 0                 Ha: diff > 0
 Pr(T < t) = 0.9920          Pr(|T| > |t|) = 0.0161          Pr(T > t) = 0.0080
```

The difference between mean BDI values in the CBT and ADM conditions is 8.2 and $t_{\text{obt}} = 2.5$ with 74 DF. Stata reports three p-values at the bottom of the output. The middle p-value is the two-tailed value, whereas the left and right values are the respective one-tailed values. Given that we are testing a two-tailed null, we will focus on the middle value: $p = 0.016$. This p-value indicates that if the null hypothesis is true, the probability of observing a difference as extreme or more extreme than 8.22 is 1.6%. Thus, we reject the null hypothesis.

5.2.2 Effect size

How big is the effect? This is a question of effect size. The simplest effect size is the difference between conditions with respect to the scale of the BDI: On average, BDI scores were 8.2 points lower in the ADM group than the CBT group. This effect size is reasonable but requires familiarity with the typical range and variability of the BDI. For example, if you knew that in normative clinical samples, the standard deviation of the BDI was 4, then an 8.2 point difference would suggest that the groups differed by just over 2 standard deviations. However, if the standard deviation was 24, then the groups differ by about one-third of a standard deviation. You would also need to know about the normative standard deviation for other outcome measures.

Given the large number of measures in the social sciences, most with distinct scaling, it is common to use effect sizes with a standardized metric. A commonly used effect size is the standardized mean difference:

$$\delta = \frac{\mu_1 - \mu_2}{\sigma}$$

where δ is the difference between two means scaled in terms of the standard deviation of the outcome.

A sample-based estimate of δ is Cohen's d:

$$d = \frac{\overline{y}_1 - \overline{y}_2}{s_p}$$

where s_p is the pooled standard deviation (that is, the standard deviation from both groups).

$$s_p = \sqrt{\frac{(n_1 - 1) s_1^2 + (n_2 - 1) s_2^2}{n_1 + n_2 - 2}}$$

Here n_1 and n_2 are the sample size, and s_1^2 and s_2^2 are the variances for each condition. Cohen's d measures how large a difference is between two groups in standard deviation units.

The `esize twosample` command can be used to estimate d.

```
. esize twosample bdi if tx == 1 | tx == 3, by(tx)
Effect size based on mean comparison

                              Obs per group:
                                        CBT =            38
                                         BA =            38

         Effect Size |   Estimate      [95% Conf. Interval]
       --------------+--------------------------------------
            Cohen´s d |   .5651669        .10459    1.022075
            Hedges´s g|   .5594161       .1035258    1.011675
```

`esize` follows the same general syntax as `ttest`. The output includes d as well as the confidence interval (CI) for d: $d = 0.57$, 95% CI $= [0.1, 1.02]$. Thus, the CBT group is 0.57 standard deviations higher on the BDI than the BA group. The output also includes Hedges's g, which was created because Cohen's d is slightly biased. Both statistics have the same interpretation, and the difference will typically be trivial.

5.3 Comparing three or more means

5.3.1 Analysis of variance

The aim of the Dimidjian et al. (2006) trial was to compare the effects of all treatments, not just CBT and ADM. A common method for comparing the effects of all treatments is

5.3.1 Analysis of variance

an ANOVA. Like regression (see chapters 3 and 4), ANOVA can be used for many research designs and analysis questions (see Winer et al. [1991] for a comprehensive treatment of ANOVA models and designs). To compare four treatments on a single outcome at a single time, you would use a one-way ANOVA. The categorical variable for treatment condition is called a factor, and given that there is a single factor, the model is a one-way ANOVA. Models with two factors are sometimes referred to as two-way ANOVA. Categories within a factor are levels—the four specific treatment types are levels of the treatment factor.

The null hypothesis for a one-way ANOVA is that the population means for each level are equal (PL is the placebo):

$$H_0 : \mu_{CBT} = \mu_{BA} = \mu_{ADM} = \mu_{PL}$$

The alternative hypothesis is that at least one of the means is different:

$$H_a : \mu_{CBT} \neq \mu_{BA} = \mu_{ADM} = \mu_{PL}$$

Here is a more general form of the hypotheses for a factor with k levels:

$$H_0: \mu_1 = \mu_2 = \mu_3 = \cdots = \mu_k$$
$$H_a: \mu_1 \neq \mu_2 = \mu_3 = \cdots = \mu_k$$

In our example, the question is whether the treatments are all sampled from a single population where the treatments do not differ or whether at least one of the treatments comes from a different population. To address this question, we compare the variability among the treatment means with the variability within each treatment. The logic is that the variability within each treatment condition represents error (for example, sampling error), whereas the variability among the treatment means represents either a) systematic differences among treatments plus error, if the alternative hypothesis is true, or b) just error, if the null hypothesis is true. To make these comparisons, we use the ratio of the between-treatments variability to the within-treatment variability.

$$\frac{\text{Between-treatments variance}}{\text{Within-treatment variance}}$$

If the alternative hypothesis is true, then we expect the numerator to represent systematic treatment differences and error:

$$\frac{\text{Systematic variance} + \text{Error variance}}{\text{Error variance}}$$

Consequently, if the alternative is true, then the ratio of the variances should be greater than 1. If the null hypothesis is true, then the numerator will represent only error:

$$\frac{\text{Error variance}}{\text{Error variance}}$$

If the null hypothesis is true, then the ratio of the variances should be 1.

To compute these ratios, we must partition the total variability in outcome into the between-conditions variance and the within-condition variance. This is precisely what ANOVA does. We start by computing the total sum of squares for depression and then break the total into two parts—the between-treatments sum of squares and the within-treatment sum of squares. The total sum of squares (SS_T; that is, the variability of the observations around the grand mean) is

$$\text{SS}_T = \sum (y_{ij} - \overline{y})^2 \tag{5.4}$$

where y_{ij} is the depression value for the ith person in treatment group j and $\overline{y}$ is the grand mean depression value. The between-treatments sum of squares (SS_B; that is, the variability of the group means around the grand mean) is

$$\text{SS}_B = n \sum (\overline{y}_j - \overline{y})^2 \tag{5.5}$$

where $\overline{y}_j$ is the mean for treatment group j and n is the number of patients within treatment groups.[2] The within-treatment sum of squares (SS_W; that is, the variability of the observations around the group means) is

$$\text{SS}_W = \sum (y_{ij} - \overline{y}_j)^2 \tag{5.6}$$

The relationship between the sums of squares is

$$\text{SS}_T = \text{SS}_B + \text{SS}_W$$

The DF for this model are also partitioned into between- and within-treatment values. The total DF for the model is [we lose 1 DF because we estimate $\overline{y}$ in (5.4)]

$$\text{DF}_T = N - 1$$

The between-treatments DF is [we lose 1 DF because we estimate $\overline{y}$ in (5.5)]

$$\text{DF}_B = k - 1$$

where k is the number of treatment conditions (4 in this example). The within-treatment DF is [we lose 4 DF because we estimate $\overline{y}_j$ in (5.6)]

$$\text{DF}_W = N - k$$

The relationship between the DFs is

$$\text{DF}_T = \text{DF}_B + \text{DF}_W$$

---

[2]. This formula can be generalized for situations where the sample size is not equal across groups. However, we will leave it to Stata to do the heavy lifting for us in that instance.

5.3.1 Analysis of variance

Recall that we want to compute the ratio of the between-treatments variance to the within-treatment variance. To produce these variances, divide the sums of squares by the appropriate DF (these variances are called mean squares).

$$\mathrm{MS}_B = \frac{\mathrm{SS}_B}{\mathrm{DF}_B}$$

$$\mathrm{MS}_W = \frac{\mathrm{SS}_W}{\mathrm{DF}_W}$$

Finally, we can construct the ratio of the variances, which follows an F distribution with DF equal to the DF_B and DF_W:

$$F = \frac{\mathrm{MS}_B}{\mathrm{MS}_W}$$

To estimate a one-way ANOVA in Stata, use the `oneway` command:

```
. oneway bdi tx
                        Analysis of Variance
    Source              SS         df      MS            F      Prob > F
Between groups     1986.57566       3   662.191886      3.72     0.0128
Within groups      26315.7911     148   177.8094

    Total          28302.3668     151   187.432893
Bartlett's test for equal variances:  chi2(3) =   19.3778  Prob>chi2 = 0.000
```

Alternatively, we can use the `anova` command, which is a more general command that allows more complicated ANOVA models in addition to the one-way ANOVA.

```
. anova bdi tx
                  Number of obs =     152     R-squared     = 0.0702
                  Root MSE      = 13.3345     Adj R-squared = 0.0513

         Source |  Partial SS     df       MS           F      Prob>F
         -------+----------------------------------------------------
          Model |   1986.5757      3    662.19189      3.72    0.0128
             tx |   1986.5757      3    662.19189      3.72    0.0128
       Residual |   26315.791    148    177.8094
         -------+----------------------------------------------------
          Total |   28302.367    151    187.43289
```

The output indicates that $F(3, 148) = 3.72$, $p = 0.013$, for the test of treatment effects. If $\alpha = 0.05$, then we reject the null hypothesis that all the treatments are equal.

We are not done yet. The main problem with this analysis is that it does not indicate which treatments are different from the others. As can be seen in figure 5.2, some treatments appear different from the others. However, the information from the F test only tells us that the data are not consistent with all the means being equal. Any time an F test has more than two DF in the numerator, we need to conduct additional tests for specific group differences. These specific group tests are often of the most scientific interest (Wampold, Davis, and Good 1990). An F test with two or more DF is called an omnibus test because it covers many group comparisons. Any test, such as a t test or an F test, with a single degree of freedom in the numerator is called a focal test.

```
. margins tx
Adjusted predictions                              Number of obs   =        152
Expression    : Linear prediction, predict()
```

	Margin	Delta-method Std. Err.	t	P>|t|	[95% Conf. Interval]	
tx						
CBT	21.7232	2.163145	10.04	0.000	17.44856	25.99784
BA	16.84499	2.163145	7.79	0.000	12.57035	21.11962
ADM	13.50326	2.163145	6.24	0.000	9.228624	17.7779
Placebo	22.26988	2.163145	10.30	0.000	17.99524	26.54452

```
. marginsplot, recast(bar) scheme(lean2) plotopts(fcolor(gs12))
>   title("") ytitle("Average depression")
   Variables that uniquely identify margins: tx
```

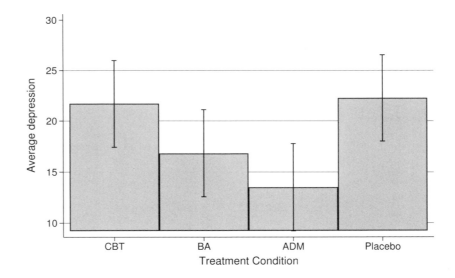

Figure 5.2. Means and 95% CI for each treatment

5.3.2 Multiple comparisons

We could use six t tests to compare all the treatments, one for each pairwise combination of the four treatments. The problem with this approach, which is known as the multiple

5.3.2 Multiple comparisons

comparison problem (Winer et al. 1991, chap. 3), is that running lots of tests can inflate the probability that we will make a type I error. In the Dimidjian et al. (2006) trial, the six comparisons are

1. BA versus CBT
2. BA versus ADM
3. BA versus placebo
4. CBT versus ADM
5. CBT versus placebo
6. ADM versus placebo

Assume the null hypothesis is true for #1, and you use a t test to compare BA and CBT. If $\alpha = 0.05$ for this test, the probability of not making a type I error is 0.95 $(1 - \alpha)$. Now assume that the null hypothesis is true for all six comparisons. What is the probability of not making a type I error for all six comparisons—that is, what is the joint probability? The probability for each comparison is $1 - \alpha$, and the joint probability is the product of all six. Thus, the probability of not making a type I error for one or more of the six comparisons, given $\alpha = 0.05$, is $(1 - \alpha)^6 = (1 - 0.05)^6 = 0.74$. Therefore, the probability of making a type I error for one or more of the six comparisons is $1 - (1 - \alpha)^6 = 1 - (1 - 0.05)^6 = 0.26$ or 26% (Winer et al. 1991).

Winer et al. (1991, 154) refer to this joint probability as α_{joint} and provide a formula for it:

$$\alpha_{\text{joint}} \leq 1 - (1 - \alpha_{\text{ind}})^m$$

where m is the number of comparisons and α_{ind} is the type I error rate for the individual comparisons. Figure 5.3 shows the relationship between the number of comparisons α and α_{joint}. It may be unusual to have enough groups to have 20 comparisons, but it is not uncommon to have multiple outcome variables in a given study. In that case, the number of comparisons made in the study can easily be 20. Consequently, α inflation can easily be a problem.

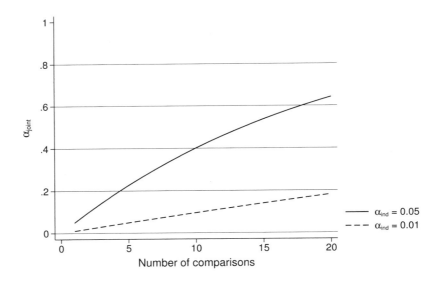

Figure 5.3. Relationship between type I error rate (α) and number of comparisons (m)

There are two primary methods for protecting against α inflation. First, limit the number of tests you run to those that are predicted prior to collecting the data. These comparisons are often referred to as planned comparisons given that they are planned prior to the study. Second, use statistical techniques that have built-in protection for multiple comparisons.

Planned comparisons

Suppose your primary interest in the treatment study is to compare i) CBT and BA, ii) CBT and ADM, and iii) CBT and placebo. These comparisons were decided upon prior to conducting the study and are the only direct comparisons we are going to make.[3] Consequently, you can consider these comparisons to be planned comparisons. To make the comparison, we use linear combinations of the four means. We denote a linear combination (ψ) as (Winer et al. 1991)

$$\psi = c_1 \overline{y}_{\text{CBT}} + c_2 \overline{y}_{\text{BA}} + c_3 \overline{y}_{\text{ADM}} + c_4 \overline{y}_{\text{PL}} \tag{5.7}$$

where c_1, c_2, c_3, and c_4 are called contrast weights, which are numeric values and must sum to 0. Whenever we use a linear combination, we must supply a value for all four values of c. For example, if one of the means is not included in the comparison we want to make, we multiply the mean we want to exclude by 0.

---

3. It is unlikely that we would only be interested in these three comparisons, but we will go with that for now to help underscore the idea.

5.3.2 Multiple comparisons

The three comparisons (Winer et al. 1991) are as follows:

1. The difference between CBT and BA: $\psi_1 = 1\overline{y}_{\text{CBT}} + -1\overline{y}_{\text{BA}} + 0\overline{y}_{\text{ADM}} + 0\overline{y}_{\text{PL}}$. Simplifying, this reduces to $\psi_1 = \overline{y}_{\text{CBT}} - \overline{y}_{\text{BA}}$. Note that the contrast weights are $1, -1, 0$, and 0, which sum to 0.
2. The difference between CBT and ADM: $\psi_2 = 1\overline{y}_{\text{CBT}} + 0\overline{y}_{\text{BA}} + -1\overline{y}_{\text{ADM}} + 0\overline{y}_{\text{PL}}$. Simplifying, this reduces to $\psi_2 = \overline{y}_{\text{CBT}} - \overline{y}_{\text{ADM}}$.
3. The difference between CBT and placebo: $\psi_3 = 1\overline{y}_{\text{CBT}} + 0\overline{y}_{\text{BA}} + 0\overline{y}_{\text{ADM}} + -1\overline{y}_{\text{PL}}$. Simplifying, this reduces to $\psi_3 = \overline{y}_{\text{CBT}} - \overline{y}_{\text{PL}}$.

The null hypothesis for each of these contrasts is that there is no difference between the two conditions in the population. We can test these nulls with an F test. The sum of squares for the comparison (SS_C) is computed as

$$\text{SS}_C = \frac{\psi^2}{n \sum c_j^2}$$

where ψ^2 is the difference squared, $\sum c_j^2$ is the sum of the squared contrast weights, and n is the number of subjects per condition. Given that the comparisons involve the difference between two means, there is 1 degree of freedom in the numerator. The residual sum of squares, SS_W, is from the one-way ANOVA. That is, the within-condition variability for the planned comparison pools the within-treatment variability from all conditions, which allows you to draw strength and precision from all the conditions. The DF is equal to the residual DF from the ANOVA. The mean squares and F ratio are computed as previously:

$$\text{MS}_C = \frac{\text{SS}_C}{\text{DF}_C}$$

$$\text{MS}_W = \frac{\text{SS}_W}{\text{DF}_W}$$

$$F = \frac{\text{MS}_C}{\text{MS}_W}$$

Two methods for estimating planned contrasts in Stata are the `test` and `lincom` postestimation commands. Because these are postestimation commands, we must first run `anova` and then specify the linear combinations.

```
. use http://www.stata-press.com/data/pspus/bdi_week8, clear
. anova bdi tx
                      Number of obs =      152    R-squared     =  0.0702
                      Root MSE      =  13.3345    Adj R-squared =  0.0513

             Source | Partial SS        df        MS         F     Prob>F
            --------+-------------------------------------------------------
              Model | 1986.5757          3    662.19189     3.72   0.0128

                 tx | 1986.5757          3    662.19189     3.72   0.0128

           Residual | 26315.791        148    177.8094

              Total | 28302.367        151    187.43289

. margins tx, coeflegend post
Adjusted predictions                               Number of obs   =      152
Expression   : Linear prediction, predict()

                 |    Margin   Legend
            -----+------------------------
               tx|
              CBT|   21.7232   _b[1bn.tx]
               BA|  16.84499   _b[2.tx]
              ADM|  13.50326   _b[3.tx]
          Placebo|  22.26988   _b[4.tx]

. test 1*1.tx + -1*2.tx + 0*3.tx + 0*4.tx = 0
 ( 1)  1bn.tx - 2.tx = 0
       F(  1,   148) =    2.54
            Prob > F =    0.1129
. lincom 1*1.tx + -1*2.tx + 0*3.tx + 0*4.tx
 ( 1)  1bn.tx - 2.tx = 0

                 |   Coef.    Std. Err.     t    P>|t|   [95% Conf. Interval]
            -----+----------------------------------------------------------
             (1) | 4.878212   3.059148    1.59   0.113   -1.16704    10.92346
```

This output also includes a call to margins that uses the coeflegend and post options. Following a one-way ANOVA, margins produces the cell means for the factor, which in this case are the means for each treatment. Both test and lincom require that you know the name Stata uses internally for the treatment means. The combination of the coeflegend and post options asks Stata to print the names (coeflegend) and make them available for test (post).

The test command requires that you write the contrast in the form of a null hypothesis. In the output above, the test syntax examines the CBT (1.tx) and BA (2.tx) difference and uses the contrast weights defined above. The output for test shows that it is testing the null hypothesis that the difference between CBT and BA is 0. The difference between these two conditions is not statistically significant, as indicated by an F test: $F(1, 148) = 2.54$, $p = 0.11$.

5.3.2 Multiple comparisons

The `lincom` command requires that you write the contrast directly as in (5.7). The `lincom` output shows that it is testing the same null hypothesis as `test` was. The output also provides an estimate of the difference between the CBT and BA conditions, along with a t test and 95% CI for the difference. The difference between conditions was $\overline{y}_\Delta = 4.9$, $p = 0.11$. Note that the p-value for this t test is identical to the F test in the `test` output. An F test comparing two conditions (with a single degree of freedom in the numerator) is equal to t^2 from a t test of the same two conditions.

Neither `test` nor `lincom` requires that we specify means that have a 0 as the contrast weight. Thus, to compare two conditions, write the comparison directly. Examples for the three comparisons described above are

```
. test 1.tx - 2.tx = 0
 ( 1)  1bn.tx - 2.tx = 0
       F(  1,   148) =    2.54
            Prob > F =    0.1129
. test 1.tx - 3.tx = 0
 ( 1)  1bn.tx - 3.tx = 0
       F(  1,   148) =    7.22
            Prob > F =    0.0080
. test 1.tx - 4.tx = 0
 ( 1)  1bn.tx - 4.tx = 0
       F(  1,   148) =    0.03
            Prob > F =    0.8584
. lincom 1.tx - 2.tx
 ( 1)  1bn.tx - 2.tx = 0
```

	Coef.	Std. Err.	t	P>|t|	[95% Conf. Interval]	
(1)	4.878212	3.059148	1.59	0.113	-1.16704	10.92346

```
. lincom 1.tx - 3.tx
 ( 1)  1bn.tx - 3.tx = 0
```

	Coef.	Std. Err.	t	P>|t|	[95% Conf. Interval]	
(1)	8.219935	3.059148	2.69	0.008	2.174683	14.26519

```
. lincom 1.tx - 4.tx
 ( 1)  1bn.tx - 4.tx = 0
```

	Coef.	Std. Err.	t	P>|t|	[95% Conf. Interval]	
(1)	-.5466792	3.059148	-0.18	0.858	-6.591931	5.498573

These tests indicate that there is a significant difference between the CBT and ADM conditions, but not between CBT and BA or CBT and placebo.

Direct adjustment for multiple comparisons

Studies often involve more than two or three comparisons, and we often want to explore the statistical significance of those comparisons. With many comparisons, standard t and F tests will not be appropriate because of α-inflation. For example, we would need six t tests to make all possible comparisons among the four conditions in the depression study. There are several methods for making comparisons among means that were designed to protect against inflated type I error rates. This chapter will cover two specific methods: Tukey's honestly significant difference (HSD) test and the Scheffé test.

When discussing these methods for protecting against inflated type I errors, we can distinguish between what is called the experimentwise error rate (α_{EW}) and the familywise error rate (α_{FW}) (Keppel and Wickens 2004; Winer et al. 1991). The experimentwise error rate refers to all possible comparisons in a study, including all pairwise comparisons among means as well as more complex comparisons involving groups of means. The Scheffé test can be used to maintain the experimentwise error rate at α_{EW}. The familywise error rate refers to subsets of comparisons from an experiment, such as all pairwise comparisons among means. These subsets are called families, and a given experiment can consist of multiple families. Tukey's HSD test can be used to maintain the familywise error rate at α_{FW}.

The Scheffé test works by adjusting the critical value we use in the null hypothesis test. In Stata, the Scheffé test is accessed using the `pwcompare` (pairwise comparisons) command, which is available for use after `anova` (among other commands). Stata will test the pairwise comparisons using a t test (see [R] **pwcompare** for details on how Stata computes the t test). The adjusted critical value for the t test is (Keppel and Wickens 2004, 129)

$$t_{\text{Scheffé}} = \sqrt{(a-1)\, F_{\alpha_{EW}}(\text{DF}_B, \text{DF}_W)} \tag{5.8}$$

where a is the number of levels of the factor (for example, treatment conditions) in your study and $F_{\alpha_{EW}}$ is the unadjusted critical F given DF_B and DF_W. Note that DF_B and DF_W are the DF from the one-way ANOVA (3 and 148, respectively, in the depression treatment example). Thus, $t_{\text{Scheffé}}$ is

```
. display "Scheffe critical value is " sqrt(3*invFtail(3, 146, 0.05))
Scheffe critical value is 2.8283781
```

Following `anova`, the `pwcompare` output with the Scheffé adjustment for the depression treatment example is

```
. pwcompare tx, mcompare(scheffe) effects
Pairwise comparisons of marginal linear predictions
Margins     : asbalanced
```

	Number of Comparisons
tx	6

5.3.2 Multiple comparisons

	Contrast	Std. Err.	Scheffe t	Scheffe P>|t|	Scheffe [95% Conf. Interval]	
tx						
BA vs CBT	-4.878212	3.059148	-1.59	0.470	-13.52927	3.772845
ADM vs CBT	-8.219935	3.059148	-2.69	0.070	-16.87099	.4311225
Placebo vs CBT	.5466792	3.059148	0.18	0.998	-8.104378	9.197737
ADM vs BA	-3.341723	3.059148	-1.09	0.755	-11.99278	5.309335
Placebo vs BA	5.424892	3.059148	1.77	0.373	-3.226166	14.07595
Placebo vs ADM	8.766614	3.059148	2.87	0.046	.1155567	17.41767

The call to `pwcompare` includes two options: `mcompare(scheffe)`, which requests the Scheffé adjustment, and `effects`, which requests that Stata print the p-values and CIs for comparison. Only one comparison had an absolute value that exceeds $t_{\text{Scheffé}}$, and thus only one comparison—ADM versus placebo—is statistically significant given $\alpha = 0.05$.

As can be seen in (5.8), the Scheffé correction is directly related to the omnibus F test. Keppel and Wickens (2004) note the consequence of this relationship:

> When the omnibus F is significant, there is at least one contrast ψ that is significant by Scheffé criterion; when the omnibus F is not significant, no contrast is significant by Scheffé's criterion. (p. 129)

This seems reasonable—if we cannot reject the null hypothesis for the overall F test, it seems unlikely that probing for pairwise differences will produce useful results. The Scheffé adjustment is used most commonly in exploratory settings, where we examine unplanned relationships.

The Tukey HSD test can be used to maintain the familywise error rate (α_{FW}) and is most commonly used when one wants to make all possible pairwise comparisons among means. Like the Scheffé test, Tukey's HSD test requires an adjustment to the critical value. The adjusted critical value is

$$D_{\text{Tukey}} = q_a \sqrt{\text{MS}_W / n}$$

where MS_W is the mean square within from the one-way ANOVA, n is the sample size per condition, and q_a is Tukey's Studentized range statistic. In Stata, q_a can be found using the `invtukeyprob()` function, which requires that we know how many means are in the study (a), the within-group DF from the one-way ANOVA (DF_W), and α.

D_{Tukey} is

```
. display invtukeyprob(4, 148, 1 - 0.05)*sqrt(177.8094/38)
7.9490748
```

Treatment conditions that have means that differ by at least |7.95| will be significant by Tukey's criterion. The pwcompare output with the Tukey HSD adjustment for the depression example is

```
. pwcompare tx, mcompare(tukey) effects
Pairwise comparisons of marginal linear predictions
Margins      : asbalanced
```

	Number of Comparisons
tx	6

	Contrast	Std. Err.	Tukey t	Tukey P>|t|	Tukey [95% Conf. Interval]	
tx						
BA vs CBT	-4.878212	3.059148	-1.59	0.385	-12.82729	3.070862
ADM vs CBT	-8.219935	3.059148	-2.69	0.040	-16.16901	-.2708602
Placebo vs CBT	.5466792	3.059148	0.18	0.998	-7.402396	8.495754
ADM vs BA	-3.341723	3.059148	-1.09	0.695	-11.2908	4.607352
Placebo vs BA	5.424892	3.059148	1.77	0.290	-2.524183	13.37397
Placebo vs ADM	8.766614	3.059148	2.87	0.024	.8175394	16.71569

The two significant differences are the ADM versus CBT comparison and the ADM versus placebo comparison, both of which had mean differences that exceeded |7.95|.

Table 5.1 provides a side-by-side comparison of the p-values for an unadjusted, Tukey, and Scheffé test. The Tukey test is more conservative than the unadjusted, and the Scheffé test is more conservative than the Tukey. This is consistent with the Tukey HSD test maintaining the familywise error rate and the Scheffé test maintaining the experimentwise error rate.

5.4 Summary

Table 5.1. Comparing the effects of the unadjusted, Tukey HSD, and Scheffé p-values

Contrast	Difference	Standard error	t	p-unadjusted	p-Tukey	p-Scheffé
BA versus CBT	−4.89	3.06	−1.59	0.11	0.39	0.47
ADM versus CBT	−8.22	3.06	−2.69	0.01	0.04	0.07
Placebo versus CBT	0.55	3.06	0.18	0.86	0.99	0.99
ADM versus BA	−3.34	3.06	−1.09	0.28	0.70	0.76
Placebo versus BA	5.42	3.06	1.77	0.08	0.29	0.37
Placebo versus ADM	8.77	3.06	2.87	0.01	0.02	0.05

5.4 Summary

The primary goal of this chapter was to introduce and illustrate how to compare means for either two groups (t test) or three or more groups (one-way ANOVA). Additionally, this chapter covered the impact of multiple comparisons on the type I error rate and discussed three ways to protect against inflated type I errors: a) planned comparisons, b) Tukey's HSD test (to control the familywise error rate), and c) the Scheffé test (to control the experimentwise error rate). Readers seeking additional coverage of t tests and one-way ANOVA can consult Keppel and Wickens (2004) and Maxwell, Delaney, and Kelley (2018). Chapter 6 covers ANOVA with two or more between-subjects factors, and chapter 7 covers ANOVA-style models for studies with within-subject factors.

6 Factorial ANOVA

The analysis of variance (ANOVA) models in chapter 5 are used for designs with a single outcome and a single factor, but many theories combine two or more factors. For example, Baldwin et al. (2013) studied whether self-persuasion promotes smoking cessation. Self-persuasion involves developing your own reasons for quitting smoking rather than reading arguments against smoking or watching anti-smoking ads. Baldwin et al. (2013) randomized participants, who smoked at least one cigarette a day, to either write their own reasons for quitting (self-condition) or read about reasons for quitting (other-condition). Baldwin et al. (2013) also wondered whether self-persuasion may be helpful because the reasons we generate for ourselves for wanting to quit smoking typically match our beliefs. That is, it is possible that reading reasons to quit that match our beliefs is just as effective as generating those reasons ourselves. To separate the source of a message from persuasiveness of a message, participants in the self- and other-conditions were further randomized to either write or read about messages they had previously rated as persuasive (matched condition) or unpersuasive (mismatched condition).

This study is known as a 2×2 factorial design. The 2×2 means that the study includes two factors (source of message and content of message), each with two levels (self versus other for source and matched versus mismatched for content). The advantage of this type of design is that it allows us to study the effects of each factor and how the factors interact. For example, does the source of a message against smoking lead to less smoking? Does the content of a message against smoking lead to less smoking? Does the effect of source depend upon the specific content of the message (and vice versa)? Because factorial designs simultaneously examine these types of questions, they are also efficient, requiring fewer participants per cell than one-factor designs (Keppel and Wickens 2004).

Factorial designs can use factors with more than two levels, such as a 3×2 design (McConnell et al. 2011), and they can include more than two factors, such as a $3 \times 3 \times 2 \times 2$ design (Ebbesen and Konecni 1975). The factorial designs discussed in this chapter are fully between-subjects designs, which means that each participant is only observed once. The Baldwin et al. (2013) study was between-subjects because participants were either in the self or in the other source condition and either in the matched or in the mismatched content condition. Factorial designs can also include all within-subject factors or a combination of within- and between-subjects factors. A factor is within-subject if participants are observed in more than one of the levels. In a longitudinal study, time is a common within-subject factor. Chapter 7 discusses within-subject designs as well as designs that mix within- and between-subjects factors.

In this chapter, you will learn the following:

1. What a multifactorial research design is.
2. Terminology and notation for factorial ANOVA, including the difference between main effects and interactions.
3. How marginal means and cell means are used to construct the sums of squares for two-factor and three-factor designs.
4. How to use ANOVA to analyze multifactorial designs.
5. How to interpret the results of a factorial ANOVA.
6. How to decompose an interaction via simple effects.
7. Effect sizes for factorial ANOVA.

Stata commands featured in this chapter

- `table`: for producing cell and marginal means
- `anova`: for fitting factorial models
- `margins`: for visualizing the results
- `contrast`: for computing simple effects
- `estat esize`: for computing effect sizes

6.1 Data for this chapter

In the United States, a person accused of a crime is afforded the right to a trial by a jury of his or her peers. A common question is how objective the juries are in making decisions. Are juries impacted by characteristics of the crime or the defendant? A series of studies have investigated the effects of the physical attractiveness of the defendant on juries' decisions. Initial studies suggested that juries tended to be more lenient to attractive defendants than to unattractive defendants (Darby and Jeffers 1988; Efran 1974; Stewart 1985). Other studies have examined how gender, type of crime, and country of origin interact with attractiveness on juries' decisions (Wuensch et al. 1993).

This chapter uses a simulated dataset inspired by these studies. The data come from a $2 \times 2 \times 2$ experiment with the following factors:

- `attract`. Attractiveness of the defendant, with two levels: attractive or unattractive.
- `sex`. Sex of the jury member, with two levels: male or female.
- `crime`. Type of crime, with two levels: burglary or swindle.

6.1 Data for this chapter

The type of crime is varied because there is some evidence that attractiveness can be more important to juries when an interpersonal relationship is part of the crime—swindling or conning a person requires an interpersonal connection, and thieves can use their attractiveness to win trust prior to the theft. This is in contrast to burglary, where the thief is not seen. Consequently, juries may consider the attractiveness of the defendant to be problematic when considering a swindle but not a burglary (Wuensch et al. 1993).

The design has eight cells, with each cell representing the cross-section of the three factors (see table 6.1). The participants, 120 men and 120 women, were randomly assigned to the levels of the `attract` and `crime` factors. Although these are simulated data, participants in this type of study typically are given a description and picture of the defendant; the picture varies across the levels of the `attract` factor. The details of the alleged crime are included, and the nature of the crime varies across the `crime` levels. Finally, defendant characteristics, such as age, are held constant across conditions. After reading about the defendant, participants rate how many years the defendant should serve in jail (Wuensch et al. 1993).

Table 6.1. Design of the $2 \times 2 \times 2$ jury study

	Men		Women	
	Burglary	Swindle	Burglary	Swindle
Attractive	$n=30$	$n=30$	$n=30$	$n=30$
Unattractive	$n=30$	$n=30$	$n=30$	$n=30$

The dataset `attractive.dta` contains five variables:

```
. use http://www.stata-press.com/data/pspus/attractive
. codebook, compact
Variable   Obs Unique     Mean  Min  Max  Label
id         240   240     120.5    1  240  Subject ID
years      240     8  3.645833    1    8  Sentence in Years
sex        240     2        .5    0    1  Gender
attract    240     2        .5    0    1  Defendant
crime      240     2        .5    0    1  Type of Crime
```

The dependent variable is `years`. The variables `sex`, `attract`, and `crime` are indicator variables that define the cell each participant is in.

Notation

It is common to label factors as A, B, C, and so on. In this example, the `attract` factor is A, `sex` is B, and `crime` is C. The outcome in a two-way factorial design is denoted as y_{ijk}, where y is the outcome for the ith person in level j of factor A and level k of factor B. An outcome for a man rating an unattractive defendant is y_{iUM}, where U is for unattractive and M is for man. In a three-way design, the outcome is denoted as y_{ijkl}, where y is the outcome for the ith person in level j of factor A, level k of factor B, and level l of factor C. This general notation is useful because it can be used to describe studies that differ with respect to the number of factors, how many levels each factor has, and the specific names of the factor. Further, many textbooks use the A, B, and C notation when describing factorial ANOVA methods. Thus, being familiar with this general notation is useful as you learn the methods.

6.2 Factorial design with two factors

Initially, I will average over crime and treat this dataset as a 2×2 design. This allows us to unpack ideas about main effects and interactions while dealing with just two factors. Interactions with three or more factors are challenging, requiring care when interpreting and requiring comfort with two-way interactions. Thus, I believe it is best to start by mastering the ins and outs of two-factor designs.

6.2.1 Examining and visualizing the data

This 2 × 2 design has four cells representing the intersection of attract and sex. Use table to compute descriptive statistics for each of the cells. Although we have used table for frequency counts of categorical variables, it can also create tables of descriptive statistics, such as the mean, standard deviation, median, and sample size for the four cells.

```
. table attract sex, c(mean years sd years p50 years n years) format(%9.2f)
```

	Gender	
Defendant	Men	Women
Unattractive	4.77	3.22
	1.33	1.06
	5.00	3.00
	60	60
Attractive	3.03	3.57
	1.04	1.03
	3.00	3.00
	60	60

Including attract and sex indicates that we want a two-way table. Because table attract sex with no options produces the sample size for each cell, we have used the contents() option with the arguments mean years, sd years, p50 years, and n years. These arguments produce the mean, standard deviation, median, and sample size for years, respectively, for each cell. The format(%9.2f) option rounds the cell results to two decimal places.

6.2.1 Examining and visualizing the data

The help documentation for `table` mentions that this output is busy and somewhat difficult to read. The output does not include labels for the statistics, and thus readers must use the syntax to determine what each number is referring to. Another option for a two-way table is to use `table`'s `by()` option:

```
. table attract, by(sex) c(mean years sd years p50 years n years) format(%9.2f)
```

Gender and Defendant	mean(years)	sd(years)	med(years)	N(years)
Men				
Unattractive	4.77	1.33	5.00	60
Attractive	3.03	1.04	3.00	60
Women				
Unattractive	3.22	1.06	3.00	60
Attractive	3.57	1.03	3.00	60

The column titles tell the reader exactly what statistics are included in the table. The only disadvantage of this display is that it does not follow the structure of a two-way table where the factors intersect (that is, one factor as the rows and one factor as the columns).

The mean and median both indicate that the most severe sentence is given by men to defendants who are unattractive. Otherwise, the conditions are roughly equivalent with respect to `years`. The standard deviation also indicates that the variability in each cell is about the same, with the most variability in the men and the unattractive condition.

Two common choices for visually inspecting the data are box plots and dot plots with confidence intervals. To create the box plot, the data need to be in wide format. Using `reshape wide` does the trick:

```
. reshape wide years, i(id) j(attract)
(note: j = 0 1)
```

Data	long	->	wide
Number of obs.	240	->	240
Number of variables	5	->	5
j variable (2 values)	attract	->	(dropped)
xij variables:			
	years	->	years0 years1

```
. graph box years0 years1, over(sex)
>       ytitle("Years")
>       legend(label(1 "Unattractive") label(2 "Attractive"))
>       scheme(lean2)
```

Figure 6.1 reflects the pattern we saw with `table`, namely, that men rating unattractive defendants produced the harshest sentences and the other conditions were all roughly equal.

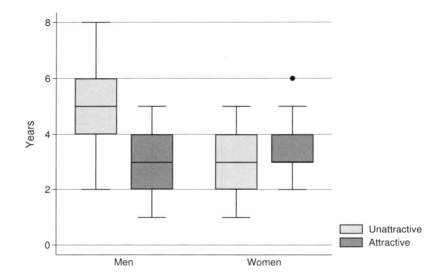

Figure 6.1. Box plot for `years` for the combination of `attract` and `sex`

There are multiple ways to create a dot plot with the means, but my usual choice is to use `anova` combined with `marginsplot`.

```
. egen twofact = group(attract sex)
. quietly anova years twofact
. margins twofact
Adjusted predictions                            Number of obs     =        240
Expression    : Linear prediction, predict()
```

	Margin	Delta-method Std. Err.	t	P>|t|	[95% Conf. Interval]	
twofact						
1	4.766667	.1449682	32.88	0.000	4.48107	5.052264
2	3.216667	.1449682	22.19	0.000	2.93107	3.502264
3	3.033333	.1449682	20.92	0.000	2.747736	3.31893
4	3.566667	.1449682	24.60	0.000	3.28107	3.852264

6.2.1 Examining and visualizing the data

```
. marginsplot, plotopts(connect(none) msize(large)) scheme(lean2)
>       xlabel(1 `" "Men" "Unattractive" "´
>               2 `" "Women" "Unattractive" "´
>               3 `" "Men" "Attractive" "´
>               4 `" "Women" "Attractive" "´)
>       plotregion(margin(20 20 0 0))
>       ylabel(0(1)6) xtitle("") ytitle("Sentence in years")
>       title("") legend(off)
>       addplot(scatter years twofact, jitter(3) msymbol(D) mcolor(gs10%30) below)
  Variables that uniquely identify margins: twofact
. drop twofact
```

This plot uses the original data rather than the wide data we created above. The first step is to create a new variable that represents the four cells in the design by using `egen` with the `group()` function. Specifically, `egen twofact = group(attract sex)` creates a new variable called `twofact` that takes on the values 1, 2, 3, and 4, which are the groups formed by the combination of `attract` and `sex` (that is, the cells of the 2×2 design). The next is to run a one-way ANOVA with `years` as the outcome and `twofact` as the factor so that we can use `margins`. The `margins twofact` command produces the means for each condition along with a standard error and confidence interval for each mean. Finally, `marginsplot` plots the means and confidence intervals (see figure 6.2).

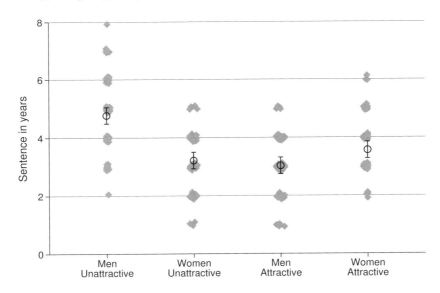

Figure 6.2. Dot plot for `years` for the combination of `attract` and `sex`

I added a few options to `marginsplot` that you may not have seen before. First, I suppressed the line that connects the dots across cells. Second, I changed the labels for the x axis to be the cell names rather than just 1, 2, 3, or 4. The strange-looking quotes for the cell labels are required to get the cell names to span two lines (Cox 2004). Third, I changed the margins for the plot by using `plotregion()` to prevent the label for the last group from running off the side of the plot (you can leave that option off to see what happens). Finally, I used `addplot()` to add a scatterplot of the data so that we can see the distribution of responses. The scatterplot includes the `jitter()` option to spread the data out and reduce the amount of overplotting.

6.2.2 Main effects

The comparisons made thus far focused on comparing specific cells of the 2×2 design. For example, the box and dot plots showed that the cell with the highest average sentence was for men rating unattractive defendants, whereas the lowest average sentence was the cell for men rating attractive defendants. We can also compare averages of cells. We could compare men versus women by averaging over type of defendant (`attract`). Likewise, we could compare attractive versus unattractive by averaging over gender (`sex`). These types of comparisons are called main effects in factorial ANOVA.

Look at figure 6.2 again. If you compare the average of the men's ratings (that is, the average of their dots) with the average of the women's ratings, you see that men, on average, provide more-severe sentences than women. Likewise, if you compare the average rating for attractive defendants with the average rating for unattractive defendants, you see that unattractive defendants receive more-severe sentences than attractive defendants.

These average ratings are called marginal means. A marginal mean is the mean of the outcome for a specific level of a factor averaging over the levels of the other factor. To be specific, the marginal mean for men is the average of the means for the men/attractive cell and the men/unattractive cell: $(4.77 + 3.03)/2 = 3.9$. Marginal means are denoted using "dot" notation (Maxwell, Delaney, and Kelley 2018). For example, the marginal mean for the jth level of A averaging over B is denoted as $\overline{y}_{j\cdot}$. The marginal mean for the kth level of B averaging over A is denoted as $\overline{y}_{\cdot k}$. Thus, the marginal mean for unattractive defendants averaging over juror sex is denoted as $\overline{y}_{U\cdot}$.

Marginal means can be added to two-way tables created by the `table` command by using the `row` and `col` options.

```
. table attract sex, c(mean years n years) format(%9.2f) row col
```

	Gender		
Defendant	Men	Women	Total
Unattractive	4.77	3.22	3.99
	60	60	120
Attractive	3.03	3.57	3.30
	60	60	120
Total	3.90	3.39	3.65
	120	120	240

6.2.3 Interactions

The table shows that a 2 × 2 design has four marginal means. The marginal means confirm what glancing at figure 6.2 suggested: men provide more-severe sentences than women averaging over type of defendant, and unattractive defendants receive more-severe sentences than attractive defendants averaging over sex.

Testing the null hypothesis

The null hypothesis for the main effect for `attract` is that the marginal mean for unattractive defendants is equal to the marginal mean for attractive defendants.

$$H_0: \mu_{U.} = \mu_{A.}$$

Said another way, the null is that the difference in the marginal means is 0.

$$H_0: \mu_{U.} - \mu_{A.} = 0$$

The null hypothesis for the main effect for `sex` is

$$H_0: \mu_{.M} = \mu_{.W} \quad \text{or} \quad H_0: \mu_{.M} - \mu_{.W} = 0$$

The null hypotheses can also be written in the A, B, and C notation for factors. For example, the `attract` factor is the A factor, with levels A_1 and A_2. The `sex` factor is the B factor, with levels B_1 and B_2. The null hypotheses for A and B in the general notation are

$$H_0: \mu_{A_1.} = \mu_{A_2.}$$
$$H_0: \mu_{.B_1} = \mu_{.B_2}$$

6.2.3 Interactions

In addition to the main effects for each factor, we can also consider how the factors interact with one another—how the effects of one factor influence the effects of the other. Interactions are one of the primary reasons for using a factorial design. In the jury study, the main effect for `sex` told us whether men and women differ with respect to the severity of their judgments, averaging over the type of defendant. But what if men and women differ more when rating unattractive defendants than when rating attractive defendants? That is, do `sex` and `attract` interact? Generally, interactions mean that the effect of one factor depends upon specific levels of another factor. If there is an interaction, averaging over `attract` would hide important information about how men and women evaluate defendants. Therefore, it is important to consider both main effects and interactions.

The general interpretation of an interaction is similar to what we saw in regression (see chapter 4). However, main effects are unique to ANOVA. Consider the following regression:

$$y = b_0 + b_1 X_1 + b_2 X_2 + b_3 X_1 X_2$$

The slope for the interaction, b_3, indexes how much the relationship between y and X_1 depends upon the specific levels of X_2.[1] The slope for X_1 is sometimes wrongly interpreted as the effect of X_1 holding constant X_2. This is wrong because there is an interaction in the model. Instead, b_1 is interpreted as the relationship between X_1 and y, when $X_2 = 0$. Likewise, b_2 is interpreted as the relationship between X_2 and y, when $X_1 = 0$.

It may seem odd that I am focusing on the interpretation of regression coefficients in an ANOVA chapter. The bottom line is that many researchers incorrectly interpret regression coefficients from models with interactions. I believe a primary reason is that researchers learn the concept of main effects in ANOVA and incorrectly apply the ANOVA interpretation to regression models.

The main effects in ANOVA are tested by comparing the marginal means. In contrast, interactions involve the cell means. This makes sense given that marginal means average over one or more factors, whereas cell means involve the intersection of two or more factors.

One way to state the null hypothesis for the interaction between sex and attract is that the difference in sentence time (years) between men and women when rating unattractive defendants is equal to the difference between men and women when rating attractive defendants.

$$H_0: \mu_{UM} - \mu_{UW} = \mu_{AM} - \mu_{AW}$$

We can also rearrange and set this equation to 0.

$$H_0: \mu_{UM} - \mu_{UW} - \mu_{AM} + \mu_{AW} = 0 \tag{6.1}$$

As in regression, interactions in ANOVA are complementary. Consequently, the null hypothesis for the interaction is that the difference in years between attractive and unattractive defendants when rated by men is the same as the difference between attractive and unattractive defendants when rated by women.

$$H_0: \mu_{UM} - \mu_{AM} = \mu_{UW} - \mu_{AW}$$

Rearranging and setting the equation to 0, we get

$$H_0: \mu_{UM} - \mu_{AM} - \mu_{UW} + \mu_{AW} = 0 \tag{6.2}$$

Equations (6.1) and (6.2) are identical, underscoring the fact that interactions are complementary.

6.2.4 Partitioning the variance

Just as we did in one-way ANOVA, we partition the total variance in factorial ANOVA into two parts:

$$SS_{Total} = SS_{Between} + SS_{Within}$$

---

[1]. Because interactions are complementary, it also indexes how much the relationship between y and X_2 depends upon the specific levels of X_1.

6.2.4 Partitioning the variance

The SS$_{\text{Total}}$ involves the deviation of each score from the grand mean [compare with (5.4)]. For a single subject, the total deviation is

$$y_{ijk} - \overline{y}_{..} \tag{6.3}$$

where y_{ijk} is the value of years for the ith person in level j of attract and level k of sex, and $\overline{y}_{..}$ is the grand mean. The grand mean includes two dots in its notation because we average A and B to compute it. Likewise, the SS$_{\text{Within}}$ involves the deviation of observations from the relevant cell mean. For a single subject, that deviation is

$$\epsilon_{ijk} = \left(y_{ijk} - \overline{y}_{jk}\right) \tag{6.4}$$

where $\overline{y}_{jk}$ is the cell mean for level j of attract and level k of sex. Finally, the SS$_{\text{Between}}$ involves the deviation of the cell means from the grand mean. For a single subject, that deviation is

$$\overline{y}_{jk} - \overline{y}_{..}$$

Unlike one-way ANOVA models, the between-cells deviation gets partitioned into three parts: i) the main effect of the A factor (attract), ii) the main effect of the B factor (sex), and iii) the interaction between A and B (attract#sex).

For a single subject, the main effect for A is estimated as

$$\alpha_j = \left(\overline{y}_{j.} - \overline{y}_{..}\right) \tag{6.5}$$

where α_j is the effect of the jth level of A averaged over the levels of B and $\overline{y}_{j.}$ is the marginal mean for the jth level of A. Thus, in the defendant dataset, there would be two unique values of α because there are two levels of attract:

$$\widehat{\alpha}_U = \overline{y}_{U.} - \overline{y}_{..} = 3.99 - 3.65 = 0.34$$
$$\widehat{\alpha}_A = \overline{y}_{A.} - \overline{y}_{..} = 3.3 - 3.65 = -0.35$$

The main effect for B is estimated as

$$\beta_k = \left(\overline{y}_{.k} - \overline{y}_{..}\right) \tag{6.6}$$

where β_k is the effect of the kth level of B averaged over the levels of A and $\overline{y}_{.k}$ is the marginal mean for the kth level of B. Thus, in the defendant dataset, there would be two unique values of β because there are two levels of sex:

$$\widehat{\beta}_M = \overline{y}_{.M} - \overline{y}_{..} = 3.90 - 3.65 = 0.25$$
$$\widehat{\beta}_W = \overline{y}_{.W} - \overline{y}_{..} = 3.39 - 3.65 = -0.26$$

The interaction is the part of the deviation of cell means from the grand mean that remains after the main effects are removed (Keppel and Wickens 2004, eq. 11.2). The interaction effect for the intersection of the jth level of A and the kth level of B is denoted as $\alpha\beta_{jk}$ and is computed as

$$\begin{aligned}
\alpha\beta_{jk} &= (\overline{y}_{jk} - \overline{y}_{..}) - \alpha_j - \beta_k \\
&= (\overline{y}_{jk} - \overline{y}_{..}) - (\overline{y}_{j.} - \overline{y}_{..}) - (\overline{y}_{.k} - \overline{y}_{..}) \\
&= \overline{y}_{jk} - \overline{y}_{..} - \overline{y}_{j.} + \overline{y}_{..} - \overline{y}_{.k} + \overline{y}_{..} \\
\alpha\beta_{jk} &= \overline{y}_{jk} - \overline{y}_{j.} - \overline{y}_{.k} + \overline{y}_{..}
\end{aligned} \tag{6.7}$$

The four interaction effects in the defendant study are

$$\begin{aligned}
\widehat{\alpha\beta}_{UM} &= 4.77 - 3.99 - 3.9 + 3.65 &= 0.53 \\
\widehat{\alpha\beta}_{UW} &= 3.22 - 3.99 - 3.39 + 3.65 &= -0.51 \\
\widehat{\alpha\beta}_{AM} &= 3.03 - 3.3 - 3.9 + 3.65 &= -0.52 \\
\widehat{\alpha\beta}_{AW} &= 3.57 - 3.3 - 3.39 + 3.65 &= 0.53
\end{aligned}$$

Taken together, the preceding discussion shows that each participant's total deviation is divided into four parts in a 2×2 design: the main effect for A (`attract`), the main effect for B (`sex`), the interaction of A and B (`attract#sex`), and the residual or within-cell deviation. If we combine (6.3), (6.4), (6.5), (6.6), and (6.7), we see that the total deviation can be written as

$$y_{ijk} - \overline{y}_{..} = \alpha_j + \beta_k + \alpha\beta_{jk} + \epsilon_{ijk}$$

Or if we write the total deviation in terms of the observations, marginal means, cell means, and grand mean, we get (Keppel and Wickens 2004, eq. 11.4)

$$y_{ijk} - \overline{y}_{..} = (\overline{y}_{j.} - \overline{y}_{..}) + (\overline{y}_{.k} - \overline{y}_{..}) + (\overline{y}_{jk} - \overline{y}_{j.} - \overline{y}_{.k} + \overline{y}_{..}) + (y_{ijk} - \overline{y}_{jk})$$

We can square each deviation and sum across all participants to obtain the sum of squares (SS):

$$\sum (y_{ijk} - \overline{y}_{..})^2 = \sum (\overline{y}_{j.} - \overline{y}_{..})^2 + \sum (\overline{y}_{.k} - \overline{y}_{..})^2 + \sum (\overline{y}_{jk} - \overline{y}_{j.} - \overline{y}_{.k} + \overline{y}_{..})^2 \\
+ \sum (y_{ijk} - \overline{y}_{jk})^2 \tag{6.8}$$

Equation (6.8) can be written in words as

$$\text{SS}_{\text{Total}} = \text{SS}_A + \text{SS}_B + \text{SS}_{AB} + \text{SS}_{\text{Within}}$$

These sums of squares are a fundamental part of the source table for a factorial ANOVA and are produced when using the `anova` command.

6.2.5 2 x 2 source table

Table 6.2 is a complete source table for a 2×2 design. The pattern for the degrees of freedom (DF) is similar to what we saw in one-way ANOVA (see chapter 5). Keppel and Wickens (2004) provide a description of how DF is defined:

6.2.5 2 × 2 source table

> The number of degrees of freedom for a sum of squares equals the number of different observations used to calculate it minus the number of constraints imposed on those observations. (p. 219)

The DF for the A factor is $a-1$, where a is the number of levels for A and also refers to the number of marginal means for that factor. The marginal means represent the "different observations" in the description above. The constraint refers to the grand mean ($\overline{y}_{..}$) that is subtracted from each marginal mean. The grand mean is a constraint because it is estimated from the data (that is, it is a data-based proxy for a population mean), as are the marginal means. The marginal means and the grand mean are therefore related to one another. Specifically, when cell sizes are equal, the average of the marginal means for A must equal the grand mean. For example, if you had two numbers, x and 4, and I said the average of x and 4 was 6, you would know that the value of x was 8. In this example, x and 4 are like the marginal means and 6 is the grand mean.

The DF for the main effect of the B factor follows the exact same pattern as the main effect for A. The DF for the interaction is equal to the DF for A times the DF for B. Finally, the DF for SS_{Within} is equal to the number of participants per cell (n) minus 1 times the levels of A and the levels of B. For each cell, there is a single constraint on SS_{Within} because the cell mean is removed from each observation. We can rewrite the DF for SS_{Within} as $abn - ab$. There are abn observations in the dataset and there are ab cell means, which means that there are ab constraints on SS_{Within} (Keppel and Wickens 2004).

Table 6.2. Factorial ANOVA source table for a 2 × 2 design

Source	SS	DF	MS	F
A	$\sum (\overline{y}_{j.} - \overline{y}_{..})^2$	$a-1$	SS_A/DF_A	$MS_A/MS_{S/AB}$
B	$\sum (\overline{y}_{.k} - \overline{y}_{..})^2$	$b-1$	SS_B/DF_B	$MS_B/MS_{S/AB}$
A×B	$\sum (\overline{y}_{jk} - \overline{y}_{j.} - \overline{y}_{.k} + \overline{y}_{..})^2$	$(a-1)(b-1)$	SS_{AB}/DF_{AB}	$MS_{AB}/MS_{S/AB}$
S/A×B	$\sum (y_{ijk} - \overline{y}_{jk})^2$	$ab(n-1)$	$SS_{S/AB}/DF_{S/AB}$	

The mean squares (MS) are computed as they were in one-way ANOVA. Divide the SS for a given line in the source table by the DF for that line. Finally, F is computed by dividing the MS for a given line by the MS within. We do not always create F in this way. If a design includes "random factors", then F is constructed differently. Random factors and related ANOVA designs are beyond the scope of this book. Interested readers can consult Keppel and Wickens (2004), Judd, Westfall, and Kenny (2012), Kenny and Judd (1986), and Westfall, Judd, and Kenny (2015) for a discussion of these designs and some specific examples of when random factors are present in psychology and related fields.

6.2.6 Using anova to estimate a factorial ANOVA

We now have sufficient background to understand the output from anova for a factorial model. The full model, including both the main effects and the interaction, is estimated with anova as

```
. anova years attract sex attract#sex
                 Number of obs =      240    R-squared     = 0.2686
                 Root MSE      =  1.12292    Adj R-squared = 0.2594

      Source |  Partial SS      df       MS           F     Prob>F
  -----------+----------------------------------------------------
       Model |   109.3125        3     36.4375       28.90   0.0000
             |
     attract |  28.704167        1   28.704167       22.76   0.0000
         sex |  15.504167        1   15.504167       12.30   0.0005
 attract#sex |  65.104167        1   65.104167       51.63   0.0000
             |
    Residual |  297.58333      236   1.2609463
  -----------+----------------------------------------------------
       Total |  406.89583      239    1.702493
```

The output includes the full source table. The main effect for attract is statistically significant, $F(1, 236) = 22.8$, $p < 0.001$, consistent with differences between attractive and unattractive defendants averaging over sex. The F value does not provide information regarding whether attractive or unattractive defendants received harsher sentences. Rather, we require plots or descriptive statistics such as those we used previously (those suggested that when averaging over men and women, unattractive defendants received harsher sentences than attractive defendants). We can also use margins to assist with interpretation (see section 5.3.2).

The main effect for sex is statistically significant, $F(1, 236) = 12.3$, $p < 0.001$, consistent with differences between men and women when averaging over attract. The previous plots and descriptive statistics suggested that men rated defendants more harshly than women when averaging over attract.

The main effects do not tell the whole story and maybe should not even be interpreted because there is a significant interaction between attract and sex: $F(1, 236) = 51.6$, $p < 0.001$. Recall what interactions indicate: the relationship between one variable depends upon the specific levels of another variable. If there is an interaction, it does not make much sense to describe the effects of one variable averaged over all levels of another. The significant interaction implies that the difference between attractive and unattractive defendants depends upon whether they are rated by men or women.

Figures 6.1 and 6.2 suggest that men show a difference between attractive and unattractive defendants. There is a slight difference for women. What is clear is that the difference is larger for men than women. Using margins after anova can help unpack the interaction effect more specifically and can help us create better graphs for visualizing the interaction.

6.2.6 Using anova to estimate a factorial ANOVA

The `margins` command for plotting the interaction is straightforward:

```
. margins attract#sex
Adjusted predictions                              Number of obs   =       240
Expression   : Linear prediction, predict()
```

	Margin	Delta-method Std. Err.	t	P>|t|	[95% Conf. Interval]	
attract#sex						
Unattractive # Men	4.766667	.1449682	32.88	0.000	4.48107	5.052264
Unattractive # Women	3.216667	.1449682	22.19	0.000	2.93107	3.502264
Attractive # Men	3.033333	.1449682	20.92	0.000	2.747736	3.31893
Attractive # Women	3.566667	.1449682	24.60	0.000	3.28107	3.852264

```
. marginsplot, scheme(lean2) plotregion(margin(20 20 0 0))
>         ylabel(0(1)6) ytitle("Sentence in years")
>         title("Defendant type by sex interaction")
  Variables that uniquely identify margins: attract sex
```

The `margins` command produces the means and 95% confidence intervals for each cell, and `marginsplot` produces a dot plot of the results (see figure 6.3).

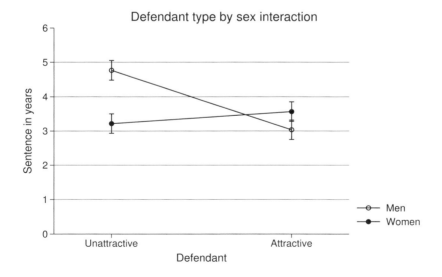

Figure 6.3. Dot plot of the `attract` by `sex` interaction

The option `ylabel(0(1)6)` defines the range of the y axis. Graphing interactions is an exercise in restraint. At the right scale, any interaction can look big, and reporting results that look big can make findings seem important. In this example, the default y-axis range began at 2.5 and ran to 5.5. This made some of the differences, such as the difference between men and women for unattractive defendants, appear quite large. Of course, if a goal of the analysis is to show that there is not an interaction or to show that a statistically significant interaction is not all that meaningful because it is small, we could go too far in the other direction and make the scale of the y axis too large. In the end, there is no one right answer to what the scale of the axes should be. I contend that the goal is to communicate the results in a way that is transparent and defensible.

Figure 6.3 demonstrates that the difference between men and women is largest when a defendant is unattractive, with men providing harsher sentences than women. Women provided slightly higher ratings than men when defendants were attractive. Figure 6.3 features `attract` on the y axis and `sex` as different lines and dots. This was the default construction based on the order of the variables in the `margins` command. Had we instead typed `margins sex#attract`, our graph would have `sex` on the y axis and `attract` as different lines and dots. The `xdimension()` option of `marginsplot` can also be used to control which variable is plotted on the y axis, regardless of the order in `margins`. As with the scale of the y axis, the choice of what to put on the y axis is up to you.

It is possible to use `margins` to produce the main effects as well as the interaction. As stated above, I do not believe it is useful to interpret main effects when there is a significant interaction because the interaction implies multiple effects rather than a single main effect. Nevertheless, if you believe there is a compelling reason to interpret the main effect and the interaction, or if the interaction is not significant, you can use `margins` to do so. If you want to just compute and graph the main effect for `attract`, the command is `margins attract`. If you want to compute and graph the main effect for `attract` as well as the interaction, the command is `margins attract attract#sex`. Either command can be followed by `marginsplot`.

6.2.7 Simple effects

The statistically significant interaction indicates that the difference between men and women varied as a function type of defendant. Figure 6.3 indicates that men rated defendants more harshly than women when the defendant was unattractive. Women rated defendants more harshly than men when the defendant was attractive, though the sex difference was much smaller than with unattractive defendants.[2] The sex differences at specific levels of `attract` are called simple effects. More generally, simple effects are the effects of a factor at one level of another factor.

A significant interaction does not indicate which simple effects are statistically significant. That is, we do not know whether the sex difference for unattractive defendants is significant, the sex difference for attractive defendants is significant, or both are

---

2. Although we discuss the sex differences varying across type of defendant, because interactions are complementary, we could also discuss differences between defendants varying across sex. The concepts in this section can be applied to either difference.

6.2.7 Simple effects

significant. We can conduct a significance test for simple effects using the `contrast` postestimation command.

There are four simple effects we will test. The first two are the sex differences. One null hypothesis is that the difference between men's ratings and women's ratings for attractive defendants is 0:

$$H_0: \mu_{AM} - \mu_{AW} = 0$$

The second null is that the difference between men's ratings and women's ratings for unattractive defendants is 0:

$$H_0: \mu_{UM} - \mu_{UW} = 0$$

The `contrast` command to test these hypotheses is

```
. contrast sex@attract, effects mcompare(scheffe)
Contrasts of marginal linear predictions
Margins      : asbalanced
```

	df	F	P>F	Scheffe P>F
sex@attract				
Unattractive	1	57.16	0.0000	0.0000
Attractive	1	6.77	0.0099	0.0356
Joint	2	31.96	0.0000	
Denominator	236			

Note: Scheffe-adjusted p-values are reported for tests on individual contrasts only.

	Number of Comparisons
sex@attract	2

	Contrast	Std. Err.	Scheffe t	Scheffe P>|t|	[95% Conf. Interval]	
sex@attract (Women vs base)						
Unattractive	-1.55	.205016	-7.56	0.000	-2.055029	-1.044971
(Women vs base)						
Attractive	.5333333	.205016	2.60	0.036	.0283042	1.038362

The `sex@attract` argument computes the simple effects of `sex` at each level of `attract`. The `effects` option tells Stata to print the differences and the t test for each difference. The `mcompare(scheffe)` option adjusts the *p*-values and confidence intervals for multiple comparisons by using the Scheffé method (see section 5.3.2).

The output shows that women are compared to the base, which in this case is men. Therefore, a positive value indicates that women sentenced defendants to more years than men, and a negative value means the opposite. The results indicate that women sentenced unattractive defendants to 1.6 fewer years than men. This difference was statistically significant, $t(236) = -7.56, p < 0.001$. The results also indicate that women sentenced attractive defendants to 0.5 more years than men, $t(236) = 2.6$, $p = 0.036$. These are the differences between the relevant cell means. However, the standard error for the differences as well as the DF for these tests are drawn from the ANOVA model rather than a simple t test on the cell means.

The third and fourth simple effects from this interaction involve the differences between unattractive and attractive defendants for specific levels of `sex`. One null hypothesis is that the difference between men's sentences for unattractive and attractive defendants is 0:

$$H_0: \mu_{UM} - \mu_{AM} = 0$$

The final null hypothesis is that the difference between women's sentences for unattractive and attractive defendants is 0:

$$H_0: \mu_{UW} - \mu_{AW} = 0$$

```
. contrast attract@sex, effects mcompare(scheffe)
Contrasts of marginal linear predictions
Margins      : asbalanced
```

	df	F	P>F	Scheffe P>F
attract@sex				
Men	1	71.48	0.0000	0.0000
Women	1	2.91	0.0891	0.2350
Joint	2	37.20	0.0000	
Denominator	236			

Note: Scheffe-adjusted p-values are reported for tests on individual contrasts only.

	Number of Comparisons
attract@sex	2

	Contrast	Std. Err.	Scheffe t	Scheffe P>|t|	[95% Conf. Interval]	
attract@sex (Attractive vs base)						
Men	-1.733333	.205016	-8.45	0.000	-2.238362	-1.228304
(Attractive vs base)						
Women	.35	.205016	1.71	0.235	-.1550291	.8550291

There was a significant difference between attractive and unattractive defendants for men, $t(236) = -8.45$, $p < 0.001$. However, the difference was not significant for women, $t(236) = 1.71$, $p = 0.235$.

6.2.8 Effect size

Like in one-way ANOVA (see chapter 5), commonly used effect sizes include η^2, ω^2, and Cohen's d. Cohen's d can be used to describe the size of the difference between two cell means, such as those that make up a simple effect. The concepts are identical to what we discussed in section 5.2.2 and so will not be discussed further here. Both η^2 and ω^2 are used to describe the size of the effect for main effects and interactions.

The most commonly used effect size in ANOVA models is η^2. In a one-way design, η^2 is the proportion of variance in the outcome accounted for by the factor. It is computed as

$$\eta^2 = \frac{\text{SS}_{\text{Between}}}{\text{SS}_{\text{Total}}}$$

In a factorial design, it is common to compute partial η^2, which is interpreted as the proportion of variance in the outcome accounted for by a factor once the variances of the other factors are removed. Recall that the sums of squares for a two-way factorial design are SS_A, SS_B, SS_{AB}, and $\text{SS}_{\text{Within}}$. When computing partial η^2, we use the SS for the factor of interest and the $\text{SS}_{\text{Within}}$. Thus, partial η^2 for the main effect of A is computed as [Warner 2013, eq. 13.27]

$$\eta^2_A = \frac{\text{SS}_A}{\text{SS}_A + \text{SS}_{\text{Within}}} \tag{6.9}$$

Because we are removing the variability associated with B and the A by B interaction, (6.9) does not involve either SS_B or SS_{AB}. Partial η^2 for the other effects are

$$\eta^2_B = \frac{\text{SS}_B}{\text{SS}_B + \text{SS}_{\text{Within}}}$$

$$\eta^2_{AB} = \frac{\text{SS}_{AB}}{\text{SS}_{AB} + \text{SS}_{\text{Within}}}$$

Use the `estat esize` postestimation command to compute partial η^2 following anova. By default, `estat esize` computes η^2.

```
. estat esize
Effect sizes for linear models
```

Source	Eta-Squared	df	[95% Conf. Interval]	
Model	.2686498	3	.1713351	.3488323
attract	.087972	1	.0310077	.1615966
sex	.0495202	1	.009584	.1123387
attract#sex	.1795049	1	.099018	.2641386

Note: Eta-Squared values for individual model terms are partial.

Thus, the main effect for `attract` accounts for 8.8% of the variability in `years` after partialing out the other effects; the main effect for `sex` accounts for 5.0%; and the interaction accounts for 18.0%. Stata also computes η^2 for the whole model, which is interpreted as the proportion of variance in the outcome accounted for by all factors (it is the same as R^2 from a regression model). In this example, it was 26.9%. I have not seen η^2 for the full model reported in conjunction with a factorial ANOVA model.

An alternative to η^2 is to use ω^2. ω^2 is the estimated population effect size, whereas η^2 is a sample effect size. Further, η^2 overestimates the population effect size because even if a factor has no relationship to the outcome in the population ($\omega^2 = 0$), a sample of data will show a relationship. It may be small, but the relationship will be there. Therefore, η^2 will nearly always be nonzero (Keppel and Wickens 2004). In contrast, ω^2 adjusts for chance relationships and thus is an unbiased estimator of the population effect size. Consequently, if you want to make inferences about the population effect size—which more often than not is what researchers want to do—use ω^2.

In a one-way design, ω^2 is the proportion of variance in the outcome accounted for by the factor in the population. An estimate of ω^2 is computed as

$$\widehat{\omega}^2 = \frac{(a-1)(F-1)}{(a-1)(F-1) + an} \tag{6.10}$$

where a is the number of levels of the factor, n is the sample size per level of a, and F is the F-ratio for the between-groups effect. The adjustment for chance is accomplished via the subtraction of 1 from F. If there is no relationship between a factor and the outcome, then $F = 1$; the only variability between groups is chance variability. That chance variability is subtracted from F in (6.10) (Keppel and Wickens 2004).

In a factorial design, it is common to estimate partial ω^2, which is interpreted as the proportion of variance in the outcome accounted for by a factor once the variances of the other factors are removed. Partial ω^2 for the main effect of A only includes the F-ratio for A in the denominator because we exclude the variance for the main effect of B and the interaction of A and B. Partial ω^2 for the main effect of A is computed as [Keppel and Wickens 2004, eq. 11.22]

$$\widehat{\omega}_A^2 = \frac{\mathrm{DF}_A\,(F_A - 1)}{\mathrm{DF}_A\,(F_A - 1) + abn}$$

where a is the number of levels of A, b is the number of levels of B, n is cell size, DF_A is the DF for the main effect of A, and F_A is the F-ratio for the main effect of A. Partial ω^2 for the other two effects are

$$\widehat{\omega}_B^2 = \frac{\mathrm{DF}_B\,(F_B - 1)}{\mathrm{DF}_B\,(F_B - 1) + abn}$$

$$\widehat{\omega}_{AB}^2 = \frac{\mathrm{DF}_{AB}\,(F_{AB} - 1)}{\mathrm{DF}_{AB}\,(F_{AB} - 1) + abn}$$

To compute partial ω^2, use `estat esize` with the `omega` option.

```
. estat esize, omega
Effect sizes for linear models
```

Source	Omega-Squared	df
Model	.2585518	3
attract	.0837837	1
sex	.0453103	1
attract#sex	.1754183	1

```
Note: Omega-Squared values for individual model terms are partial.
```

We see that the main effect for `attract` accounts for 8.4% of the variability in the outcome after partialing out the other effects; the main effect for `sex` accounts for 4.5% of the variability; and the interaction accounts for 17.5% of the variability. Each of these estimates is smaller than the corresponding η^2 estimates. The difference is not large (which is typical); however, in the long run, the ω^2 values will be closer to the population effect size than η^2.

6.3 Factorial design with three factors

The two-way analysis suggests that the sentences men and women gave to defendants differed the most when defendants were unattractive. Specifically, men provided more severe sentences than women when defendants were unattractive ($\overline{y}_{UW} - \overline{y}_{UM} = -1.6$ years). In contrast, women provided more severe sentences than men when defendants were attractive, though the difference was smaller ($\overline{y}_{AW} - \overline{y}_{AM} = 0.5$ years).

These analyses have ignored a third factor—the type of crime the defendant is charged with. Specifically, `crime` has two levels: swindle and burglary. As discussed in section 6.1, `crime` could interact with `attract` because a swindle requires a personal relationship between the criminal and the victim whereas a burglary does not. Thus, attractive defendants may be judged more harshly than unattractive defendants when they are convicted of a swindle as compared with a burglary (Wuensch et al. 1993). Furthermore, given the sex difference in judgments, it is also possible that `sex` might interact with `crime` as well.

6.3.1 Examining and visualizing the data

As with the two-way model, it is useful to start by examining the data across the cells. Use `table` to create a three-dimension table of cell means. Section 6.2.1 demonstrates how to add additional summary statistics to the `table` output.

```
. table sex attract crime, c(mean years) format(%9.2f)
```

	Type of Crime and Defendant			
	Burglary		Swindle	
Gender	Unattractive	Attractive	Unattractive	Attractive
Men	5.37	2.97	4.17	3.10
Women	3.20	3.57	3.23	3.57

Although the three-way table provides exactly what we need, the way the information is displayed can be a bit difficult to follow, especially if there are more than two levels to any given factor. The `by()` option can help simplify the presentation. For example, removing `sex` from the `table` varlist and adding `by(sex)` as an option produces two tables showing the sex-specific cell means for `attract` and `crime`.

```
. table attract crime, by(sex) c(mean years) format(%9.2f)
```

Gender and Defendant	Type of Crime	
	Burglary	Swindle
Men		
Unattractive	5.37	4.17
Attractive	2.97	3.10
Women		
Unattractive	3.20	3.23
Attractive	3.57	3.57

Which form of `table` to use is a matter of preference. However, I find that using the `by()` option makes the table more digestible, especially when factors have three or more levels.

6.3.1 Examining and visualizing the data

We can create figure 6.4, which is similar to figure 6.1 except now we add panels for the levels of `crime`. Panels are created with the `by()` option:

```
. reshape wide years, i(id) j(attract)
(note: j = 0 1)

Data                                 long    ->   wide
-----------------------------------------------------------------------------
Number of obs.                        480    ->    240
Number of variables                     5    ->      5
j variable (2 values)              attract   ->   (dropped)
xij variables:
                                     years   ->   years0 years1
-----------------------------------------------------------------------------

. graph box years0 years1, over(sex) by(crime, note(""))
>         ytitle("Years")
>         legend(label(1 "Unattractive") label(2 "Attractive"))
>         scheme(lean2)
```

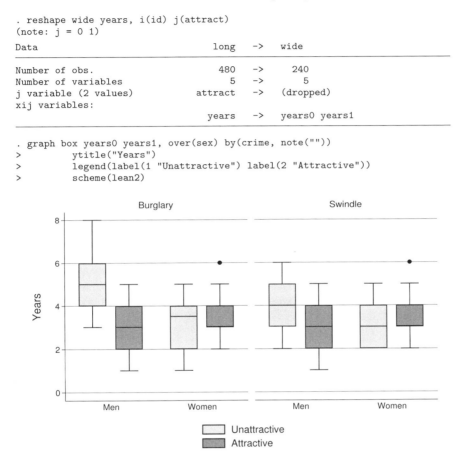

Figure 6.4. Box plot for `years` for the combination of `attract`, `sex`, and `crime`

Which factor you choose to use in the `over()` option and which factor you use in the `by()` option is simply a matter of preference and clarity. Test out different ways of reshaping the data and altering what you use in `over()` and `by()` to see how the box plot changes, and determine whether your conclusions change at all as a consequence.

The summary statistics and figure 6.4 are not consistent with the hypothesis that attractive defendants accused of swindles receive harsher punishments than unattractive defendants. Unattractive defendants accused of burglary receive the harshest punishments when judged by men only. Thus, the descriptive data are consistent with a three-way interaction between `sex`, `attract`, and `crime`, although it is not consistent with the hypothesized relationships.

6.3.2 Marginal means

Two-way designs have cell means, marginal means, and a grand mean. Three-way designs also have cell means and a grand mean. The marginal means, however, have two levels: a) marginal means that are averaged over two factors and b) marginal means that are averaged over one factor.

Keppel and Wickens (2004) refer to marginal means averaged over one factor as two-way marginal means because they represent the means created at the intersection of two variables. In a three-way design, there are three sets of two-way marginal means:

1. $A \times B$ averaging over C
2. $A \times C$ averaging over B
3. $B \times C$ averaging over A

The two-way marginal means are denoted using "dot" notation. For example, the two-way marginal mean for the intersection of the jth level of A and the kth level of B averaging over C is denoted as $\bar{y}_{jk.}$. The dot following j and k indicates that we have averaged over all levels of C. The two-way marginal mean for unattractive defendants rated by men averaging over type of crime is denoted as $\bar{y}_{UM.}$ or $\bar{y}_{A_1 B_1 .}$.

We can use `table` to produce these two-way marginal means. The key is to exclude the factor that we average over when creating the table. Note that the `sex` by `attract` two-way marginal means are the same as the marginal means discussed in section 6.2.2. In the two-way analysis, we did not include the `crime` factor, which simply means that we averaged over that factor. The three two-way marginal means are

```
. table attract sex, c(mean years) format(%9.2f)
```

	Gender	
Defendant	Men	Women
Unattractive	4.77	3.22
Attractive	3.03	3.57

```
. table attract crime, c(mean years) format(%9.2f)
```

	Type of Crime	
Defendant	Burglary	Swindle
Unattractive	4.28	3.70
Attractive	3.27	3.33

```
. table sex crime, c(mean years) format(%9.2f)
```

	Type of Crime	
Gender	Burglary	Swindle
Men	4.17	3.63
Women	3.38	3.40

6.3.3 Main effects and interactions

Keppel and Wickens (2004) refer to the marginal means averaging over two factors as one-way marginal means because these means are for a single factor. There are three sets of one-way marginal means in a three-way design:

1. A averaging over B and C
2. B averaging over A and C
3. C averaging over A and B

The one-way marginal means use the dot notation. For example, the one-way marginal mean for the jth level of the A factor is denoted as $\overline{y}_{j..}$. The two dots indicate that we have averaged over all levels of B and C. The one-way marginal mean for unattractive defendants averaging over sex and crime is denoted as $\overline{y}_{U..}$ or $\overline{y}_{A_1..}$.

```
. table attract, c(mean years) format(%9.2f)
```

Defendant	mean(years)
Unattractive	3.99
Attractive	3.30

```
. table sex, c(mean years) format(%9.2f)
```

Gender	mean(years)
Men	3.90
Women	3.39

```
. table crime, c(mean years) format(%9.2f)
```

Type of Crime	mean(years)
Burglary	3.78
Swindle	3.52

6.3.3 Main effects and interactions

Armed with the cell means, two-way marginal means, one-way marginal means, and the grand mean, we can now go through how the main effects and the interactions are defined in a three-way model. The effects in three-way models are direct extensions of the effects in two-way models, but with more complexity and more parts. The three-way model has three main effects, three two-way interactions, and one three-way interaction.

The main effects describe the variability of the one-way marginal means. The null hypothesis is that there is no difference among the one-way marginal means. For example, the main effect for attract is

$$H_0: \mu_{A..} = \mu_{U..}$$

where $\mu_{A..}$ is the population mean for the attractive defendants averaging over **sex** and **crime**, and $\mu_{U..}$ is the population mean for the unattractive defendants averaging over **sex** and **crime**. The null hypotheses for the main effects for **sex** and **crime** are, respectively,

$$H_0: \mu_{.M.} = \mu_{.W.} \quad \text{and} \quad H_0: \mu_{..B} = \mu_{..S}$$

The SS column of table 6.3 shows how the variability due to the main effect is computed using the one-way marginal means and the grand mean.

The two-way interactions involve the two-way marginal means. The null hypothesis for the **attract** by **crime** interaction (that is, $A \times C$) is that the difference in the sentence given to unattractive and attractive defendants who are convicted of a burglary is equal to the difference in the sentence given to unattractive and attractive defendants who are convicted of a swindle.

$$H_0: \mu_{U.B} - \mu_{A.B} = \mu_{U.S} - \mu_{A.S}$$

The null hypotheses for the **attract** by **sex** interactions and the **sex** by **crime** interactions are, respectively,

$$H_0: \mu_{UM.} - \mu_{AM.} = \mu_{UW.} - \mu_{AW.} \qquad (6.11)$$
$$H_0: \mu_{.MB} - \mu_{.WB} = \mu_{.MS} - \mu_{.WS}$$

Table 6.3 shows how the interactions are computed.

6.3.3 Main effects and interactions

Table 6.3. Factorial ANOVA source table for a $2 \times 2 \times 2$ design

Source	SS	DF	MS	F
A	$\sum \left(\overline{y}_{j..} - \overline{y}\right)^2$	$a-1$	SS_A/DF_A	$MS_A/MS_{S/AB}$
B	$\sum \left(\overline{y}_{.k.} - \overline{y}\right)^2$	$b-1$	SS_B/DF_B	$MS_B/MS_{S/AB}$
C	$\sum \left(\overline{y}_{..l} - \overline{y}\right)^2$	$c-1$	SS_C/DF_C	$MS_C/MS_{S/AB}$
A×B	$\sum \left(\overline{y}_{jk.} - \overline{y}_{j..} - \overline{y}_{.k.} + \overline{y}\right)^2$	$(a-1)(b-1)$	SS_{AB}/DF_{AB}	$MS_{AB}/MS_{S/AB}$
A×C	$\sum \left(\overline{y}_{j.l} - \overline{y}_{j..} - \overline{y}_{..l} + \overline{y}\right)^2$	$(a-1)(c-1)$	SS_{AC}/DF_{AC}	$MS_{AC}/MS_{S/AB}$
B×C	$\sum \left(\overline{y}_{.kl} - \overline{y}_{.k.} - \overline{y}_{..l} + \overline{y}\right)^2$	$(b-1)(c-1)$	SS_{BC}/DF_{BC}	$MS_{BC}/MS_{S/AB}$
A×B×C	$\sum \left(\overline{y}_{jkl} - \overline{y}_{jk.} - \overline{y}_{j.l} - \overline{y}_{.kl} + \overline{y}_{j..} + \overline{y}_{.k.} + \overline{y}_{..l} - \overline{y}\right)^2$	$(a-1)(b-1)(c-1)$	SS_{ABC}/DF_{ABC}	$MS_{ABC}/MS_{S/AB}$
S/A×B	$\sum \left(y_{ijkl} - \overline{y}_{jkl}\right)^2$	$ab(n-1)$	$SS_{S/AB}/DF_{S/AB}$	

6.3.4 Three-way interaction

The three-way interaction is the trickiest part of a three-way design to understand. In the case of a two-way interaction, an interaction means that the effect of a factor varies as a function of the specific levels of another factor. A three-way interaction follows the same pattern. Specifically, a three-way interaction means that the size of a two-way interaction varies as a function of the specific levels of another factor. It might even mean that the direction of an interaction varies as a function of the specific levels of another factor. In other words, a three-way interaction is the moderation of a two-way interaction—an interaction with an interaction. The multiple levels of interactions makes interpreting three-way interactions challenging.

In the jury example, a three-way interaction means that the size of the interaction between the attractiveness of the defendant and the juror's sex depends upon the specific type of crime committed. The null hypothesis is that the attract by sex interaction does not depend on crime. Recall that the null hypothesis for the two-way attract by sex interaction is written as [see (6.11)]

$$H_0: \mu_{UM.} - \mu_{AM.} = \mu_{UW.} - \mu_{AW.}$$

This says that the difference between unattractive and attractive defendants for men is equal to the difference between unattractive and attractive defendants for women. The means in this null hypothesis are the two-way marginal means—they average over crime. That is, the two-way interaction does not reference any specific levels of crime.

In the three-way interaction, rather than average over crime, we test the two-way interaction at specific levels of crime. Consequently, the interaction is built using the cell means because the cell means are at the intersection of attract, sex, and crime. The null hypothesis for the three-way interaction is written as

$$H_0: \overbrace{(\mu_{UMB} - \mu_{AMB}) - (\mu_{UWB} - \mu_{AWB})}^{\text{attract} \times \text{sex when crime} = \text{burglary}} = \overbrace{(\mu_{UMS} - \mu_{AMS}) - (\mu_{UWS} - \mu_{AWS})}^{\text{attract} \times \text{sex when crime} = \text{swindle}} \qquad (6.12)$$

The left-hand term in (6.12) is the two-way interaction between attract and sex when the crime is a burglary. The right-hand term is the attract by sex interaction when the crime is a swindle. Because the interactions are equal to one another, we say that the two-way interaction between attract and sex does not depend upon crime, if the null is not rejected.

Although we have discussed the three-way interaction in terms of the attract by sex interaction varying over crime, the three-way interaction can also be formulated in two additional ways:

6.3.5 Fitting the model with anova

1. `attract` by `crime` varying over levels of `sex`
2. `sex` by `crime` varying over levels of `attract`

Because interactions are complementary, none of the formulations is more correct than another. The null hypotheses for these additional formulations are

1. $H_0: \overbrace{(\mu_{UMB} - \mu_{AMB}) - (\mu_{UMS} - \mu_{AMS})}^{\text{attract} \times \text{crime when sex} = \text{men}} = \overbrace{(\mu_{UWB} - \mu_{AWB}) - (\mu_{UWS} - \mu_{AWS})}^{\text{attract} \times \text{crime when sex} = \text{women}}$

2. $H_0: \overbrace{(\mu_{UMB} - \mu_{UWB}) - (\mu_{UMS} - \mu_{UWS})}^{\text{sex} \times \text{crime when attract} = \text{unattractive}} = \overbrace{(\mu_{AMB} - \mu_{AWB}) - (\mu_{AMS} - \mu_{AWS})}^{\text{sex} \times \text{crime when attract} = \text{attractive}}$

Given that these formulations are equivalent, it is challenging (impossible) to interpret interactions with only an F-value and significance test. Additional data and analyses are needed (see section 6.3.6).

6.3.5 Fitting the model with anova

The results of the three-way model are

```
. anova years attract sex crime attract#sex attract#crime sex#crime
> attract#sex#crime
```

	Number of obs =	240	R-squared	=	0.3224
	Root MSE =	1.09012	Adj R-squared =		0.3020

Source	Partial SS	df	MS	F	Prob>F
Model	131.19583	7	18.742262	15.77	0.0000
attract	28.704167	1	28.704167	24.15	0.0000
sex	15.504167	1	15.504167	13.05	0.0004
crime	4.0041667	1	4.0041667	3.37	0.0677
attract#sex	65.104167	1	65.104167	54.78	0.0000
attract#crime	6.3375	1	6.3375	5.33	0.0218
sex#crime	4.5375	1	4.5375	3.82	0.0519
attract#sex#crime	7.0041667	1	7.0041667	5.89	0.0160
Residual	275.7	232	1.1883621		
Total	406.89583	239	1.702493		

The syntax includes terms for the two-way and three-way interactions. An alternative syntax for this model is `anova years attract##sex##crime`. The double pound signs (##) indicate that we want all main effects and interactions among the factors.

The main effects for `attract` [$F(1, 232) = 28.7, p < 0.001$] and `sex` [$F(1, 232) = 15.5, p < 0.001$] are significant, as they were in the two-way model. Likewise, the `attract` by `sex` interaction is significant [$F(1, 232) = 65.1, p < 0.001$]. The main effect for `crime` is not [$F(1, 232) = 4.0, p = 0.07$], suggesting that there is no difference between burglary and swindle averaging over defendant attractiveness and juror sex.

Both the two-way and the three-way interactions suggest that `crime` is associated with the outcome. The `sex` by `crime` interaction does not quite reach statistical significance $[F(1, 232) = 4.5, p = 0.052]$. The `attract` by `crime` interaction is significant $[F(1, 232) = 6.3, p = 0.02]$, suggesting that the difference between attractive and unattractive defendants varied across type of crime. Finally, the three-way interaction between `attract`, `sex`, and `crime` is significant $[F(1, 232) = 7.0, p = 0.02]$.

Because there is a three-way interaction in the model, interpreting the main effects and two-way interactions is tricky. The three-way interaction indicates that the `attract` by `crime` interaction is different for male and female jurors. It is possible that the `attract` by `crime` interaction is present for men but not women. In that situation, drawing conclusions that apply to both men and women about the two-way interaction, which is averaged over men and women, is problematic. Consequently, I strongly recommend unpacking the three-way interaction and understanding the pattern of results that produced it. Once you have a good grasp on what is going on with the three-way interaction, you can then work to understand whether the main effects and two-way interactions provide additional information that is useful.

6.3.6 Interpreting the interaction

Three-way interactions are tricky because there are so many parts. Consequently, using `contrast` to interpret the interaction requires care. Mitchell (2012a) notes the following:

> In the case of a three-way interaction, there are even more ways that such interactions can be dissected. The best practice would be to use your research questions to develop an analysis plan that describes how the interactions will be dissected. This plan would also consider the number of statistical tests that will be performed and consider whether a strategy is required to control the overall type I error rate. By contrast, an undesirable practice would be to include an interaction because it happened to be significant, dissect it every which way possible until an unpredicted significant test is obtained, and make no adjustment for the number of unplanned statistical tests that were performed. (pp. 249–250)

I simply want to echo these words—three-way interactions create many opportunities for analytical nonsense and fiddling. If there is one thing that the debate over rigor and replication in psychology and science has highlighted, it is that while analytic fiddling might produce publications, it does not produce good inferences (Simmons, Nelson, and Simonsohn 2011).

6.3.6 Interpreting the interaction

A good first step when interpreting a three-way interaction is to visualize the interaction by using `margins`.

```
. quietly anova years attract sex crime attract#sex attract#crime sex#crime
> attract#sex#crime
. margins attract#sex#crime
Adjusted predictions                            Number of obs     =        240
Expression     : Linear prediction, predict()
```

	Margin	Delta-method Std. Err.	t	P>|t|	[95% Conf. Interval]	
attract#sex# crime						
Unattractive # Men # Burglary	5.366667	.1990278	26.96	0.000	4.974534	5.7588
Unattractive # Men # Swindle	4.166667	.1990278	20.94	0.000	3.774534	4.5588
Unattractive # Women # Burglary	3.2	.1990278	16.08	0.000	2.807867	3.592133
Unattractive # Women # Swindle	3.233333	.1990278	16.25	0.000	2.8412	3.625466
Attractive # Men # Burglary	2.966667	.1990278	14.91	0.000	2.574534	3.3588
Attractive # Men # Swindle	3.1	.1990278	15.58	0.000	2.707867	3.492133
Attractive # Women # Burglary	3.566667	.1990278	17.92	0.000	3.174534	3.9588
Attractive # Women # Swindle	3.566667	.1990278	17.92	0.000	3.174534	3.9588

```
. marginsplot, by(crime) byopts(title("")) scheme(lean2)
>         plotregion(margin(10 10 0 0))
>         ylabel(0(1)10)
   Variables that uniquely identify margins: attract sex crime
```

Adding the `by()` option to `marginsplot` creates the two separate panels for levels of `crime` (see figure 6.5). We asked for the predicted means for the `attract#sex#crime` interaction. By default, `marginsplot` places the first factor listed in the interaction—`attract` in this case—on the x axis, and the second factor is identified by different types of dots.

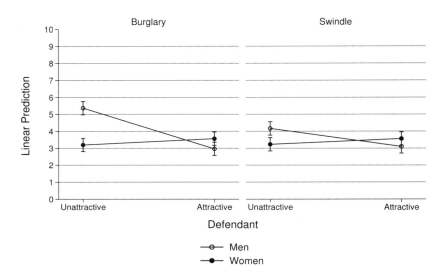

Figure 6.5. Plot of the `attract` by `sex` by `crime` interaction with `attract` on the x axis

Given that the research question is framed as the moderators of judgments made about attractive and unattractive defendants, I prefer to place `sex` on the x axis. This will allow us to examine the difference in the outcome by looking at the vertical distance between dots that are above or below one another rather than looking at the vertical distance between dots that are far to the right or left of one another. To make this change, add `xdimension(sex)` to `marginsplot`. This places `sex` on the x axis.

```
. marginsplot, xdimension(sex) by(crime) byopts(title(""))
>         scheme(lean2)
>         plotregion(margin(10 10 0 0))
>         ylabel(0(1)10)
   Variables that uniquely identify margins: attract sex crime
```

Note that figures 6.5 and 6.6 look very similar. This is just a coincidence and unique to this particular dataset.[3] The point is that now we are able to easily make the visual comparisons between the expected values for unattractive and attractive defendants for the combinations of `sex` and `crime`.

---

3. Remember, these are simulated data.

6.3.6 Interpreting the interaction

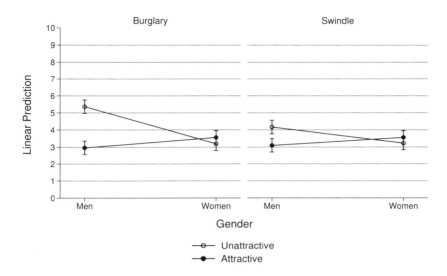

Figure 6.6. Plot of the `attract` by `sex` by `crime` interaction with `sex` on the x axis

The F test for the three-way interaction tests whether the two-way interactions represented in figure 6.6 are significantly different from one another. We do not know which of the interactions are significantly different from 0. It is possible that the `attract` by `sex` interaction is statistically significant only when the crime is burglary (that is, the left-hand side of figure 6.6), only when the crime is swindle (that is, the right-hand side of figure 6.6), or both. These interactions are known as simple interactions, which in the case of a three-way design means two-way interactions at specific levels of a third factor.

The null for the simple interaction between `attract` and `sex` when the crime is burglary is

$$H_0: \mu_{UMB} - \mu_{AMB} = \mu_{UWB} - \mu_{AWB}$$

Note that we are not averaging over `crime` in the simple interaction. We are examining the two-way interaction at a specific level of `crime`. The null for the simple interaction between `attract` and `sex` when the crime is swindle is

$$H_0: \mu_{UMS} - \mu_{AMS} = \mu_{UWS} - \mu_{AWS}$$

You can test simple interactions by using `contrast` after `anova`. The `contrast` syntax for testing the `attract` by `sex` interaction when the crime is swindle is

```
. contrast attract#sex@crime, mcompare(scheffe) noeffects
Contrasts of marginal linear predictions
Margins      : asbalanced
```

	df	F	P>F	Scheffe P>F
attract#sex@crime				
Burglary	1	48.31	0.0000	0.0000
Swindle	1	12.37	0.0005	0.0024
Joint	2	30.34	0.0000	
Denominator	232			

Note: Scheffe-adjusted p-values are reported for tests on individual contrasts only.

The syntax `attract#sex@crime` indicates that you want to evaluate the `attract` by `sex` interaction at the different levels of `crime`. `contrast` produces F tests for three values: the two simple interactions and a joint test, which tests the null hypothesis that both simple interactions are 0. I cannot think of a research question I have encountered where the joint test is relevant, so I ignore it. Both the `attract` by `sex` interaction when `crime` is burglary [$F(1, 232) = 48.31$, $p < 0.001$] and the `attract` by `sex` interaction when `crime` is swindle [$F(1, 232) = 12.37$, $p = 0.0024$] are statistically significant. Thus, the simple interaction tests indicate that there are significant interactions between `attract` and `sex` at both levels of `crime`. Further, the significant three-way interaction indicates that the simple interactions differ from one another.

In section 6.2.7, I discussed the use of simple effects to decompose interactions. Specifically, I compared the levels of `attract` at specific levels of `sex`. Simple effects in three-way designs are similar, except you compare levels of a factor at specific levels of an interaction. For example, you can compare the specific levels of `attract` for male jurors when the crime is a burglary. In a $2 \times 2 \times 2$ design, there are four simple effects per factor. The simple effects for `attract` and the corresponding null hypotheses are

1. Unattractive versus attractive when `sex` = men and `crime` = burglary.
   $H_0: \mu_{UMB} - \mu_{AMB} = 0$
2. Unattractive versus attractive when `sex` = women and `crime` = burglary.
   $H_0: \mu_{UWB} - \mu_{AWB} = 0$
3. Unattractive versus attractive when `sex` = men and `crime` = swindle.
   $H_0: \mu_{UMS} - \mu_{AMS} = 0$
4. Unattractive versus attractive when `sex` = women and `crime` = swindle.
   $H_0: \mu_{UWS} - \mu_{AWS} = 0$

We can use `contrast` to estimate the simple effects.

```
. contrast attract@sex#crime, mcompare(scheffe) pveffects nowald
Contrasts of marginal linear predictions
Margins      : asbalanced

                   |  Number of
                   | Comparisons
-------------------+------------
 attract@sex#crime |       4

                                                               Scheffe
                                     Contrast   Std. Err.     t     P>|t|
-----------------------------------+-------------------------------------
                 attract@sex#crime
 (Attractive vs base) Men#Burglary |      -2.4   .2814678  -8.53    0.000
  (Attractive vs base)  Men#Swindle| -1.066667   .2814678  -3.79    0.007
(Attractive vs base) Women#Burglary|  .3666667   .2814678   1.30    0.791
 (Attractive vs base) Women#Swindle|  .3333333   .2814678   1.18    0.843
```

The syntax `attract@sex#crime` indicates that we want to compare the levels of `attract` at all combinations of the levels of `sex` and `crime`. We correct for multiple comparisons using the Scheffé adjustment. The `pveffects` option prints the estimate and p-value for each of the simple effects, and the `nowald` option suppresses the F tests (they are redundant with the contrasts and do not provide any useful information in this situation).

The output indicates that men sentenced attractive defendants to 2.4 fewer years than unattractive defendants when the crime was a burglary ($p < 0.001$). Men sentenced attractive defendants to 1.1 fewer years than unattractive defendants when the crime was a swindle ($p = 0.007$). In contrast, the differences between the women's sentences given to unattractive and attractive defendants were small and not significant for both burglary (0.37, $p = 0.79$) and swindle (0.33, $p = 0.84$).

We have decomposed the three-way interaction by focusing on the simple interaction of `attract` by `sex` at levels of `crime`. We also could examine the `sex` by `crime` interaction at levels of `attract` or the `attract` by `crime` interaction at levels of `sex`. Likewise, we used one of many possibilities for the simple effects. As noted previously, which simple interactions and simple effects you focus on should be driven by research questions and not by which one provides significant effects.

6.3.7 A note about effect size

Effect size in three-way designs is the same as it was in two-way designs. The postestimation command `estat esize` works in the same way as discussed in section 6.2.8. For example, the ω^2 results for the three-way design are

```
. estat esize, omega
```
Effect sizes for linear models

Source	Omega-Squared	df
Model	.3011077	7
attract	.0900408	1
sex	.0489609	1
crime	.0100245	1
attract#sex	.1868924	1
attract#crime	.0181803	1
sex#crime	.0119006	1
attract#sex#crime	.0204859	1

Note: Omega-Squared values for individual model terms are partial.

6.4 Conclusion

This chapter covered methods for analyzing two- and three-way factorial designs, focused on fully between-subjects designs. For designs involving within-subject factors, see chapter 7 as well as the references discussed in that chapter. I discussed designs with factors that had two levels, but the analysis strategies I covered generalize to factors with more than two levels. Further details about these additional types of designs can be found in Keppel and Wickens (2004) and Maxwell, Delaney, and Kelley (2018). Additional Stata resources for analyzing factorial designs can be found in Mitchell (2012a) and Mitchell (2015).

7 Repeated-measures models

Researchers often design studies with repeated measures. Examples of repeated measures include experiments where participants complete multiple conditions; intervention studies where participants are measured before, during, and after treatment; developmental studies where participants are tracked over months or years; and intensive longitudinal studies where participants are measured many times over a short period (for example, hourly for a week). The statistical methods in chapters 3–6 can be used with cross-sectional data only (that is, where participants are observed only once). This chapter covers methods for modeling repeated measures.

The most common use of repeated-measures designs is to address research questions that involve the passage of time (Keppel and Wickens 2004). For example, developmental studies often investigate how a psychological construct unfolds over time. Trying to understand how self-identity develops by collecting data cross-sectionally, where we only observe someone at one timepoint but sample people at different ages, is fraught with inferential problems. Suppose we want to compare self-identity at age 15 and age 16. If we compare a sample of 15-year-olds to a sample of 16-year-olds, it is not possible to separate between-people differences from within-person (or developmental) differences. Instead, we need to compare each participant at age 15 and age 16 because that is a within-person comparison.

In addition to the inferential benefits, repeated-measures designs can also be more statistically powerful (see chapter 8 for a discussion of power). That is, repeated observations on the same person tend to be correlated. Using models that account for this correlation can increase the precision of the estimates in our study (Keppel and Wickens 2004).

In this chapter, we analyze repeated-measures designs using mixed models, which use maximum likelihood, rather than analysis of variance (ANOVA), which uses least squares. This is an atypical approach to introducing repeated-measures models, especially in psychology. Many textbooks teach repeated-measures ANOVA as the primary or only method for analyzing repeated-measures data (Keppel and Wickens 2004). Mixed models can do everything repeated-measures ANOVA can do. Further, mixed models are more flexible and expandable as problems become more complicated (Littel, Stroup, and Freund 2002; Littell et al. 2007). Mixed models will be the basis for modeling not only highly structured repeated-measures designs that we cover in this chapter, but also for nested designs (see chapter 9) and longitudinal studies (see chapter 10). Stata's `mixed` command is well suited for fitting a huge variety of models.

Two key benefits of using mixed models rather than ANOVA are

1. Mixed models are better at handling missing data. In ANOVA or multivariate versions of ANOVA for repeated measures, participants missing any data are dropped from the analysis. Thus, in a study with four repeated measures, if a person did not complete one of the measures or there was a recording error, that person is dropped from the analysis. This is known as listwise deletion and is a limitation of least-squares methods. In contrast, mixed models use the available data. Participants only completing one trial still contribute to the analysis but only contribute a single data point. This is an advantage of maximum likelihood methods. There's a catch, however. The missing data must be ignorable. That is, they must occur randomly and not be systematically related to variables not included in the model. For example, if participants with the poorest attention abilities are the only people with missing data, then the analysis from a mixed model will be biased. If the missing data are not ignorable, then the listwise deletion used in ANOVA methods will also produce a biased result (Enders 2010).

2. It can be difficult, if not impossible, to include covariates that are measured at the between-subjects level in repeated-measures ANOVA (Littell, Stroup, and Freund 2002). For example, consider a repeated-measures treatment study for depression where participants complete a baseline depression measure prior to treatment and then are randomized to two treatment conditions. It is common to include baseline depression as a covariate in a model to control for any differences in baseline symptoms. This is straightforward to do using a mixed model but nearly impossible for repeated-measures ANOVA.

In this chapter, you will learn the following:

1. The difference between wide and long format datasets.
2. How repeated-measures data violate the assumptions of between-subjects models.
3. How to fit a repeated-measures model by using a mixed-effects model for both one-way designs and multifactorial designs.
4. Options for the covariance structure of the residuals and how to choose between them.
5. Why degrees of freedom is tricky in repeated-measures models.
6. How to allow the residuals to vary across the levels of variable in the model.

Stata commands featured in this chapter

- `reshape`: for changing data from wide to long format and vice versa
- `mixed`: for fitting repeated-measures models
- `contrast`: for computing F tests following `mixed`
- `estat wcorrelation`: for printing the within-subject correlation matrix after `mixed`
- `estat ic`: for printing fit statistics after `mixed`
- `pwcompare`: for making pairwise comparisons after `mixed` and adjusting for multiple comparisons
- `margins` and `marginsplot`: for visualizing the results

7.1 Data for this chapter

The data for this chapter come from a study aimed at investigating the effects of treadmill or cycling desks on attention and other cognitive tasks.[1] To combat the negative effects of sitting at a desk all day, companies developed exercise desks that allow people to move at a slow pace—either walking or cycling—while still being able to work at a computer and do other desk-related tasks. This study aimed to find out whether a walking or cycling desk was better for attention.[2]

The attention task is known as the Paced Auditory Serial Addition Test (Roman et al. 1991). In this task, participants are presented with the numbers 2 and 3; they are asked to provide the sum of the two numbers (5). They are then presented with the number 4 and asked to add this new number with the last of the original pair: $3+4 = 7$. This goes on for 50 pairs, and the number of correct responses (among other things) is recorded. Participants complete four trials of 50 pairs (the pairs are not the same from trial to trial). Each trial gets progressively harder because the delay between the second number in a pair and the next number gets shorter. The delays for trials 1–4 are 2.4, 2.0, 1.6, and 1.2 seconds. Therefore, the number of correct answers is expected to go down from trial to trial.

Repeated-measures data are stored in either wide or long format. I have provided both formats for this chapter, and I will show you how to turn wide data to long data and vice versa by using `reshape`.

---

1. I would like to thank Michael Larson for sharing this dataset. It is currently unpublished.
2. Participants also completed the attention task prior to walking or cycling to provide a baseline estimate. Thus, the main effect for time provides a within-subject test of whether attention is better sitting or moving. The model to test that requires additional concepts beyond the scope of this chapter, and thus the baseline values are ignored in this chapter.

Wide format means there is one line in the dataset for every participant, and there are as many variables as there are repeated measures. The variables in the wide format dataset are (`pasatwide.dta`)

- `part`: participant number
- `cond`: condition (1 = cycling, 0 = treadmill)
- `cor_pasat1`–`cor_pasat4`: number of correct PASAT responses for trials 1–4.

The variables in the long format dataset are (`pasatlong.dta`)

- `part`: participant number
- `cond`: condition (1 = cycling, 0 = treadmill)
- `trial`: trial number
- `cor_pasat`: number of correct PASAT responses

The wide format for the PASAT data looks like this:

```
. use http://www.stata-press.com/data/pspus/pasatwide
. list in 1/3, clean
        part        cond    cor_pa~1   cor_pa~2   cor_pa~3   cor_pa~4
  1.     102     Cycling          49         42         42         41
  2.     103    Treadmill         49         47         44         41
  3.     104    Treadmill         47         43         39         33
```

As described above, each participant has a single line in the dataset and four variables representing the PASAT scores. Wide format data is common, especially when collecting data. For example, it's sometimes easier to set up web-based surveys to write data to a new variable for a specific person rather than add a new line to a dataset. Further, if you are going to manually enter data, entering all timepoints or trials on a single line is time saving. Wide data can be useful when exploring data descriptively and graphically. With wide data, summarizing the PASAT scores is straightforward:

```
. summarize cor_pasat1-cor_pasat4
    Variable |       Obs        Mean    Std. Dev.       Min        Max
-------------+--------------------------------------------------------
  cor_pasat1 |       102    45.83333    4.129304         30         49
  cor_pasat2 |       102    41.58824    5.737411         22         49
  cor_pasat3 |       102    37.83333    6.467104         16         49
  cor_pasat4 |       102    31.18627    6.919238         14         48
```

7.1 Data for this chapter

A box plot demonstrates the impact of trials (see figure 7.1):

```
. graph box cor_pasat1-cor_pasat4, ytitle("Number of correct answers")
> scheme(lean2) legend(rows(1) position(6))
```

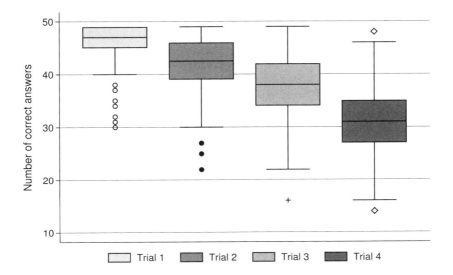

Figure 7.1. Number of correct PASAT answers over four trials

Fitting repeated-measures models in Stata, using `mixed` or using `anova`, requires the data to be in long format. Long format means there is one outcome variable (as opposed to four in the wide format) and each participant has as many rows in the dataset as he or she has observations. Thus, for the PASAT data, there is just one outcome variable—cor_pasat—and each participant has up to four observations representing the four trials. The `reshape` command can change wide data into long format and vice-versa.[3]

To change `pasatwide.dta` to long format, the command is

```
. reshape long cor_pasat, i(part) j(trial)
(note: j = 1 2 3 4)
Data                              wide   ->   long

Number of obs.                     102   ->    408
Number of variables                  6   ->      4
j variable (4 values)                    ->   trial
xij variables:
    cor_pasat1 cor_pasat2 ... cor_pasat4 ->   cor_pasat
```

---

[3] All major statistics software packages have the ability to reshape data from wide to long and vice-versa. However, `reshape` is, in my opinion, the best implemented method. It is flexible and simple to reshape complex datasets that include a combination of variables that are constants for a specific person (for example, age at baseline) and variables that vary across repeated measures.

```
. list in 1/12, clean

      part   trial       cond   cor_pa~t
  1.   102       1    Cycling         49
  2.   102       2    Cycling         42
  3.   102       3    Cycling         42
  4.   102       4    Cycling         41
  5.   103       1   Treadmill        49
  6.   103       2   Treadmill        47
  7.   103       3   Treadmill        44
  8.   103       4   Treadmill        41
  9.   104       1   Treadmill        47
 10.   104       2   Treadmill        43
 11.   104       3   Treadmill        39
 12.   104       4   Treadmill        33
```

When using **reshape**, we specify either **reshape long** (to go from wide to long) or **reshape wide** (to go from long to wide). Following the specification of either **long** or **wide**, we indicate which variable we want to reshape. In the case of reshaping to a long dataset, we specify the stub of the variable name that we will reshape. In this case, we have four variables—cor_pasat1, cor_pasat2, cor_pasat3, and cor_pasat4—and the stub is cor_pasat.

The i() option designates the variable that identifies the participants (or whatever unit you have repeatedly observed). In this example, we said i(part) because **part** is the participant identification number. The j() option creates a new variable that indexes the repeated measures. When going from wide to long format, we decide the name of this new variable. In the PASAT data, participants were observed repeatedly over trials, so we said j(trial). If we name the wide variables with a number at the end as we did with cor_pasat,[4] then Stata will pull those numbers into the new variable **trial**.

If we need to return the dataset to wide format, we type **reshape wide**. Given that we have already specified the participant ID and the trial ID (that is, the variable that indexes the repeated measures), Stata can put the data back into wide format. See [D] **reshape** for more details about **reshape**, including how to move from long to wide when the data are stored in long format.

7.2 Basic model

The first model we fit with the PASAT data is a fully within-subject model. The only predictor will be **trial**, and we observe each participant across all four trials. Missing data could mean that some participants are not observed at each of the four trials, but the factor is still within-subject. The outcome is the number of correct answers on the PASAT (cor_pasat). The model can be written as a regression model:

$$\text{cor\_pasat}_{ij} = b_0 + b_1 t_2 + b_2 t_3 + b_3 t_4 + \epsilon_{ij} \tag{7.1}$$

---

4. I strongly recommend that you name your variables like we did in the PASAT data. If you do, you will find that using **reshape** is far easier.

7.2 Basic model

The outcome is the number of correct PASAT answers for trial i and person j. Trial is dummy coded (see chapter 4) with trial 1 set as the reference category. Thus, b_0 is the mean number of correct PASAT answers; b_1, b_2, and b_3 are the difference between trials 2-4 and trial 1, respectively; and ϵ_{ij} is the residual for trial i and person j. In a regression or ANOVA model for between-subjects data, we assume that the residuals are independent of one another. That is, we assume that the residual for any one observation is uncorrelated with the residual for any other observation. This is sensible when comparing residuals in between-subjects data or when comparing residuals across participants in within-subject data. For example, imagine yourself in the PASAT study with just a single timepoint. There's no reason to think that your residual—how you differ from the overall mean—is systematically related to any other residual in the dataset because the residuals all come from different people.

The assumption of independence is not sensible in repeated-measures data because we are no longer only considering residuals across participants; we are also considering residuals observed on the same participant. For example, in the PASAT data, it may be sensible to assume that the residual for the first person at the first trial, ϵ_{11}, is uncorrelated with the residual for the second person at the first trial, ϵ_{12} (or any other person at any other trial). These are between-subjects comparisons just as we saw in a regression or ANOVA model. However, it is not sensible to assume that the residual for the first person at the first trial, ϵ_{11}, is uncorrelated with the residual for the first person at the second trial, ϵ_{21} (or any other trial for the first person). Indeed, we would expect anything unique to a specific person at one trial to be present to some degree at subsequent trials. This will create a correlation among the residuals for any specific person.

A within-subject model that is correctly constructed will account for the correlations among residuals in some fashion. To understand how this works, note that each participant has four observations on cor_pasat. All four observations on the outcome can be placed into a vector (Rabe-Hesketh and Skrondal 2012b). For example, we create a vector for the first participant, $\mathbf{y}_1$, that contains that participant's four observations.

$$\mathbf{y}_1 = \begin{bmatrix} y_{11} \\ y_{21} \\ y_{31} \\ y_{41} \end{bmatrix}$$

We can view these in Stata:

```
. use http://www.stata-press.com/data/pspus/pasatlong, clear
. list part trial cor_pasat in 1/4, clean
        part    trial   cor_pa~t
  1.    102     1        49
  2.    102     2        42
  3.    102     3        42
  4.    102     4        41
```

Just like the outcome variables, we can also place the residuals for each trial into a vector. The vector of residuals for the first participant, ϵ_1, that contains the residuals at each trial is

$$\epsilon_1 = \begin{bmatrix} \epsilon_{11} \\ \epsilon_{21} \\ \epsilon_{31} \\ \epsilon_{41} \end{bmatrix} \qquad (7.2)$$

Because the residuals in (7.2) are for a single participant, each has the same value for the j index (that is, 1). Furthermore, because the residuals are for a single participant, they will likely be correlated.

A key part of repeated-measures analyses is determining the structure of the correlations/covariances. Specifically, we need to construct a covariance matrix for each participant, which consists of the residual variance for trials on the diagonal and the covariance among the trials on the off-diagonal. The covariance matrix for participant j is Σ_j, and the covariance matrix for the first participant is

$$\Sigma_1 = \begin{bmatrix} \sigma_1^2 & & & \\ \sigma_{21} & \sigma_2^2 & & \\ \sigma_{31} & \sigma_{32} & \sigma_3^2 & \\ \sigma_{41} & \sigma_{42} & \sigma_{43} & \sigma_4^2 \end{bmatrix}$$

Typically, some constraints are placed on the elements of Σ. For example, it is common to assume that variances on the diagonal are all equal (that is, the residual variability is constant across trials): $\sigma_1^2 = \sigma_2^2 = \sigma_3^2 = \sigma_4^2 = \sigma^2$. Likewise, we can also add constraints to the covariances. See section 7.3.1 for a discussion of common constraints on the variances and covariances.

Σ is also constrained to be equal across participants. That is, we do not estimate a unique value for Σ for all participants but instead assume Σ is constant across participants. In regression, we estimated a single value of the residual variance that was the same across all participants. We do the same here, except now we estimate a covariance matrix rather than just a variance. This constraint across participants can be relaxed, and we can estimate unique values of Σ for subsets of participants, such as men and women, or unique values for different treatment groups (see section 7.5). Estimating unique values of Σ for subsets of participants becomes especially important as data structures and designs become more complex (Baldwin et al. 2011).

In sum, the basic repeated-measures model is similar to a regression or ANOVA model for between-subjects data except with respect to the residuals. Likewise, in between-subjects data, all residuals are assumed independent of one another. In repeated-measures data, residuals from different participants are still assumed independent. However, residuals within the same participant are not independent. The specific structure of the covariance/correlation among the residuals within a participant is a key feature of a repeated-measures analysis.

7.3 Using mixed to fit a repeated-measures model

We can use `mixed` to fit the model in (7.1). We will start with a model that treats the PASAT data as a between-subjects dataset rather than a repeated-measures dataset. That is, we will assume that all residuals are independent of one another. This allows us to compare the results to the regression and ANOVA models we have covered in previous chapters. Furthermore, the `mixed` syntax can become quite complicated, so it also allows us to build up the model step by step.

The `mixed` command to estimate (7.1) with independent residuals is

```
. mixed cor_pasat i.trial || part:, noconstant reml dfmethod(residual)
> residuals(independent) nolog
Note: all random-effects equations are empty; model is linear regression
Mixed-effects REML regression                   Number of obs     =        408
DF method: Residual                             DF:       min =     404.00
                                                          avg =     404.00
                                                          max =     404.00

                                                F(3,   404.00)    =     112.72
Log restricted-likelihood = -1300.2023          Prob > F          =     0.0000
```

cor_pasat	Coef.	Std. Err.	t	P>|t|	[95% Conf. Interval]	
trial						
2	-4.245098	.827435	-5.13	0.000	-5.871714	-2.618482
3	-8	.827435	-9.67	0.000	-9.626616	-6.373384
4	-14.64706	.827435	-17.70	0.000	-16.27367	-13.02044
_cons	45.83333	.5850849	78.34	0.000	44.68314	46.98352

Random-effects Parameters	Estimate	Std. Err.	[95% Conf. Interval]	
var(Residual)	34.91708	2.444685	30.43981	40.0529

Stata confirms in the output that our model is equivalent to a linear regression. The initial syntax for `mixed` is the same as `regress` or `anova`—the dependent variable comes first and then any predictors. Although everything after the predictors is not necessary if we want to fit a model with independent residuals, I have included these details here because we need it for the other models in this chapter. Following the predictors are two "pipes", ||, followed by `part:, noconstant`. This is called the random-effects portion of the syntax (see chapters 9 and 10 for more details). The variable following || is the one in the dataset that identifies who was observed multiple times. Because we observed participants multiple times, we use the participant ID. If we repeatedly observed mice, we would use mouse ID; if we repeatedly observed schools, we would use school ID. The `noconstant` option prevents `mixed` from including random effects in the model. For the models covered in this chapter, `noconstant` must be included or the models will not converge. The `reml` option tells Stata to use restricted maximum likelihood estimation methods. This will be required for models where we want to use the Satterthwaite method to adjust degrees of freedom for statistical tests. Additionally, dfmethod(residual) tells Stata to use t tests for the significance tests and to compute the degrees of freedom for the t tests by using the residual method ($N - k - 1$, where k is the number of predictors). By default, Stata uses z tests for significance tests. Using dfmethod(residual) makes the tests in the mixed output identical to a model fit using regress cor_pasat i.trial.

7.3.1 Covariance structures

The final part of the mixed syntax is the residuals() option, which is used to indicate how we believe the residuals are related to one another. The independent option requires that mixed constrain Σ to have the following form:

$$\Sigma = \begin{bmatrix} \sigma^2 & & & \\ 0 & \sigma^2 & & \\ 0 & 0 & \sigma^2 & \\ 0 & 0 & 0 & \sigma^2 \end{bmatrix} \tag{7.3}$$

A single error variance common to all participants is estimated, and all covariances among residuals are fixed to 0. A regression model fit using regress cor_pasat i.trial produces identical parameter estimates and standard errors.

The coefficients b_1, b_2, and b_3 are the differences in number of correct PASAT answers between trials 2, 3, and 4 and trial 1, respectively. For example, $b_3 = -14.6$, which indicates that, on average, participants had about 15 fewer correct responses at trial 4 than at trial 1. The intercept indicates that the average number of correct responses at trial 1 was $b_0 = 45.8$. The t tests provide the significance test for the coefficients.

The t tests test the null hypothesis for the pairwise comparisons among the trials. However, we also often want to know about the joint null hypothesis that the means for all trials are equal: $H_0: \mu_1 = \mu_2 = \mu_3 = \mu_4$ (see chapter 5). This null hypothesis is tested with an F test and can be produced using the contrast postestimation command.

```
. contrast trial, small
Contrasts of marginal linear predictions
Margins      : asbalanced
```

	df	ddf	F	P>F
cor_pasat trial	3	404.00	112.72	0.0000

We reject the null hypothesis that the mean number of correct PASAT responses is equal across trials because $F(3, 404) = 112.7$, $p < 0.001$. The option small was added to produce an F test rather than the default χ^2 test.[5] In large samples, the F test and χ^2 test will produce nearly identical results. However, in small samples, they will typically

---

5. Note that this option will only work if the dfmethod() option was used in the mixed command. Otherwise, contrast will default to a χ^2 test.

7.3.1 Covariance structures

differ just as a t test and z test differ in small samples. The F test is identical to a one-way ANOVA for this dataset.

```
. anova cor_pasat i.trial
                  Number of obs  =      408    R-squared      =  0.4556
                  Root MSE       =  5.90907    Adj R-squared  =  0.4516

         Source |  Partial SS       df         MS           F      Prob>F
        --------+----------------------------------------------------------
          Model |  11807.537         3      3935.8456     112.72   0.0000

          trial |  11807.537         3      3935.8456     112.72   0.0000

       Residual |  14106.5         404      34.917079
        --------+----------------------------------------------------------
          Total |  25914.037       407      63.670852
```

Repeated-measures analyses will differ from traditional between-subjects analyses once we specifically model the correlation among the residuals. More often than not, modeling the correlation among the residuals will increase the precision of the analysis and thus its statistical power (see chapter 8). Given that the independent covariance structure (7.3) is not appropriate, the challenge moving forward is selecting an appropriate structure for the dataset and research design. I review four choices for the structure of the residuals: compound symmetry, autoregressive, Toeplitz, and unstructured. Stata has additional choices to consider that may be appropriate for other situations. Fortunately, the principles of understanding and selecting a covariance structure can easily be generalized to the other available structures.

Compound symmetry (exchangeable)

The first option for Σ is compound symmetry, which Stata calls exchangeable. A compound symmetry matrix has the following form:

$$\Sigma = \sigma^2 \begin{bmatrix} 1 & & & \\ \rho & 1 & & \\ \rho & \rho & 1 & \\ \rho & \rho & \rho & 1 \end{bmatrix} \quad (7.4)$$

There is a common variance across all trials (that is, repeated measures) as well as a common correlation/covariance between the trials. Equation (7.4) is written as a correlation matrix. The common correlation is denoted by ρ, and the common variance is outside the matrix and is denoted as σ^2.

To use compound symmetry for the residuals, the `mixed` command is

```
. mixed cor_pasat i.trial || part:, noconstant residuals(exchangeable) reml
>     dfmethod(satterthwaite, oim) nolog
```

Mixed-effects REML regression				Number of obs	=	408
Group variable: part				Number of groups	=	102
				Obs per group:		
				min	=	4
				avg	=	4.0
				max	=	4
DF method: Satterthwaite				DF: min	=	160.14
				avg	=	267.29
				max	=	303.00
				F(3, 303.00)	=	392.00
Log restricted-likelihood = -1169.1182				Prob > F	=	0.0000

cor_pasat	Coef.	Std. Err.	t	P>|t|	[95% Conf. Interval]	
trial						
2	-4.245098	.443699	-9.57	0.000	-5.11822	-3.371977
3	-8	.443699	-18.03	0.000	-8.873122	-7.126878
4	-14.64706	.443699	-33.01	0.000	-15.52018	-13.77394
_cons	45.83333	.5850849	78.34	0.000	44.67786	46.98881

Random-effects Parameters	Estimate	Std. Err.	[95% Conf. Interval]	
part: (empty)				
Residual: Exchangeable				
var(e)	34.91708	3.902122	28.0487	43.46734
cov(e)	24.87677	3.859256	17.31277	32.44077

LR test vs. linear model: chi2(1) = 262.17   Prob > chi2 = 0.0000

Note: The reported degrees of freedom assumes the null hypothesis is not on
the boundary of the parameter space. If this is not true, then the
reported test is conservative.

Note: The observed information matrix is used to compute Satterthwaite degrees
of freedom.

There are two differences between the code for compound symmetry and the code for independent models: now we use `residuals(exchangeable)` to use compound symmetry, and we use `dfmethod(satterthwaite, oim)` to change the calculation of the degrees of freedom to the Satterthwaite approximation (Littell et al. 2007; Satterthwaite 1946). Section 7.3.2 discusses why I use the Satterthwaite degrees of freedom for these models. For now, just know that it is a reasonable choice and that we can still see how the choice about the structure of Σ affects the available degrees of freedom.

The bottom of the `mixed` output displays the estimate of σ^2, `var(e)` = 34.9, and the common covariance, `cov(e)` = 24.9. Adding the option `stddeviations` to `mixed` will make Stata report the common correlation from (7.4). We can also use the `estat wcorrelation` postestimation command, which stands for within-participant correlation, to print the residual standard deviation and Σ.

7.3.1 Covariance structures

```
. estat wcorrelation
Standard deviations and correlations for part = 102:
Standard deviations:
         obs |     1       2       3       4
     --------+--------------------------------
          sd | 5.909   5.909   5.909   5.909
Correlations:
         obs |     1       2       3       4
     --------+--------------------------------
           1 | 1.000
           2 | 0.712   1.000
           3 | 0.712   0.712   1.000
           4 | 0.712   0.712   0.712   1.000
```

Thus, the common standard deviation is $\widehat{\sigma} = 5.9$ and the common correlation is $\widehat{\rho} = 0.7$.

The coefficient estimates from the compound symmetry model are identical to the estimates from the independence models. However, the standard errors are about half as big in the compound symmetry model as the independence model. Ignoring the correlation among the residuals reduces precision—the independence model does not take advantage of the fact that we are controlling for person-specific variation in a repeated-measures design.

Table 7.1 provides the denominator degrees of freedom (DDF) and F for `trial` as well as the Akaike information criterion (AIC) and Bayesian information criterion (BIC) for the five covariance structures considered in this chapter. Details regarding the computation of the AIC and BIC can be found in [R] **estat ic** (StataCorp 2017a), and both can be obtained following `mixed` by using the `estat ic` postestimation command. The AIC and BIC both suggest (smaller values are better) that the compound symmetry model fits the data better than the independence model, as expected. The F is also bigger, which is a reflection of the increased precision that occurs by accounting for the person-specific correlation.

Table 7.1. Test statistics and fit indices for five covariance structures applied to the PASAT data

Model	DDF	F	AIC	BIC
Independent	404.00	112.72	2610.405	2630.461
Exchangeable	303.00	392.00	2350.236	2374.304
Autoregressive	298.98	213.35	2317.399	2341.467
Toeplitz	226.34	153.45	2430.546	2454.613
Unstructured	101.00	242.83	2286.038	2342.196

Interestingly, the DDF is smaller in the compound symmetry model as compared with the independence model. This may come as a surprise because the number of observations has not changed. Unfortunately, what counts as the sample in a repeated-measures design, and any multilevel design (see chapter 9 and 10), is controversial (Faes et al.

2009). Is the sample size the number of participants, the number of observations, or somewhere in between?

Imagine a situation where the compound symmetry assumption is appropriate. Suppose that we observed participants four times, just like in the PASAT data, but the correlation among the residuals was $\rho = 1$. In that case, we obtain no unique information from the repeated measures on a participant—it is fully redundant. Thus, the effective sample size, the sample size that provides unique information, is equal to the number of participants. Now suppose that the common correlation among residuals was $\rho = 0$, just like a between-subjects design. Now all observations provide unique information, and the effective sample size is equal to the number of observations.

In most cases, including the PASAT example, the truth is somewhere in between because the common correlation will be positive. The closer the common correlation is to 0, the more unique pieces of information each observation provides and the more DDF we have. This comes with a cost, however. Precision goes down here because we are not able to control for as much person-specific variability as when the common correlation is large. The closer the common correlation is to 1, the less unique information is provided by each observation and the fewer DDF we have. Precision goes up. Because the common correlation is $\widehat{\rho} = 0.7$, we lose degrees of freedom but gain precision in the compound symmetry model.

First-order autoregressive

A problem with the compound symmetry structure is the common correlation among the trials. This is reasonable with only two trials because there is just one correlation. However, with three or more trials, the common correlation assumption can be problematic.[6] For example, it makes sense that correlations between observations that are close in time (for example, trial 1 and trial 2 or trial 2 and trial 3) will often be more strongly correlated than observations that have a large spacing (for example, trial 1 and trial 3 or trial 2 and trial 4).

The first-order autoregressive structure for Σ directly accounts for the fact that correlations should get smaller the further apart the repeated measures get. The first-order autoregressive structure is

$$\Sigma = \sigma^2 \begin{bmatrix} 1 & & & \\ \rho & 1 & & \\ \rho^2 & \rho & 1 & \\ \rho^3 & \rho^2 & \rho & 1 \end{bmatrix} \tag{7.5}$$

Two parameters are estimated: a common variance across trials, σ^2, and a correlation, ρ. The correlation is assumed equal for adjacent trials—the correlation between ϵ_1 and

---

6. In fact, in repeated-measures ANOVA, there are numerous corrections, which are automatically done in Stata when performing a repeated-measures analysis, that adjust for the fact that the compound symmetry assumption (as well as a related assumption called sphericity) is rarely met (Winer, Brown, and Michels 1991).

7.3.1 Covariance structures

ϵ_2 is equal to the correlation between ϵ_2 and ϵ_3 and the correlation between ϵ_3 and ϵ_4. However, the correlation is squared (and thus gets smaller) for correlations that are two trials apart. Finally, for trials that are three trials apart, the correlation is cubed and thus gets even smaller.

The exponent for the correlation represents the distance between trials. Therefore, if there were many trials, such that there could be a distance of 10 between trials, then the correlation would be raised to the tenth power and would get close to 0. It makes sense because trials that are quite far apart are likely to be unrelated. However, the decay in the correlation imposed by the autoregressive structure can be a problem if there is strong person-specific variance present in any measurement, regardless of distance among trials.

To use the first-order autoregressive structure, the `mixed` syntax is

```
. mixed cor_pasat i.trial || part:, noconstant residuals(ar 1, t(trial)) reml
>         dfmethod(satterthwaite, oim) nolog
```

```
Mixed-effects REML regression                   Number of obs    =       408
Group variable: part                            Number of groups =       102

                                                Obs per group:
                                                           min =         4
                                                           avg =       4.0
                                                           max =         4
DF method: Satterthwaite                        DF:        min =    179.08
                                                           avg =    306.39
                                                           max =    391.20

                                                F(3,   298.98)  =    213.35
Log restricted-likelihood = -1152.6995          Prob > F        =    0.0000
```

cor_pasat	Coef.	Std. Err.	t	P>|t|	[95% Conf. Interval]	
trial						
2	-4.245098	.3832798	-11.08	0.000	-4.999348	-3.490848
3	-8	.5111674	-15.65	0.000	-9.005302	-6.994698
4	-14.64706	.5919144	-24.75	0.000	-15.81079	-13.48333
_cons	45.83333	.5760723	79.56	0.000	44.69657	46.9701

Random-effects Parameters	Estimate	Std. Err.	[95% Conf. Interval]	
part: (empty)				
Residual: AR(1)				
rho	.7786661	.0265772	.7209673	.8256427
var(e)	33.84965	3.577198	27.51693	41.63977

```
LR test vs. linear model: chi2(1) = 295.01               Prob > chi2 = 0.0000
```

Note: The reported degrees of freedom assumes the null hypothesis is not on the boundary of the parameter space. If this is not true, then the reported test is conservative.

Note: The observed information matrix is used to compute Satterthwaite degrees of freedom.

The only change is `residuals(ar 1, t(trial))`. The `ar 1` specifies the first-order autoregressive structure, and the `t(trial)` tells `mixed` what variable indexes the repeated measures (`trial` in this case). The `t()` option is required with an autoregressive structure, and the trials must be evenly spaced across participants.

It is possible to use a second-order or higher autoregressive structure. In the first-order structure, the correlation is a function of the preceding value. In a second-order structure, the correlation is a function of the preceding two values. In a third-order structure, the correlation is a function of the preceding three values, and so on. The disadvantage of doing this is that you have to estimate additional parameters. Also, the structure of the residuals is not as simple as (7.5).

For the first-order model, $\hat{\Sigma}$ is

```
. estat wcorrelation
Standard deviations and correlations for part = 102:
Standard deviations:
         trial |    1        2        3        4
            sd |  5.818    5.818    5.818    5.818
Correlations:
         trial |    1        2        3        4
             1 |  1.000
             2 |  0.779    1.000
             3 |  0.606    0.779    1.000
             4 |  0.472    0.606    0.779    1.000
```

The correlation is a little larger for adjacent trials than in the compound symmetry model, and the decay in the correlation is apparent. Remember, however, the decay is imposed by the model and is a function of the relationship between adjacent trials. It is certainly possible that the decay implied by the model is too severe or not severe enough.

As seen in table 7.1, the DDF and F-value are a bit smaller in the autoregressive model than the compound symmetry model. Furthermore, the AIC and BIC both suggest that the autoregressive model fits the data better than the compound symmetry model. Together, these points suggest that the common correlation in the compound symmetry model, while an improvement over the independent model, does not capture the relationship among the trials as well as the autoregressive model where the correlation decays over time.

7.3.1 Covariance structures

Toeplitz

The Toeplitz structure is also commonly used. Like the autoregressive structure, Toeplitz can be first order, second order, and so on. A first-order structure is

$$\Sigma = \sigma^2 \begin{bmatrix} 1 & & & \\ \rho_1 & 1 & & \\ 0 & \rho_1 & 1 & \\ 0 & 0 & \rho_1 & 1 \end{bmatrix}$$

Two parameters are estimated: a common variance across trials, σ^2, and a common correlation among adjacent trials, ρ_1. Correlations among trials that are two or more trials apart are fixed to 0. A second-order Toeplitz model has three parameters, with the additional parameter being a correlation among trials that are two trials apart, ρ_2; trials that are three trials apart have a 0 correlation.

To use the Toeplitz structure in `mixed`, the command is

```
. mixed cor_pasat i.trial || part:, noconstant residuals(toeplitz 1, t(trial))
> reml dfmethod(satterthwaite, oim) nolog
Mixed-effects REML regression                   Number of obs    =        408
Group variable: part                            Number of groups =        102
                                                Obs per group:
                                                             min =          4
                                                             avg =        4.0
                                                             max =          4
DF method: Satterthwaite                        DF:          min =     335.50
                                                             avg =     352.62
                                                             max =     403.98
                                                F(3,  226.34)    =     153.45
Log restricted-likelihood = -1209.2728          Prob > F         =     0.0000
```

cor_pasat	Coef.	Std. Err.	t	P>|t|	[95% Conf. Interval]	
trial						
2	-4.245098	.5402825	-7.86	0.000	-5.307214	-3.182982
3	-8	.7625441	-10.49	0.000	-9.49997	-6.50003
4	-14.64706	.7625441	-19.21	0.000	-16.14703	-13.14709
_cons	45.83333	.5392001	85.00	0.000	44.77269	46.89397

Random-effects Parameters	Estimate	Std. Err.	[95% Conf. Interval]
part: (empty)			
Residual: Toeplitz(1)			
cov1	14.76799	1.407666	12.00901 17.52696
var(e)	29.65515	2.289655	25.49056 34.50014

LR test vs. linear model: chi2(1) = 181.86   Prob > chi2 = 0.0000

Note: The reported degrees of freedom assumes the null hypothesis is not on the boundary of the parameter space. If this is not true, then the reported test is conservative.

Note: The observed information matrix is used to compute Satterthwaite degrees of freedom.

The only change from the previous model is `residuals(toeplitz 1, t(trial))`. The `toeplitz 1` specifies the first-order Toeplitz structure, and the `t(trial)` tells `mixed` which variable indexes the repeated measures. The `t()` option is required with a Toeplitz structure, and the trials must be evenly spaced across participants.

The output includes the common variance and a common covariance. If you want to see the correlation, you can either use the `stddeviations` option for `mixed` or print the correlation matrix:

```
. estat wcorrelation
Standard deviations and correlations for part = 102:
Standard deviations:
     trial |     1       2       3       4
        sd |  5.446   5.446   5.446   5.446
Correlations:
     trial |     1       2       3       4
         1 |  1.000
         2 |  0.498   1.000
         3 |  0.000   0.498   1.000
         4 |  0.000   0.000   0.498   1.000
```

The correlation between adjacent trials is $\hat{\rho} = 0.5$, which is smaller than in either the compound symmetry or the first-order autoregressive structure. Thus, fixing the correlations to 0 for trials that are two or more trials apart affects the estimate of the adjacent correlation as well.

As seen in table 7.1, the DDF and F-value are smaller in the Toeplitz model than the autoregressive model. The AIC and BIC both suggest that the autoregressive model fits the data better than the Toeplitz model. This suggests that assumptions of the first-order Toeplitz model are not as reasonable as the autoregressive model. In fact, the Toeplitz model does not fit the data as well as the compound symmetry model.

7.3.1 Covariance structures

One possibility is to fit a second- or third-order Toeplitz model.

```
. quietly mixed cor_pasat i.trial || part:, noconstant
> residuals(toeplitz 2, t(trial) ) reml
> dfmethod(satterthwaite, oim) nolog

. quietly estimates store toep2

. estat ic
Akaike's information criterion and Bayesian information criterion
```

Model	Obs	ll(null)	ll(model)	df	AIC	BIC
toep2	408	.	-1165.684	7	2345.368	2373.446

Note: N=Obs used in calculating BIC; see [R] BIC note.

```
. quietly mixed cor_pasat i.trial || part:, noconstant
> residuals(toeplitz 3, t(trial) ) reml
> dfmethod(satterthwaite, oim) nolog

. quietly estimates store toep3

. estat ic
Akaike's information criterion and Bayesian information criterion
```

Model	Obs	ll(null)	ll(model)	df	AIC	BIC
toep3	408	.	-1149.592	8	2315.184	2347.274

Note: N=Obs used in calculating BIC; see [R] BIC note.

```
. lrtest toep1 toep2
Likelihood-ratio test                       LR chi2(1) =      87.18
(Assumption: toep1 nested in toep2)         Prob > chi2 =     0.0000
Note: LR tests based on REML are valid only when the fixed-effects
      specification is identical for both models.

. lrtest toep2 toep3
Likelihood-ratio test                       LR chi2(1) =      32.18
(Assumption: toep2 nested in toep3)         Prob > chi2 =     0.0000
Note: LR tests based on REML are valid only when the fixed-effects
      specification is identical for both models.
```

A likelihood-ratio test suggests that the second-order model fits the data significantly better than the first-order model. However, the AIC and BIC both suggest that the additional parameters do not improve fit over the autoregressive model. A likelihood-ratio test suggests that the third-order model fits the data significantly better than the second-order model. The AIC also suggests that the third-order model fits better than the autoregressive model. However, the BIC suggests that the autoregressive model fits better than the Toeplitz model. This likely indicates that the four parameters of the third-order model improve fit as compared with the two parameters of the first-order autoregressive model, but they do so at the cost of complexity.

Unstructured

The final option we consider is to estimate an unstructured matrix:

$$\Sigma = \begin{bmatrix} \sigma_1^2 & & & \\ \sigma_{12} & \sigma_2^2 & & \\ \sigma_{13} & \sigma_{23} & \sigma_3^2 & \\ \sigma_{14} & \sigma_{24} & \sigma_{34} & \sigma_4^2 \end{bmatrix}$$

The unstructured matrix includes four variances and six covariances for a four-trial repeated-measures design. No constraints are placed on the elements. The advantage of this is that any misfit that occurs in the compound symmetry, autoregressive, or Toeplitz structures because matrices involve constraining variances and covariances will not be present in the unstructured matrix. The disadvantage is that the complexity of the model increases. The autoregressive model describes the correlation among the residuals with 2 parameters whereas the unstructured requires 10. That complexity creates potential problems as we consider replication in new datasets. For example, variances may differ across trials simply because of sampling error, and the unstructured matrix will model those differences nonetheless. In the next sample, or even most samples, the variances may just not be that different.

To use an unstructured matrix, the code is

```
. mixed cor_pasat i.trial || part:, noconstant
> residuals(unstructured, t(trial)) reml dfmethod(satterthwaite, oim) nolog
Mixed-effects REML regression                   Number of obs    =         408
Group variable: part                            Number of groups =         102

                                                Obs per group:
                                                              min =           4
                                                              avg =         4.0
                                                              max =           4

DF method: Satterthwaite                        DF:           min =      101.00
                                                              avg =      101.00
                                                              max =      101.00

                                                F(3,   101.00)   =      242.83
Log restricted-likelihood = -1129.0192           Prob > F        =      0.0000
```

cor_pasat	Coef.	Std. Err.	t	P>|t|	[95% Conf. Interval]	
trial						
2	-4.245098	.3755372	-11.30	0.000	-4.990063	-3.500133
3	-8	.4704426	-17.01	0.000	-8.933232	-7.066768
4	-14.64706	.5502594	-26.62	0.000	-15.73863	-13.55549
_cons	45.83333	.4088619	112.10	0.000	45.02226	46.64441

7.3.1 Covariance structures

Random-effects Parameters	Estimate	Std. Err.	[95% Conf. Interval]	
part: (empty)				
Residual: Unstructured				
var(e1)	17.05114	2.399422	12.94116	22.46642
var(e2)	32.91786	4.632139	24.98344	43.37217
var(e3)	41.82342	5.885304	31.74244	55.106
var(e4)	47.87585	6.737003	36.33599	63.08063
cov(e1,e2)	17.79207	2.948114	12.01387	23.57026
cov(e1,e3)	18.15015	3.212814	11.85315	24.44715
cov(e1,e4)	17.02144	3.309209	10.53551	23.50737
cov(e2,e3)	29.88117	4.740342	20.59028	39.17207
cov(e2,e4)	29.53289	4.92326	19.88348	39.1823
cov(e3,e4)	36.88283	5.770001	25.57384	48.19182

LR test vs. linear model: chi2(9) = 342.37   Prob > chi2 = 0.0000

Note: The reported degrees of freedom assumes the null hypothesis is not on the boundary of the parameter space. If this is not true, then the reported test is conservative.

Note: The observed information matrix is used to compute Satterthwaite degrees of freedom.

The only change from the previous model is `residuals(unstructured, t(trial))`. The `t()` option is required with an unstructured matrix, and the trials must be evenly spaced across participants.

The residual variances grow in size across trials. All previous structures we have considered have forced those variances to be equal, so this is an area where the unstructured matrix improves on previous models. The correlations among the residuals are

```
. estat wcorrelation
Standard deviations and correlations for part = 102:
Standard deviations:
    trial |    1      2      3      4
       sd | 4.129  5.737  6.467  6.919
Correlations:
    trial |    1      2      3      4
        1 | 1.000
        2 | 0.751  1.000
        3 | 0.680  0.805  1.000
        4 | 0.596  0.744  0.824  1.000
```

These are quite similar to the autoregressive estimates.

Table 7.1 shows that the degrees of freedom is smaller for the unstructured model than all others. It also has the second largest F-value. The high F is a function of the fact that the unstructured model captures the person-specific correlations well. Furthermore, because it lacks constraints across parameters, it captures the "nuances" of the person-specific relationships in the data (with the caveat that some of the nuances may be dataset specific). The F-value for the unstructured model is not as big as the compound symmetry model because the latter likely overestimates the residual correlations, especially for those residuals two or more trials apart. The small degrees of freedom is a function of the model complexity. That is, it takes information from the data to estimate all the variances and covariances, and this is reflected in the smallest denominator degrees of freedom of any model.

The AIC indicates that the unstructured matrix fits the data the best, whereas the BIC indicates that the autoregressive model fits the best. We saw this pattern when comparing the autoregressive model to the third-order Toeplitz model. The BIC is sensitive to the additional complexity of the unstructured model as compared with the relative simplicity of the autoregressive model, just as the BIC was sensitive to the additional model complexity of the third-order Toeplitz model as compared with the autoregressive model.

Given the discrepancies, the question of whether to prefer the unstructured or the autoregressive model is a function of whether we believe the differences among the variances from trial to trial are meaningful or noise. If they are meaningful, then the constraints of the autoregressive model are too severe. However, if they are noise, then the autoregressive model is a good approximation. The increased variability across participants in trials is common using the PASAT because the PASAT is more difficult for later trials, which adds variability among participants.

So which model do we choose? As seen in table 7.1, the structure of Σ has a large effect on DDF, F, and fit indices, so this is not a trivial decision. All options except the independent structure take advantage of the person-specific variability in the PASAT over time. Thus, fit improves and F-values go up for structures appropriate to repeated-measures designs. Because more parameters are estimated and because observations are not fully independent, the available degrees of freedom go down. This highlights a key consideration when designing a repeated-measures study. Namely, are the observations within a person going to be sufficiently correlated (that is, is there sufficient person-specific variability?) that we overcome the reduction in degrees of freedom as compared with a between-subjects design. Generally, the answer is going to be yes, but this can be a problem for repeated-measures designs where there are only two timepoints and the correlation between the residuals is less than $\rho = 0.5$.

In this example, the choice among structures comes down to either the first-order autoregressive model or the unstructured model. If parsimony is used as the criteria and if there are not concerns about the variability in the residual variance across trials, then the autoregressive model is a good choice. However, if the variability is expected to increase over trials, then the unstructured model is a good choice. In this instance, I believe the unstructured matrix is the most reasonable choice, so I proceed with the unstructured matrix.

Σ for between-subjects and repeated-measures designs

Suppose the PASAT data consisted of three participants with just one observation each. The covariance matrix of the residuals would look like this:

	ϵ_1	ϵ_2	ϵ_3
ϵ_1	σ^2		
ϵ_2	0	σ^2	
ϵ_3	0	0	σ^2

All residuals share the same variance and all covariances are 0 (that is, residuals are independent in between-subjects data). This matrix is an example of a diagonal matrix, which is a matrix with nonzero numbers on the diagonal and 0s on the off-diagonal. Now suppose the PASAT data consisted of three participants with four observations each. If the residuals within a participant had an unstructured matrix, then the covariance matrix would look like this:

	ϵ_{11}	ϵ_{21}	ϵ_{31}	ϵ_{41}	ϵ_{12}	ϵ_{22}	ϵ_{32}	ϵ_{42}	ϵ_{13}	ϵ_{23}	ϵ_{33}	ϵ_{43}
ϵ_{11}	σ_1^2											
ϵ_{21}	σ_{21}^2	σ_2^2										
ϵ_{31}	σ_{31}^2	σ_{32}^2	σ_3^2									
ϵ_{41}	σ_{41}^2	σ_{42}^2	σ_{43}^2	σ_4^2								
ϵ_{12}	0	0	0	0	σ_1^2							
ϵ_{22}	0	0	0	0	σ_{21}^2	σ_2^2						
ϵ_{32}	0	0	0	0	σ_{31}^2	σ_{32}^2	σ_3^2					
ϵ_{42}	0	0	0	0	σ_{41}^2	σ_{42}^2	σ_{43}^2	σ_4^2				
ϵ_{13}	0	0	0	0	0	0	0	0	σ_1^2			
ϵ_{23}	0	0	0	0	0	0	0	0	σ_{21}^2	σ_2^2		
ϵ_{33}	0	0	0	0	0	0	0	0	σ_{31}^2	σ_{32}^2	σ_3^2	
ϵ_{43}	0	0	0	0	0	0	0	0	σ_{41}^2	σ_{42}^2	σ_{43}^2	σ_4^2

This matrix shows that a) residuals within a participant are related, b) residuals between participants are unrelated (that is, have a covariance of 0), and c) the structure of the covariance among the residuals is constant across participants. This matrix is an example of a block diagonal matrix, which is a matrix with blocks of numbers (in this case corresponding to participants) on the diagonal and 0s everywhere else. You can consider other types of covariance structures discussed in section 7.3.1.

7.3.2 Degrees of freedom

Determining degrees of freedom in one-way ANOVA, factorial ANOVA, and regression models was straightforward. Repeated-measures designs are more difficult. In cases where the data are fully balanced and the residual variance structure is uncomplicated (that is, compound symmetry), determining the degrees of freedom is also straightforward. However, in cases where the data are not fully balanced and where the residual variance structure is complex (for example, autoregressive or unstructured), determining degrees of freedom is complex and can often only be approximated (Littell, Stroup, and Freund 2002; Littell et al. 2007; Baldwin et al. 2011). This becomes increasingly true with the addition of random effects, as we discuss in chapters 9 and 10. The challenge for degrees of freedom in repeated-measures designs is determining how many unique observations are available, which is a function of how correlated the observations are (see section 7.3.1).

The default in mixed is to use z tests and other large-sample tests to determine p-values and confidence intervals. This will work wonderfully if your sample size is large.[7] In Stata 14, degrees of freedom approximations were added to mixed for small samples. There are several options including residual, repeated, ANOVA, Satterthwaite, and Kenward–Roger. The latter two are the most flexible and can be used even as models become more complex. In this chapter, I use the Satterthwaite degrees of freedom. For many models, the Satterthwaite and Kenward–Roger degrees of freedom will be identical. I have found the Satterthwaite degrees of freedom to be stable and easy to estimate using mixed. Kenward–Roger can sometimes fail because it requires additional computation and information (Kenward and Roger 2009). Both the Satterthwaite and the Kenward–Roger degrees of freedom take into account the structure of Σ when approximating the degrees of freedom, which means that both approximate the number of unique pieces of information available (see [ME] mixed for more details on how the degrees of freedom is approximated).

As seen previously, the Satterthwaite approximation is implemented in mixed with the dfmethod() option. Likewise, if you want to use the Satterthwaite degrees of freedom in a postestimation command such as contrast (see section 7.3.1) or pwcompare (see section 7.3.3), add the option small to those commands after using dfmethod() with mixed. When using Satterthwaite degrees of freedom in mixed, you have two options: use the expected information matrix—dfmethod(satterthwaite, eim), which is the default—or use the observed information matrix—dfmethod(satterthwaite, oim). I recommend the observed information matrix (Efron and Hinkley 1978). The results tend to be more sensible, and it allows comparison to other software packages that default to the observed information matrix, such as SAS or SPSS. The details of the information matrix are beyond the scope of this book. Suffice to say that it is directly related to variance–covariance matrix of the estimates, which is where standard errors come from. Thus, the information is directly related to estimates of sampling variability.

7.3.3 Pairwise comparisons

Given that we treated trial as a factor variable in the mixed syntax, the coefficients for trial represent the difference between the trial of interest (for example, trial 2) and the

---

7. How large is sufficiently large is a matter of debate, and there is not an easy answer.

7.3.3 Pairwise comparisons

reference category (trial 1). Thus, the three coefficients provide the difference between trial 2 and trial 1, trial 3 and trial 1, and trial 4 and trial 1. Often, we are also interested in all pairwise comparisons. Fortunately, we can use the `pwcompare` postestimation command after `mixed` to obtain the estimates of the pairwise comparisons, including corrections for multiple comparisons as was discussed in chapter 5.

The pairwise comparisons are estimated as follows:

```
. mixed cor_pasat i.trial || part:, noconstant
> residuals(unstructured, t(trial))
> reml dfmethod(satterthwaite, oim) nolog
 (output omitted)
. pwcompare trial, small mcompare(scheffe) effects
Pairwise comparisons of marginal linear predictions
Margins      : asbalanced
```

	Number of Comparisons
cor_pasat trial	6

	Contrast	Std. Err.	Scheffe t	Scheffe P>|t|	Scheffe [95% Conf. Interval]	
cor_pasat trial						
2 vs 1	-4.245098	.3755372	-11.30	0.000	-5.31283	-3.177366
3 vs 1	-8	.4704426	-17.01	0.000	-9.337568	-6.662432
4 vs 1	-14.64706	.5502594	-26.62	0.000	-16.21156	-13.08255
3 vs 2	-3.754902	.3832131	-9.80	0.000	-4.844458	-2.665346
4 vs 2	-10.40196	.4615398	-22.54	0.000	-11.71422	-9.089705
4 vs 3	-6.647059	.3952365	-16.82	0.000	-7.7708	-5.523318

Here we have used the Scheffé adjustment [see chapter 5, (5.8)]. The Tukey's honestly significant difference adjustment used in chapter 5 is not available following `mixed`. In addition to the Scheffé adjustment, the Bonferroni and Šidák adjustments are available after `mixed`.

Recall that the Scheffé adjustment changes the critical t-value for the pairwise comparisons. Specifically, in this example, the critical t-value is

```
. display "Scheffe critical value is " sqrt(3*invFtail(3, 101, 0.05))
Scheffe critical value is 2.8432121
```

Given that all observed t-values (see the t's in the `pwcompare` output) are larger than 2.84, all the pairwise comparisons are statistically significant. This is not surprising given the large sample size and relatively large mean differences between trials. The confidence intervals are also wider with the Scheffé correction than with no correction.

```
. pwcompare trial, small mcompare(noadjust) effects
Pairwise comparisons of marginal linear predictions
Margins       : asbalanced
```

	Contrast	Std. Err.	t	Unadjusted P>|t|	Unadjusted [95% Conf. Interval]	
cor_pasat						
trial						
2 vs 1	-4.245098	.3755372	-11.30	0.000	-4.990063	-3.500133
3 vs 1	-8	.4704426	-17.01	0.000	-8.933232	-7.066768
4 vs 1	-14.64706	.5502594	-26.62	0.000	-15.73863	-13.55549
3 vs 2	-3.754902	.3832132	-9.80	0.000	-4.515094	-2.99471
4 vs 2	-10.40196	.4615398	-22.54	0.000	-11.31753	-9.48639
4 vs 3	-6.647059	.3952366	-16.82	0.000	-7.431102	-5.863016

7.4 Models with multiple factors

Like between-subjects designs, repeated-measures models can include multiple factors. These factors can be within-subject or between-subjects factors. Interactions can be between two or more within-subject factors, two or more between-subjects factors, or a combination of within-subject and between-subjects factors. In the PASAT data, there is a between-subjects factor, cond, which was the activity condition—cycling or treadmill desk. Thus, we can fit a model that examines the main effect of condition, the main effect of trial, and the interaction between the two.

A box plot for PASAT scores across trials for each condition helps visualize the relationships we are modeling. A box plot across trials is best created by putting trial in wide format.

```
. reshape wide cor_pasat, i(part cond) j(trial)
(note: j = 1 2 3 4)
Data                               long   ->   wide

Number of obs.                      408   ->   102
Number of variables                   7   ->   9
j variable (4 values)             trial   ->   (dropped)
xij variables:
                              cor_pasat   ->   cor_pasat1 cor_pasat2 ...
> cor_pasat4
```

7.4 Models with multiple factors

```
. graph box cor_pasat1-cor_pasat4, over(cond) scheme(lean2)
>       ytitle("Correct PASAT")
>       legend(label(1 "1") label(2 "2") label(3 "3") label(4 "4")
>                       title("Trial", size(medium)))
. reshape long
(note: j = 1 2 3 4)

Data                                wide    ->   long

Number of obs.                       102    ->    408
Number of variables                    9    ->      7
j variable (4 values)                        ->   trial
xij variables:
    cor_pasat1 cor_pasat2 ... cor_pasat4    ->   cor_pasat
```

After the graph is created, using `reshape long` puts the dataset back in the long format for analysis. Figure 7.2 depicts the relationships. There is a strong effect of `trial` across both conditions as evidenced by the decrease from trial 1 to trial 4. There is no visual evidence of a `cond` effect or an interaction between `cond` and `trial`. That is, the boxes line up similarly across conditions—for example, the box for trial 1 in the treadmill condition and the box for trial 1 in the cycling condition (the same is true for the other trials). The `trial` effect is similar for both conditions, as evidenced by the similar decrease in correct responses across trials.

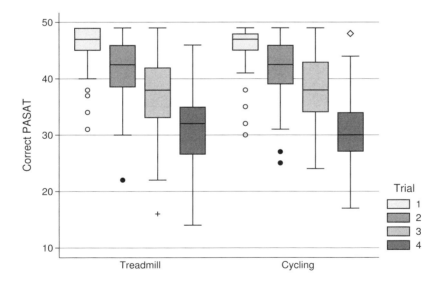

Figure 7.2. Number of correct PASAT answers over four trials and stratified by condition

To provide a statistical test of these relationships, we can fit a regression model:

$$\texttt{cor\_pasat}_{ij} = b_0 + b_1 t_{2j} + b_2 t_{3j} + b_3 t_{4j} + b_4 \texttt{cond}_j + b_5 t_{2j} \texttt{cond}_j + b_6 t_{3j} \texttt{cond}_j$$
$$+ b_7 t_{4j} \texttt{cond}_j + \epsilon_{ij}$$

As before, we specify a structure for the matrix of the residuals, Σ. Table 7.2 provides a listing of the numerator degrees of freedom, denominator degrees of freedom, F-values for both main effects and the interaction, AIC, and BIC statistics for the exchangeable, first-order autoregressive structure, first-order Toeplitz, and unstructured matrix.

Table 7.2. Test statistics and fit indices for five covariance structures applied to the PASAT data; DF = numerator degrees of freedom

Model	DF	DDF	F	AIC	BIC
Exchangeable				2350.409	2390.522
Trial	3	300	385.98		
Condition	1	100	<0.01		
Trial × Condition	3	300	0.59		
Autoregressive				2317.534	2357.647
Trial	3	296.07	210.52		
Condition	1	110.49	<0.01		
Trial × Condition	3	296.07	0.68		
Toeplitz				2430.845	2470.958
Trial	3	224.33	150.84		
Condition	1	281.50	<0.01		
Trial × Condition	3	224.33	0.26		
Unstructured				2285.829	2358.032
Trial	3	100	238.01		
Condition	1	100	<0.01		
Trial × Condition	3	100	0.76		

The results are similar to the analysis with `trial` as the only factor. The autoregressive and unstructured models fit the data best; the AIC was best for the unstructured model, and the BIC was best for the autoregressive model. We again choose the unstructured model given the fact that we expect the residual variances to increase across trials (see figure 7.2). The `mixed` syntax for the unstructured model is

7.4 Models with multiple factors

```
. mixed cor_pasat i.trial##i.cond || part:, noconstant
> residuals(unstructured, t(trial)) reml
> dfmethod(satterthwaite, oim) nolog
```

Mixed-effects REML regression	Number of obs =	408
Group variable: part	Number of groups =	102

	Obs per group:	
	min =	4
	avg =	4.0
	max =	4
DF method: Satterthwaite	DF: min =	100.00
	avg =	100.00
	max =	100.00
	F(7, 100.45) =	103.47
Log restricted-likelihood = -1124.9144	Prob > F =	0.0000

cor_pasat	Coef.	Std. Err.	t	P>|t|	[95% Conf. Interval]	
trial						
2	-4.571429	.5070308	-9.02	0.000	-5.577363	-3.565494
3	-8.5	.6337187	-13.41	0.000	-9.75728	-7.24272
4	-14.78571	.7460501	-19.82	0.000	-16.26586	-13.30557
cond						
Cycling	-.5675466	.8238277	-0.69	0.492	-2.201997	1.066904
trial#cond						
2#Cycling	.7236025	.7550149	0.96	0.340	-.7743256	2.221531
3#Cycling	1.108696	.9436647	1.17	0.243	-.7635084	2.9809
4#Cycling	.3074534	1.110936	0.28	0.783	-1.896613	2.511519
_cons	46.08929	.5532421	83.31	0.000	44.99167	47.1869

Random-effects Parameters	Estimate	Std. Err.	[95% Conf. Interval]	
part: (empty)				
Residual: Unstructured				
var(e1)	17.1403	2.424	12.99094	22.61499
var(e2)	33.24089	4.700933	25.19391	43.8581
var(e3)	42.1677	5.963351	31.95973	55.6361
var(e4)	48.33752	6.835901	36.63593	63.77661
cov(e1,e2)	17.99235	2.989097	12.13383	23.85088
cov(e1,e3)	18.40921	3.258286	12.02309	24.79534
cov(e1,e4)	17.15437	3.350768	10.58699	23.72176
cov(e2,e3)	30.15866	4.80747	20.73619	39.58113
cov(e2,e4)	29.83847	4.997052	20.04443	39.63251
cov(e3,e4)	37.2872	5.855368	25.81089	48.76351

LR test vs. linear model: chi2(9) = 341.40   Prob > chi2 = 0.0000

Note: The reported degrees of freedom assumes the null hypothesis is not on
      the boundary of the parameter space. If this is not true, then the
      reported test is conservative.

Note: The observed information matrix is used to compute Satterthwaite degrees
      of freedom.

The residual variances increase across trials, as they did before. Thus, adding `cond` and `cond#trial` did not account for the increasing residual variances. The correlations among the residuals are

```
. estat wcorrelation
Standard deviations and correlations for part = 102:
Standard deviations:
     trial |     1        2        3        4
-----------+----------------------------------
        sd |  4.140    5.765    6.494    6.953
Correlations:
     trial |     1        2        3        4
-----------+----------------------------------
         1 |  1.000
         2 |  0.754    1.000
         3 |  0.685    0.806    1.000
         4 |  0.596    0.744    0.826    1.000
```

The test of the main effects of `cond` and `trial` and their interaction is

```
. contrast trial cond trial#cond, small
Contrasts of marginal linear predictions
Margins      : asbalanced
```

	df	ddf	F	P>F
cor_pasat				
trial	3	100.00	238.01	0.0000
cond	1	100.00	0.00	0.9752
trial#cond	3	100.00	0.76	0.5204

The main effect for `trial` is statistically significant, $F(3, 100) = 238.01, p < 0.001$. The main effects for `cond` [$F(1, 100) = 0.001, p = 0.98$] and the `trial` × `cond` interaction are not significant [$F(3, 100) = 0.76, p = 0.52$]. Using `margins` can help with interpretation, and underscores the statistical tests and the box plot of the raw data (figure 7.2).

```
. quietly margins, at(trial = (1(1)4) cond = (0 1))
. marginsplot, scheme(lean2) ytitle("Expected PASAT") title("")
    Variables that uniquely identify margins: trial cond
```

As seen in figure 7.3, the expected PASAT scores are essentially identical across conditions, and the rate at which PASAT scores decrease is also identical across conditions—thus, no main effect for `cond` and no interaction.

7.5 Estimating heteroskedastic residuals

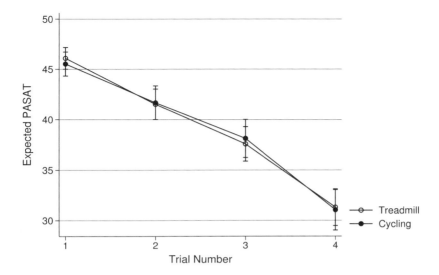

Figure 7.3. Expected PASAT scores over the four trials and stratified by condition

7.5 Estimating heteroskedastic residuals

As noted in section 7.2, the models up to this point have produced one set of values for Σ per model. That is, we assumed that all observations were best described by one set of values for Σ. We do not have to make this assumption of homoskedastic residuals. Instead, we can introduce heteroskedasticity into the model and estimate a unique value of Σ for subgroups, such as intervention condition. Heteroskedastic results are relevant any time you believe that the residuals will be influenced by a variable, but they can be particularly important in complex designs (Baldwin et al. 2011).

In the PASAT example, suppose that the treadmill desk increased the variability in PASAT responses more than the cycling desk. It is possible that all participants find the treadmill desk sufficiently different than a normal desk when it comes to sustaining attention. Some participants may find the cycling desk is close enough to sitting down at a normal desk that their responses are not any different than they would have been at a normal desk. Other participants may find that the cycling desk enhances the challenge of the PASAT as compared with sitting at a normal desk. Therefore, there may be more variability in PASAT responses in the cycling condition than the treadmill condition. This could be true even if there was not a mean difference between conditions, which the previous models suggested.

To fit a heteroskedastic model with respect to treatment condition, we add a `by()` option to the `residuals()` option.

```
. mixed cor_pasat i.trial##i.cond || part:, noconstant
> residuals(unstructured, t(trial) by(cond)) reml
> dfmethod(satterthwaite, oim) nolog
```

Mixed-effects REML regression   Number of obs   =   408
Group variable: part   Number of groups =   102

                                                 Obs per group:
                                                              min =   4
                                                              avg =   4.0
                                                              max =   4

DF method: Satterthwaite   DF: min =   55.00
                                                              avg =   74.43
                                                              max =   97.50

                                                 F(7,   92.85) =   104.94
Log restricted-likelihood = -1119.744   Prob > F   =   0.0000

cor_pasat	Coef.	Std. Err.	t	P>|t|	[95% Conf. Interval]	
trial						
2	-4.571429	.4670002	-9.79	0.000	-5.507318	-3.635539
3	-8.5	.6446585	-13.19	0.000	-9.791925	-7.208075
4	-14.78571	.7549468	-19.59	0.000	-16.29866	-13.27277
cond						
Cycling	-.5675466	.8311182	-0.68	0.496	-2.218124	1.083031
trial#cond						
2#Cycling	.7236025	.7675105	0.94	0.348	-.8015553	2.24876
3#Cycling	1.108696	.9400425	1.18	0.241	-.7569084	2.9743
4#Cycling	.3074534	1.108	0.28	0.782	-1.891587	2.506494
_cons	46.08929	.5304216	86.89	0.000	45.0263	47.15227

7.5 Estimating heteroskedastic residuals

Random-effects Parameters		Estimate	Std. Err.	[95% Conf. Interval]	
part:	(empty)				
Residual: Unstructured, by cond					
Treadmill:	var(e1)	15.75544	3.004398	10.84215	22.89527
	var(e2)	32.25412	6.150422	22.1959	46.87031
	var(e3)	48.53722	9.255355	33.40126	70.53211
	var(e4)	51.63336	9.845826	35.53179	75.03151
	cov(e1,e2)	17.89828	3.881123	10.29142	25.50515
	cov(e1,e3)	20.50996	4.642265	11.41129	29.60863
	cov(e1,e4)	17.73595	4.528614	8.860027	26.61187
	cov(e2,e3)	32.25281	6.882822	18.76272	45.74289
	cov(e2,e4)	31.43984	6.946039	17.82585	45.05383
	cov(e3,e4)	43.16321	8.91257	25.69489	60.63153
Cycling:	var(e1)	18.83287	3.970274	12.45861	28.46844
	var(e2)	34.44689	7.261918	22.78789	52.07101
	var(e3)	34.38264	7.248376	22.74538	51.97389
	var(e4)	44.30923	9.341086	29.31213	66.97935
	cov(e1,e2)	18.10727	4.658479	8.976819	27.23772
	cov(e1,e3)	15.84157	4.468232	7.083992	24.59914
	cov(e1,e4)	16.4435	4.954911	6.732053	26.15495
	cov(e2,e3)	27.59906	6.576039	14.71026	40.48786
	cov(e2,e4)	27.88119	7.154742	13.85815	41.90422
	cov(e3,e4)	30.10534	7.34801	15.70351	44.50718

LR test vs. linear model: chi2(19) = 351.74   Prob > chi2 = 0.0000

Note: The reported degrees of freedom assumes the null hypothesis is not on the boundary of the parameter space. If this is not true, then the reported test is conservative.

Note: The observed information matrix is used to compute Satterthwaite degrees of freedom.

. estimates store heterosc

. estat ic

Akaike's information criterion and Bayesian information criterion

Model	Obs	ll(null)	ll(model)	df	AIC	BIC
heterosc	408	.	-1119.744	28	2295.488	2407.803

Note: N=Obs used in calculating BIC; see [R] BIC note.

The `residuals(unstructured, t(trial) by(cond))` option tells Stata to estimate a distinct variance–covariance matrix for each condition. Note that the variable included in `by` cannot be a continuous variable; it must be a categorical variable where there are multiple participants per category. The residuals now include 20 estimates, 10 for each condition. Some of the residuals look a bit different. For example, the variance for trial 3 is $\sigma^2_{3T} = 49$ in the treadmill condition and $\sigma^2_{3C} = 34$ in the cycling condition.

We expect some differences due to sampling error. Fortunately, the homoskedastic model is nested within the heteroskedastic model, so we can test whether estimating unique values of Σ by condition significantly improves model fit. The test is done with a likelihood-ratio test by using the `lrtest` command. To use this command, save the results from the heteroskedastic and homoskedastic models with `estimates store` (as was done above for the heteroskedastic model). Then use `lrtest`.

```
. quietly mixed cor_pasat i.trial##i.cond || part:, noconstant
> residuals(unstructured, t(trial)) reml
> dfmethod(satterthwaite, oim)
. estimates store homosc
. estat ic

Akaike's information criterion and Bayesian information criterion
```

Model	Obs	ll(null)	ll(model)	df	AIC	BIC
homosc	408	.	-1124.914	18	2285.829	2358.032

Note: N=Obs used in calculating BIC; see [R] BIC note.

```
. lrtest homosc heterosc
Likelihood-ratio test                            LR chi2(10) =     10.34
(Assumption: homosc nested in heterosc)          Prob > chi2 =    0.4111
```
Note: The reported degrees of freedom assumes the null hypothesis is not on
      the boundary of the parameter space. If this is not true, then the
      reported test is conservative.
Note: LR tests based on REML are valid only when the fixed-effects
      specification is identical for both models.

The likelihood-ratio test is not significant, $\chi^2(10) = 10.34$, $p = 0.41$, which suggests that the homoskedastic and heteroskedastic models fit the data equally well. The AIC and BIC both preferred the homoskedastic model, which provides more evidence that the additional parameters are not needed.

7.6 Summary

In this chapter, I showed how to analyze repeated-measures data including models with a single within-subject factor as well as models with a mix of between-subjects and within-subject factors. We used the `mixed` command to fit the models rather than the `anova` command. The `mixed` command allowed us to thoroughly investigate what correlation structure best accounts for the relationship among the residuals. Although the `mixed` command defaults to z tests when computing p-values and constructing confidence intervals, the `dfmethod()` option allows us to use t tests and F tests with `mixed`. Specifying the `satterthwaite` option for `dfmethod()` ensures that the degrees of freedom for the t and F tests is appropriate given the factors in the model and correlation structure for the residuals. Finally, using `mixed` rather than `anova` provides more flexibility with respect to missing data.

This chapter did not cover designs with multiple within-subject factors. The `mixed` command can accommodate these designs, but doing so requires the use of random effects, which are covered in chapter 9. Additional resources for using mixed models to analyze repeated-measures data can be found in Littell, Stroup, and Freund (2002), Littell et al. (2007), and Rabe-Hesketh and Skrondal (2012b).

8 Planning studies: Power and sample-size calculations

If you have done any psychological research, you know that planning and designing studies requires determining how many people you are going to recruit for your study. Human-subjects approval requires an estimate of how many people are going to be involved. Grant applications require you to specify sample sizes so that there will be enough data to answer the research questions and you can justify your budget. Dissertation and thesis committee members want to know your sample sizes to determine whether you can successfully recruit the necessary subjects and thus complete your project.

Where do these values come from? How do you decide how many subjects to recruit? Generally, the methods fall into three categories. First, figure out how many subjects you can afford and pick a number close to the maximum. Second, determine how many subjects published studies in the research area used and use that number. Third, do formal sample-size calculations. Ideally, studies would use formal analyses for all sample-size determinations, perhaps in combination with budget concerns when those are present. It is a bad idea to rely on the published literature as a guide for what is a good sample size. Psychology studies are notorious for using unreasonably small samples (Button et al. 2013; Cohen 1962), and drawing inferences from small samples is problematic (see section 8.4). Furthermore, researchers do not consistently report how they determined the sample size they used (Larson and Carbine 2017) and whether the sample size provided sufficient precision to test their hypotheses.

This chapter will teach you how to do power analyses and sample-size calculations. Specifically, in this chapter you will learn the following:

1. Key concepts in power analyses (for example, type I and type II errors).
2. How to use simulation methods to compute power.
3. How to manually compute power.
4. How to use Stata's commands to perform power analyses and sample-size calculations.
5. How power is related to the precision of estimates and inferences.

Stata commands and functions featured in this chapter

- `power`: for computing power, performing sample-size calculations, and determining detectable difference
- `program`: for writing programs for simulations
- `simulate`: for performing simulations
- `normal()`: the cumulative distribution function for the normal distribution, which is used in power calculations

8.1 Foundational ideas

Power is most commonly defined as follows: The probability of rejecting the null hypothesis, if the null hypothesis is false and given a specific alternative or effect size. Suppose you conduct a study comparing a new insomnia drug to a placebo. You use a t test to compare the number of hours of sleep following treatment in each condition. The null hypothesis for the t test is that the number of hours slept when taking the drug does not differ from the hours slept when taking the placebo in the population. Imagine you know that the null hypothesis is false and that those who take the drug will sleep 1 hour more per night than those taking the placebo. The standard deviation for hours slept is 2 hours, so this 1-hour difference represents a half-standard-deviation difference ($\delta = 0.5$). That is, the real effect of the drug is 1 additional hour of sleep.

You design your study, collect data on 20 people taking the drug and 20 people taking the placebo, and test the difference with a t test. (I know you would do something more sophisticated if you were actually doing a clinical trial.) Power in this case is the probability that you will reject the null hypothesis given that the real effect is a 1-hour difference. If power was 40%, then, in the long run, 40% of trials with 40 participants would reject the null hypothesis. Thus, you want power to be high, so you have a high probability of rejecting the null. (As you will see below, power is important even if you are not using null hypothesis significance testing to make inferences.)

8.1.1 Null and alternative distributions

Imagine two sampling distributions (see section 3.3.7): a sampling distribution where the null hypothesis is true and a sampling distribution where the alternative hypothesis is true. The null distribution is centered at a difference of $\delta_0 = 0$, and the alternative distribution is centered at a difference of $\delta_A = 0.5$. Figure 8.1 depicts these sampling distributions as t distributions, with 38 degrees of freedom (DF; DF $= n_1 + n_2 - 2 = 20 + 20 - 2 = 38$). Focus on the null distribution. The tall, bold vertical lines represent

8.1.1 Null and alternative distributions

the critical values, which is determined by the DF and α. α is known as the type I error rate, which is the probability that you will reject the null hypothesis when the null is true.

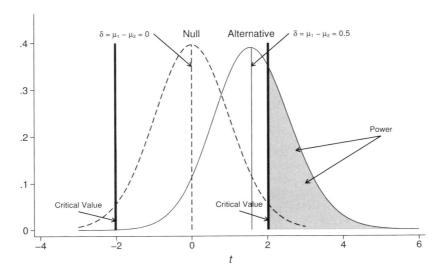

Figure 8.1. Null and alternative distributions

If the drug actually has no effect ($\delta = \mu_1 - \mu_2 = 0$), then any samples from the population will come from the null distribution—repeated samples will look just like the null distribution. You do not know whether the null is true, but the logic of null hypothesis significance testing says that if your observed results are in the tails of the null distribution (that is, the regions beyond the critical values), you say that the null is unlikely. But there is still some probability that your sample came from the null distribution, and that probability is α. Specifically, if α is 0.05, then, in the long run, 5% of the time you will falsely reject the null hypothesis—your type I error rate is 5%. If α is 0.1, your type I error rate is 10%.

Now focus on the alternative distribution. If the drug actually produces an effect of $\delta = \mu_1 - \mu_2 = 0.5$, then all samples will come from the alternative distribution. If a sample from the alternative distribution falls beyond the critical values in the null distribution, then you reject the null hypothesis. Power represents the proportion of samples from the alternative distribution that exceed the critical value of the null distribution. That is, the proportion of the alternative distribution exceeds the thick vertical bars in figure 8.1. If 40% exceed the critical value, then power is 40%. Note that figure 8.1 presents a two-tailed test. Consequently, "exceed" means greater than the critical value in the positive direction *and* less the critical value in the negative direction. Given that the alternative distribution is to one side or the other of the null distribution, power is only meaningfully going to be affected by how much the alternative exceeds the critical value in one of the directions.

When power is low, you run the risk of making a type II error, which is the probability of failing to reject the null hypothesis when the null is actually false. The type II error rate is denoted as β and is equal to $1 - \text{Power}$. Thus, if power is 40%, then the type II error rate is 60%—in the long run you will fail to reject the null hypothesis, when it is really false, 60% of the time.

Null or alternative distribution?

Step back for a second. Given our assumptions, either the null is true or the alternative is true. Thus, when you collect data on a sample, the sample is either from the null distribution or the alternative distribution. You are trying to figure out which. Power is about making sure you have sufficient precision in your study to ensure that you do not mistakenly conclude that you sampled from the null distribution, when you actually are sampling from the alternative distribution. Of course, perhaps you specified the wrong alternative distribution—the effect is a lot larger or smaller than you guessed. Such is the imprecise nature of power analyses and science generally. But do not despair. Thoughtful power analyses have many benefits, as you will see.

8.1.2 Simulating draws out of the null and alternative distributions

To solidify your understanding of power analyses, write a simulation program that can sample either from the null or alternative distribution. I have written such a program, which I have called `simpower`. This program simulates a two-condition study with a single timepoint, such as the hypothetical insomnia drug study I described above.

There are six arguments for `simpower`: the population means for groups 1 and 2 (`m1` and `m2`), the population standard deviations for groups 1 and 2 (`sd1` and `sd2`), and the sample sizes for groups 1 and 2 (`n1` and `n2`). Once the values for these arguments are determined, the program simulates a sample from the population. It produces an outcome variable for groups 1 and 2 (`y1` and `y2`) from a normal distribution with supplied means and standard deviations. It then performs a t test and saves the specified population difference (`delta`), sample means, sample standard deviations, sample sizes, t-statistics, DF, and two-sided p-value.

8.1.2 Simulating draws out of the null and alternative distributions

```
. program define simpower, rclass
  1.         version 14.1
  2.         syntax [, m1(real 0) m2(real 0)
>         sd1(real 1) sd2(real 2)
>         n1(integer 1) n2(integer 1)]
  3.         drop _all
  4.         local obs = max(`n1´, `n2´)
  5.         set obs `obs´
  6.         generate y1 = rnormal(`m1´, `sd1´) in 1/`n1´
  7.         generate y2 = rnormal(`m2´, `sd2´) in 1/`n2´
  8.         ttest y1==y2, unpaired
  9.         return scalar delta = `m1´ - `m2´
 10.         return scalar m1 = r(mu_1)
 11.         return scalar m2 = r(mu_2)
 12.         return scalar sd1 = r(sd_1)
 13.         return scalar sd2 = r(sd_2)
 14.         return scalar n1 = r(N_1)
 15.         return scalar n2 = r(N_2)
 16.         return scalar t = r(t)
 17.         return scalar df = r(df_t)
 18.         return scalar p = r(p)
 19. end
```

You can use `simulate` to perform the simulations with `simpower`. In this initial simulation, sample from the null distribution by giving both groups the same population mean (I used 0 for both groups so $\delta = 0$). This simulation assumes 20 subjects per group.

```
. simulate m1 = r(m1) m2 = r(m2) sd1 = r(sd1) sd2 = r(sd2)
>             n1 = r(n1) n2 = r(n2) t = r(t) df = r(df) p = r(p)
>             delta = r(delta), reps(1000) seed(43714) nodots: simpower,
>             m1(0) m2(0) sd1(1) sd2(1) n1(20) n2(20)
      command:  simpower, m1(0) m2(0) sd1(1) sd2(1) n1(20) n2(20)
           m1:  r(m1)
           m2:  r(m2)
          sd1:  r(sd1)
          sd2:  r(sd2)
           n1:  r(n1)
           n2:  r(n2)
            t:  r(t)
           df:  r(df)
            p:  r(p)
        delta:  r(delta)
```

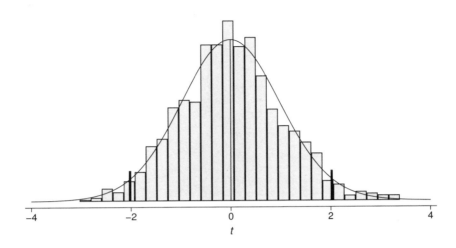

Figure 8.2. Simulated samples from the null distribution

Figure 8.2 displays a histogram of the t-values from each of the 1,000 samples as well as the null hypothesis sampling distribution (a t distribution with 38 DF). The thick black lines represent the critical values for rejecting the null hypothesis for $\alpha = 0.05$. If you count the number of times that the t-value for a sample exceeds the critical value (± 2.02), that is the type I error rate.

```
. display invt(38, .975)
2.0243942

. generate reject = t > 2.02 | t < -2.02

. tabulate reject

      reject |      Freq.     Percent        Cum.
-------------+-----------------------------------
           0 |        952       95.20       95.20
           1 |         48        4.80      100.00
-------------+-----------------------------------
       Total |      1,000      100.00
```

Out of the 1,000 samples, about 5% exceed the critical value. Thus, you would make a type I error 5% of the time. It is critical to understand that the null is true in this simulation. Thus, if you reject the null hypothesis, you will be wrong, and you will make a type I error.

Repeat the simulation, but this time make the alternative hypothesis true and make the difference between the two groups one-half of a standard deviation. If the population standard deviations for each group are 1, then giving group 1 a mean of 0.5 and group 2 a mean of 0 means that $\delta = 0.5$.

8.1.2 Simulating draws out of the null and alternative distributions

```
. simulate m1 = r(m1) m2 = r(m2) sd1 = r(sd1) sd2 = r(sd2)
>         n1 = r(n1) n2 = r(n2) t = r(t) df = r(df) p = r(p)
>         delta = r(delta), reps(1000) seed(438854) nodots: simpower,
>         m1(.5) m2(0) sd1(1) sd2(1) n1(20) n2(20)
   command:  simpower, m1(.5) m2(0) sd1(1) sd2(1) n1(20) n2(20)
        m1:  r(m1)
        m2:  r(m2)
       sd1:  r(sd1)
       sd2:  r(sd2)
        n1:  r(n1)
        n2:  r(n2)
         t:  r(t)
        df:  r(df)
         p:  r(p)
     delta:  r(delta)
```

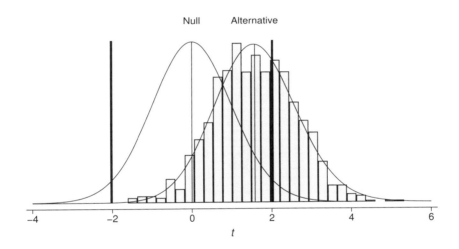

Figure 8.3. Simulated samples from the alternative distribution

Figure 8.3 presents a histogram of the samples from the alternative distribution as well as the null distribution. The tall, thick lines are the critical values used for rejecting the null hypotheses. All the samples come from the alternative distribution. Thus, even if there is a real effect of $\delta = 0.5$, you can still get tiny effects in some samples or even samples where the placebo is better than the treatment.

If you count the number of times the t-values from the simulation exceed the critical value, that is power. Because you are performing a two-tailed test, you will reject the null if it is greater than 2.02 or less than -2.02. So you need to count on both sides. Given that the real effect is positive, the number of samples that will actually be less than -2.02 will be small and sometimes even 0.

```
. generate reject_alt = t > 2.02 | t < -2.02
. tabulate reject_alt
```

reject_alt	Freq.	Percent	Cum.
0	651	65.10	65.10
1	349	34.90	100.00
Total	1,000	100.00	

Thus, with a sample size of 20 per group and a population difference between the conditions of $\delta = 0.5$ (that is, one-half of a standard deviation), 35% of the samples will lead you to reject the null hypothesis. This also means that 65% of the samples will lead to not rejecting the null hypothesis, even though they came from the alternative distribution. You basically have a 1 in 3 chance of rejecting the null, even though it is false. However, you can do better just by increasing the sample size. Let's see how that works.

Repeat the simulation where the alternative hypothesis is true, but this time increase the sample size to 50 per condition.

```
. simulate m1 = r(m1) m2 = r(m2) sd1 = r(sd1) sd2 = r(sd2)
>              n1 = r(n1) n2 = r(n2) t = r(t) df = r(df) p = r(p)
>              delta = r(delta), reps(1000) seed(438854) nodots: simpower,
>              m1(.5) m2(0) sd1(1) sd2(1) n1(50) n2(50)
     command:  simpower, m1(.5) m2(0) sd1(1) sd2(1) n1(50) n2(50)
          m1:  r(m1)
          m2:  r(m2)
         sd1:  r(sd1)
         sd2:  r(sd2)
          n1:  r(n1)
          n2:  r(n2)
           t:  r(t)
          df:  r(df)
           p:  r(p)
       delta:  r(delta)
```

8.1.2 Simulating draws out of the null and alternative distributions

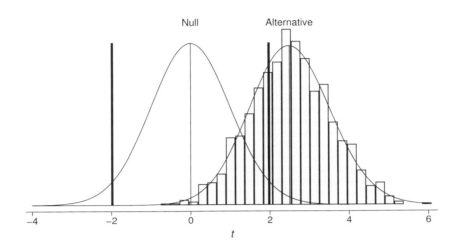

Figure 8.4. Simulated samples from the alternative distribution with 50 participants per condition

Figure 8.4 depicts the draws from this alternative distribution. Compare figures 8.3 and 8.4. Note how the number of samples that exceed the critical value is much higher in figure 8.4 than 8.3. With 50 people per condition, power is 71%.

```
. generate reject_alt_n50 = t > 1.98 | t < -1.98
. tabulate reject_alt_n50

reject_alt_
       n50 |      Freq.     Percent        Cum.
------------+-----------------------------------
          0 |        290       29.00       29.00
          1 |        710       71.00      100.00
------------+-----------------------------------
      Total |      1,000      100.00
```

By using simulation to draw samples from the alternative hypothesis population and then compute a t test, you have simulated power. Simulating power calculations is by far the most flexible method for performing power analyses (Muthén and Muthén 2002). As long as you can specify a null and an alternative population and simulate samples from each, you can perform power analyses. For example, I have used simulation to perform power analyses for multilevel models with a complex data structure (Baldwin et al. 2011), as well as when planning mediation studies and measurement studies for grants.[1]

---

[1]. An additional benefit for simulating power is that you learn a lot about a model if you have to build a simulation for it. It is worth the time to learn this stuff, I promise.

The main problem with a simulation study is that it can be challenging to manipulate certain aspects of the design to see what happens to other parts. In the examples above, you changed sample size to see what happened to power. However, what if you wanted to know what sample size would give you 80% power. You would need to iterate through multiple simulations to find that value. Fortunately, Stata has commands for power analyses and sample-size calculations for many data types and analysis options. When you design a study that uses one of these methods, these commands can prove invaluable. Before discussing the use of these commands in more detail, let's learn how to use Stata to solve the equations to compute power and sample-size requirements. This provides a deeper understanding of power and the commands themselves.

8.2 Computing power manually

We are going to learn the manual methods for computing power using a two-sample z test. Why not a t test? The z test allows us to use the normal distribution, whereas a t test requires us to use a noncentral t distribution. Dealing with the noncentral t distribution is, frankly, a bit of a pain when first learning power because you have to keep the peculiarities of that distribution in your head at the same time as learning power. Furthermore, the concepts are exactly the same, so if you master power calculations for z tests, you will have an excellent foundation for power calculations for other designs and situations.

To compute power for a two-sample z test, you need to know three quantities:

1. The effect size, which in this case is the difference between the means of each group.
$$\delta = \mu_1 - \mu_2$$

2. The standard error of the difference between the two groups. This is the standard deviation of the sampling distribution of the difference between the means. To compute this, you need to know the population variance[2] for each group, σ_1^2 and σ_2^2, and the sample size for each group, n_1 and n_2. For now, you will assume the sample sizes are equal, but you can generalize the concepts to unequal samples.
$$\sigma_D = \sqrt{\frac{\sigma_1^2}{n_1} + \frac{\sigma_2^2}{n_2}}$$

3. You need to know α, the type I error rate, which is used to determine the critical values, $z_{\alpha/2}$ (for example, ± 1.96 for $\alpha = 0.05$).

---

2. If these are unknown, you use the two-sample t test. For those calculations, you will just use the built-in Stata commands.

8.2 Computing power manually

With those three values in hand, you can compute power for a two-tailed test with the following equation:

$$\text{Power} = \Phi\left(\frac{\delta}{\sigma_D} - z_{1-\alpha/2}\right) + \Phi\left(-\frac{\delta}{\sigma_D} - z_{1-\alpha/2}\right) \tag{8.1}$$

where Φ is the cumulative distribution function for the standard normal distribution. The standard normal distribution is a normal distribution with a mean equal to 0 and a standard deviation equal to 1. A cumulative distribution function tells you how much of the distribution falls below a specific value. For example, if you want to know what proportion of the standard normal distribution falls below $z = -0.2$, you can use Φ. Stata includes a cumulative distribution function for the standard normal distribution that can be accessed using the function `normal()`. For example, if you want to know what proportion of the standard normal distribution falls below $z = -0.2$, you would type

```
. display normal(-0.2)
.42074029
```

which indicates that 42% of the normal distribution falls below $z = -0.2$. If you want to know how much falls above $z = -0.2$, you would type

```
. display 1 - normal(-0.2)
.57925971
```

The expression δ/σ_D transforms the effect size into a z-value so that you can use the normal distribution as a reference distribution. The difference between δ/σ_D and $z_{1-\alpha/2}$ tells you how many z score units the effect size and critical value are away from one another. The first part of (8.1) gives you the probability that a given effect size will exceed the critical value in the positive direction. The second part of (8.1) gives you the probability that the effect size will be smaller than the critical value in the negative direction.

Suppose you design a study to compare two treatments for depression—an active treatment [for example, cognitive behavioral therapy (CBT)] and a placebo treatment (for example, active listening). You will randomly assign 100 patients to either the CBT or the placebo condition so that there are 50 patients per condition. Your primary outcome is a measure of depressive symptoms, which you will standardize so that the outcome has a standard deviation of 1. You believe that patients receiving CBT will improve more than those receiving placebo and that the difference is half of a standard deviation on the depression measure: $\delta = 0.5$.

To use the steps listed above, you need to know the following:

- δ, which is 0.5 (see step 1).
- σ_1^2 and σ_2^2, which are both 1 (see step 2).
- n_1 and n_2, which are both 50 (see step 2).
- α, which you will set to 0.05 (see step 3).

You have specified δ. Next, compute σ_D:

$$\sigma_D = \sqrt{\frac{1}{50} + \frac{1}{50}} = 0.2$$

Then, find the two-tailed critical value for $\alpha = 0.05$, which is $z = \pm 1.96$. Finally, plug the resulting values into (8.1).

$$\text{Power} = \Phi\left(\frac{0.5}{0.2} - 1.96\right) + \Phi\left(-\frac{0.5}{0.2} - 1.96\right) = 0.71$$

Thus, power in this example is 0.71; that is, if the real effect is $\delta = 0.5$, you have a 71% chance of rejecting the null hypothesis. This value can also be computed directly in Stata.

```
. display normal(((.5/.2) - 1.96)) + normal(((-.5/.2) - 1.96))
.70540558
```

8.2 Computing power manually

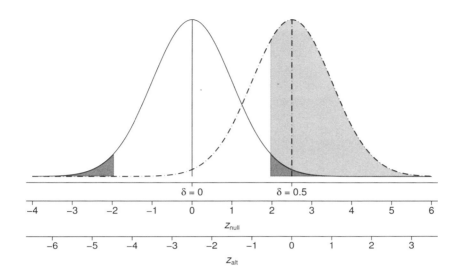

Figure 8.5. Illustrating power by using z scores for both the null (z_{null}) and the alternative (z_{alt}) distributions

Figure 8.5 helps illustrate why (8.1) works. The left-hand distribution is the null population, and the right-hand distribution is the alternative population. The null population is centered at $\delta = 0$ and the alternative at $\delta = 0.5$. The two x axes represent z scores for the null distribution (the top axis) and the alternative distribution (the bottom axis). An effect size of $\delta = 0.5$ is equal to a z score of 2.5 in the null population but is equal to a z score of 0 in the alternative population. The critical value in the null population is $z = \pm 1.96$. For effects greater than $\delta = 0$ (that is, effects greater than the null hypothesis δ), the critical value is $z = -0.54$ in the alternative population $(1.96 - 2.5 = -0.54)$. For effects less than $\delta = 0$, the critical value is $z = -4.45$ in the alternative distribution $(-1.96 - 2.5 = -4.45)$. You can see this by extending the critical values down to the z_{alt} axes on figure 8.5.

Power is defined by determining the proportion of the alternative distribution that is less than $z = -4.45$ and greater than $z = -0.54$. Recall that the cumulative distribution function for the normal curve, Φ, provides the proportion of the distribution below a particular z score. Thus, $\Phi(-4.45)$ provides the proportion of the distribution that is less than $z = -4.45$. To obtain the proportion greater than $z = -0.5$, you have two options. First, you could compute $1 - \Phi(-0.5)$. Second, you could, as the first part of (8.1) does, take advantage of the fact that the normal distribution is symmetrical; that is, $\Phi(0.5)$ is the same as $1 - \Phi(-0.5)$. You can verify this equivalence in Stata:

```
. display (1 - normal(-0.5))
.69146246

. display normal(0.5)
.69146246
```

8.3 Stata's commands

Stata's `power` command is a flexible utility for conducting power analyses. You can perform power analyses for a) one sample, b) two independent samples, c) two paired samples, d) more than two samples (repeated or independent samples), e) contingency tables, and f) survival data. Within these designs, you can also perform several power analyses for several kinds of outcomes. For example, when dealing with two independent samples, you can calculate power for a z test, t test, comparison of two proportions, comparison of two correlations, or comparison of two variances. See the *Stata Power and Sample-Size Manual* (StataCorp 2017g) for a complete description of possible analyses.

This section of the chapter will demonstrate how to use `power` to compute power and sample-size calculations for common designs in psychology. I begin by computing power for the two-sample z test from section 8.2. I then show how to compute power for a t test, correlation, one-way analysis of variance (ANOVA), and factorial ANOVA.

8.3.1 Two-sample z test

Section 8.2 computed power for a two-sample z test where the population effect size was $\delta = 0.5$, and n_1 and n_2 were both 50. A z test was appropriate because we assumed that the group variances, σ_1^2 and σ_2^2, were known. The power for this z test was 0.7. This analysis can be done in Stata using the following code:

```
. power twomeans 0.5 0, sd(1) n(100) alpha(0.05) knownsds
Estimated power for a two-sample means test
z test assuming sd1 = sd2 = sd
Ho: m2 = m1  versus  Ha: m2 != m1

Study parameters:

        alpha =    0.0500
            N =       100
  N per group =        50
        delta =   -0.5000
           m1 =    0.5000
           m2 =    0.0000
           sd =    1.0000

Estimated power:

        power =    0.7054
```

The first argument of `power` is `twomeans` to signal that the design is comparing two independent means. Following `twomeans`, we specify the means for groups 1 and 2, 0.5 and 0, respectively. We set the standard deviation for each group to 1 and the total sample size to 100. When the total sample size is used, Stata automatically divides the total evenly between the two groups.[3] The α level is also set to 0.05; this is the default and does not need to be made explicit. I prefer to make this explicit in my power calculations for transparency and clarity. Finally, we use the `knownsds` option

---
3. You can also set distinct standard deviations for each group by using the `sd1()` and `sd2()` options. The same is true for the sample size: `n1()` and `n2()`.

8.3.2   Two-sample t test   215

to signal that we want to treat the standard deviations as known, which means Stata will compute power for a z test. Stata defaults to assuming the standard deviations are sample estimates and will compute power for t tests.

The output first includes details about the study parameters for the proposed study. It is critical to ensure that parameters match what you intended, especially if you rely on Stata's defaults. The bottom of the output includes the estimated power, which is 0.7 in this example. This is the same value as in section 8.2.

8.3.2   Two-sample t test

Typically, it is not reasonable to assume that the standard deviations are known, and t tests rather z tests should be used. To compute power for the example depression-treatment study, repeat the previous power analysis but leave off the `knownsds` option.

```
. power twomeans 0.5 0, sd(1) n(100) alpha(0.05)
Estimated power for a two-sample means test
t test assuming sd1 = sd2 = sd
Ho: m2 = m1   versus  Ha: m2 != m1

Study parameters:
          alpha =    0.0500
              N =       100
    N per group =        50
          delta =   -0.5000
             m1 =    0.5000
             m2 =    0.0000
             sd =    1.0000

Estimated power:
          power =    0.6969
```

The output now indicates that power is computed for a t test.

Often, it is useful to compute power for a range of sample sizes. You can use a number list for sample size with `power`. For example, if you would like to compute power for sample sizes between 20 and 100, increasing sample size by 10 each time, the code is

```
. power twomeans 0.5 0, sd(1) n(20(10)100) alpha(0.05)
```
Estimated power for a two-sample means test
t test assuming sd1 = sd2 = sd
Ho: m2 = m1   versus   Ha: m2 != m1

alpha	power	N	N1	N2	delta	m1	m2	sd
.05	.1851	20	10	10	-.5	.5	0	1
.05	.2624	30	15	15	-.5	.5	0	1
.05	.3379	40	20	20	-.5	.5	0	1
.05	.4101	50	25	25	-.5	.5	0	1
.05	.4779	60	30	30	-.5	.5	0	1
.05	.5407	70	35	35	-.5	.5	0	1
.05	.5981	80	40	40	-.5	.5	0	1
.05	.6502	90	45	45	-.5	.5	0	1
.05	.6969	100	50	50	-.5	.5	0	1

Tables of output can be difficult to digest, and graphs of power curves can be useful to summarize this information. To do this, you can use the graph() option (see figure 8.6).

```
. power twomeans 0.5 0, sd(1) n(20(10)100) alpha(0.05) graph(scheme(lean2))
```

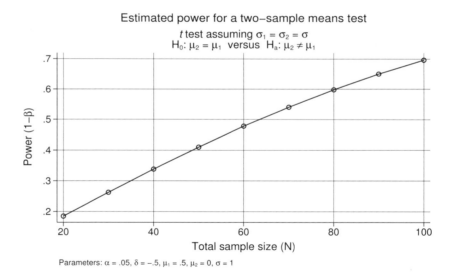

Figure 8.6. Power curve for a two-sample t test

It can be useful to think about the detectable difference for a given sample size at a specific power level. For example, suppose you are still in the design phase of your depression-treatment study. You know that you can recruit 100 patients, but you would like to know how big of a population effect you can detect with the 100 patients if you set power to 0.8. Furthermore, you want to see how big of an effect you could detect if

8.3.2 Two-sample t test

the total sample size was 60 (if the budget is tight) all the way up to 200 (if your grant gets funded).

```
. power twomeans 0, sd(1) n(60(10)200) power(0.8) alpha(0.05)
> graph(ytitle("Detectable difference ({&delta})")
> ylabel(0(.25)1) xlabel(60(20)200) scheme(lean2))
```

To get the detectable difference from `power`, omit group 1's mean and set group 2's mean. Because this example uses effects in the standardized mean difference metric, we just set the mean of group 2 to 0. The output of this analysis will be the detectable difference between group 1 and group 2 in standard deviation units. We set power to 0.8 and α to 0.05.

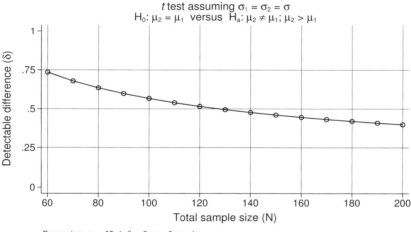

Figure 8.7. Detectable difference across sample size

Figure 8.7 shows the results of the detectable difference analysis. When N is 60, the detectable difference is $\delta = 0.75$. That is, when $N = 60$, we would have 80% power to detect a 0.75 standard deviation difference between groups. That is a large effect in treatment research and most likely implausible (Wampold and Imel 2015). For $N = 100$, the detectable difference is just above $\delta = 0.5$. $N = 130$ is required to detect $\delta = 0.5$ with 80% power, and $N = 200$ will provide 80% power to detect $\delta = 0.4$. Thus, if effect sizes are small (as are most effect sizes in psychology), you will need well over 200 participants to have 80% power.

You can adjust the analyses to compute sample size needed given a specific power value and effect size. This type of analysis is aptly named a sample-size calculation. For example, suppose you are still in the design phase of the depression-treatment study. You hope that the difference between the conditions will be $\delta = 0.5$, but differences all the way down to $\delta = 0.2$ would be clinically significant and more probable.

```
. power twomeans 0, sd(1) diff(0.2(.05)0.5) power(0.8) alpha(0.05)
> graph(xtitle("Effect size ({&delta})") scheme(lean2))
```

The code is altered to add the `diff()` option, which takes either a single number or a number list. In this case, we add a number list to examine effect sizes ranging from $\delta = 0.2$ to $\delta = 0.5$. Figure 8.8 shows the results. To detect a difference of $\delta = 0.2$ (that is, a 0.2 standard deviation difference), a study needs 800 total patients. For an effect size of $\delta = 0.4$, 200 patients are needed, and so on.

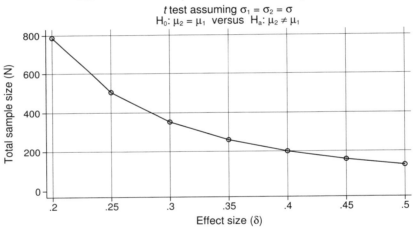

Figure 8.8. Sample sizes needed to have 80% power to detect effect sizes between $\delta = 0.2$ to $\delta = 0.5$

8.3.2 Two-sample t test

You can also vary two aspects of the design simultaneously. For example, you can compute the detectable difference for distinct values of N and power (see figure 8.9).

```
. power twomeans 0, sd(1) n(20(10)100) power(0.6(0.1)0.9) alpha(0.05)
> graph(ytitle("Detectable difference ({&delta})")
> title("Detectable difference for a two-tailed t-test ({&alpha} = 0.05)")
> subtitle("") note("") scheme(lean2))
```

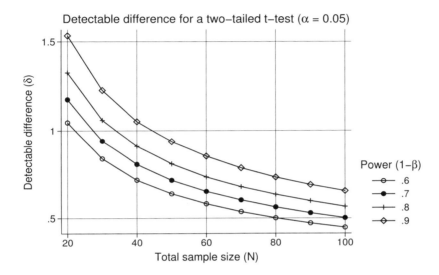

Figure 8.9. Detectable difference as a function of sample size and power

Or you could compute sample size needed as a function of effect size and power (see figure 8.10).

```
. power twomeans 0, sd(1) diff(0.2(.2)0.8) power(0.6(.1)0.9) alpha(0.05)
> graph(ytitle("Total sample size (N)")
> title("Sample size requirements as a function of power and effect size")
> subtitle("") note("") legend(title("Effect size ({&delta})"))
> scheme(lean2))
```

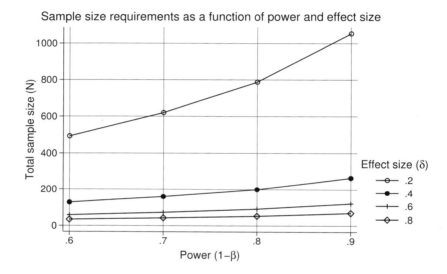

Figure 8.10. Sample size as function of power and effect size

8.3.3 Correlation

As part of the hypothetical treatment study planned in section 8.2, you measure fidelity to CBT in the treatment condition. You are interested in whether therapy cases where treatment was delivered with high fidelity will have better outcomes than cases with low fidelity. Both fidelity and outcome are continuous measures, so you will assess the relationship with a correlation. The average fidelity–outcome correlation reported in the literature is small ($r = 0.02$) and not statistically significant (Webb, DeRubeis, and Barber 2010). You want to know the power to detect a population correlation of $\rho = 0.2$, which is the smallest correlation you believe is clinically important. Possible sample sizes for the treatment condition range from 25 to 150.[4]

---

4. Given finite research budgets, you often have to prioritize some analyses over others. Because the primary aim of this type of study is to examine treatment effects, you would probably prioritize power for that aim. That is, you may only be able to plan for enough subjects to have sufficient power to detect treatment effects, even if that means you would be underpowered for the fidelity–outcome analysis.

8.3.3 Correlation

The code to estimate power to detect a correlation of $\rho = 0.2$ is

```
. power onecorrelation 0.2 0, n(25(25)150) alpha(0.05) graph(scheme(lean2))
```

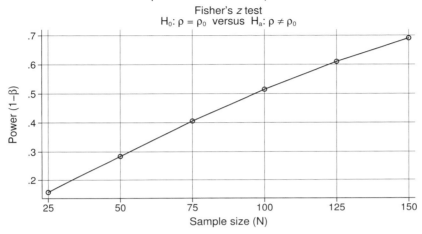

Figure 8.11. Power to detect a population correlation of $\rho = 0.2$ as a function of sample size

Figure 8.11 shows that even with $N = 150$, power does not exceed 0.7. Consequently, if the fidelity–outcome correlation was a critical aim of the study, you would need to recruit more patients.

To determine the necessary sample size, alter the code for **power** to this:

```
. power onecorrelation 0  (0.1(0.1)0.5), power(0.8) graph(scheme(lean2))
```

Stata will compute the necessary sample size to have 80% power to detect population effect sizes from $\rho = 0.1$ to $\rho = 0.5$. Figure 8.12 shows that around 200 participants are needed for a population correlation of $\rho = 0.2$, and nearly 800 are needed for a population correlation of $\rho = 0.1$.

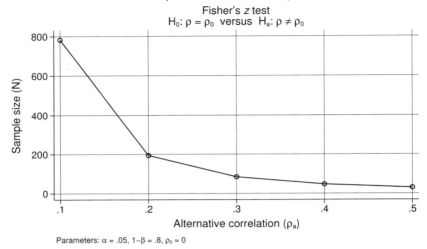

Figure 8.12. Sample size needed to have 80% power to detect a population correlation between $\rho = 0.1$ and $\rho = 0.5$

Finally, you can compute the detectable correlation as a function of sample size (see figure 8.13).

```
. power onecorrelation 0, n(25(25)150) power(0.8)
> graph(title("Detectable correlation as a function of sample size")
> ytitle("Detectable correlation ({&rho})") scheme(lean2))
```

8.3.4 One-way ANOVA

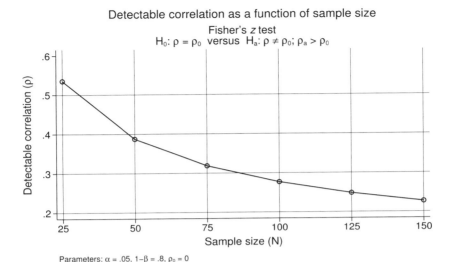

Figure 8.13. Detectable population correlation as a function of sample size

8.3.4 One-way ANOVA

Power for a one-way ANOVA analysis follows the same principles as t tests and correlations. That is, power is a function of effect size, sample size, and α. A key difference for computing power in Stata for a one-way ANOVA design is that effect size is now defined as Cohen's f (Cohen 1988). A formula for f is

$$f = \sqrt{\frac{\sigma^2_{\text{group}}}{\sigma^2_{\text{error}}}} \tag{8.2}$$

Here σ^2_{group} is the variability among the group means, and σ^2_{error} is the error variability. Given that it can be difficult to have good guesses about specific values of the group and error variability, I suggest you think about how much of the total variance in the outcome is accounted for by σ^2_{group}. Furthermore, if you standardized the outcome so that it has a variance of 1, you will have an easier time with these calculations. Specifically, you can specify σ^2_{group} as a value between 0 and 1, and σ^2_{error} as $1 - \sigma^2_{\text{group}}$. Indeed, you can reexpress (8.2) as

$$f = \sqrt{\frac{\sigma^2_{\text{group}}}{1 - \sigma^2_{\text{group}}}}$$

Suppose you extend the depression-treatment example to include three conditions: CBT, antidepressant medication, and placebo therapy. You believe that treatment condition will account for 5% of the total variance. In that case, $\sigma^2_{\text{group}} = 0.05$ and $\sigma^2_{\text{error}} = 1 - 0.05 = 0.95$. Thus, f is $\sqrt{(0.05/0.95)} = 0.23$.

To estimate power for this design in Stata, use the `power oneway` command.

```
. power oneway, n(90) varmeans(.05) varerror(.95) ngroups(3) alpha(0.05)
Estimated power for one-way ANOVA
F test for group effect
Ho: delta = 0  versus  Ha: delta != 0

Study parameters:
            alpha =    0.0500
                N =        90
      N per group =        30
            delta =    0.2294
              N_g =         3
            Var_m =    0.0500
            Var_e =    0.9500

Estimated power:
            power =    0.4671
```

In the output, σ^2_{group} is `Var_m`, σ^2_{error} is `Var_e`, and f is `delta`. Thus, when group accounts for 5% of the variance in the outcome and with 90 patients (30 per group), the power to detect an effect is 0.47.

8.3.4 One-way ANOVA

Power for the same effect size but across multiple sample sizes can be found by using a number list for sample size (see figure 8.14).

```
. power oneway, n(30(30)210) varmeans(.05) varerror(.95) ngroups(3)
> graph(scheme(lean2)) alpha(0.05)
```

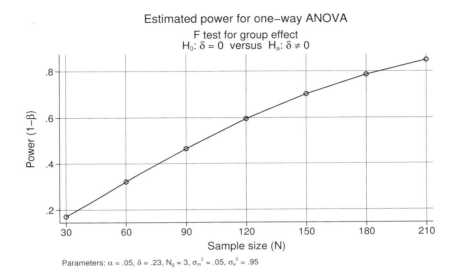

Figure 8.14. Power as a function of sample size for a one-way ANOVA

Finally, sample-size estimates are provided when the power level and effect size are specified.

```
. power oneway, ngroups(3) varmeans(.05(.05).25) varerror(.95(.05).75) parallel
> alpha(0.05) power(0.8)
Estimated sample size for one-way ANOVA
F test for group effect
Ho: delta = 0   versus   Ha: delta != 0
```

alpha	power	N	N_per_group	delta	N_g	Var_m	Var_e
.05	.8	189	63	.2294	3	.05	.95
.05	.8	90	30	.3333	3	.1	.9
.05	.8	60	20	.4201	3	.15	.85
.05	.8	42	14	.5	3	.2	.8
.05	.8	33	11	.5774	3	.25	.75

We used number lists for both `varmeans()` and `varerror()`. We also added the `parallel` option. If you do not use the `parallel` option, Stata will try to compute sample sizes for all possible combinations from the number lists. That is not desirable behavior here because of how we have defined the effect size. When Stata is computing

f, it should treat $\sigma^2_{\text{group}} = 0.05$ and $\sigma^2_{\text{error}} = 0.95$ as a pair, $\sigma^2_{\text{group}} = 0.1$ and $\sigma^2_{\text{error}} = 0.9$ as a pair, and so on. The `parallel` option correctly pairs the σ^2_{group} and σ^2_{error} values.

The output indicates that 33 patients, 11 per condition, are needed if treatment condition accounts for 25% of the outcome variance. In contrast, 189 patients, 63 per group, are needed if treatment group accounts for 5% of the outcome variance.

8.3.5 Factorial ANOVA

Power for factorial ANOVA is similar to power for one-way ANOVA, except that with factorial ANOVA you need to conduct analyses separately for main effects and interaction terms. If your study is a 2 × 3 design, then you will need to compute power for the column main effect, the row main effect, and the interaction main effect.

Suppose you conduct a treatment for depression study with a 2 × 2 design, with a psychotherapy factor and a drug factor. The levels of the psychotherapy factor are CBT or no CBT. The levels of the drug factor are active medication or placebo. You plan to recruit 50 patients per cell, for a total sample size of $N = 200$. You would like to compute power for the main effect of CBT, the main effect of drug, and the interaction.

The effect size is still f; however, f will be specific depending on which effect you are interested in. In a 2 × 2 design, there are three f values: f_A, f_B, and $f_{A \times B}$. These are computed as

$$f = \sqrt{\frac{\sigma^2_A}{\sigma^2_{\text{error}}}}$$

$$f = \sqrt{\frac{\sigma^2_B}{\sigma^2_{\text{error}}}}$$

$$f = \sqrt{\frac{\sigma^2_{A \times B}}{\sigma^2_{\text{error}}}}$$

In one-way ANOVA, the key quantity when thinking about effect size is the ratio of group variance to total variance: $\sigma^2_{\text{group}}/\sigma^2_{\text{group}} + \sigma^2_{\text{error}}$. In factorial ANOVA, the key quantity is the ratio of factor variance (or interaction variance) to factor variance plus error $\sigma^2_{\text{factor}}/\sigma^2_{\text{factor}} + \sigma^2_{\text{error}}$. Thus, the factor-specific f-value is similar to a partial correlation or partial η^2 (Cohen 1988).

If the CBT factor accounts for 3% of the factor plus error variance, then $f_{\text{CBT}} = \sqrt{0.03/0.97} = 0.18$. If the drug factor accounts for 5% of the factor plus error variance, then $f_{\text{Drug}} = \sqrt{0.05/0.95} = 0.23$. Finally, if the interaction accounts for 1% of the interaction plus error variance, then $f_{\text{CBT} \times \text{Drug}} = \sqrt{0.01/0.99} = 0.1$. Power for each of these is computed as

8.3.5 Factorial ANOVA

```
. power twoway, varrow(0.03) varerror(0.97) nrows(2) ncols(2) n(200)
Estimated power for two-way ANOVA
F test for row effect
Ho: delta = 0   versus  Ha: delta != 0

Study parameters:
        alpha =      0.0500
            N =         200
    N per cell =          50
        delta =      0.1759
          N_r =           2
          N_c =           2
        Var_r =      0.0300
        Var_e =      0.9700

Estimated power:
        power =      0.6967

. power twoway, varcolumn(0.05) varerror(0.95) nrows(2) ncols(2) n(200)
Estimated power for two-way ANOVA
F test for column effect
Ho: delta = 0   versus  Ha: delta != 0

Study parameters:
        alpha =      0.0500
            N =         200
    N per cell =          50
        delta =      0.2294
          N_r =           2
          N_c =           2
        Var_c =      0.0500
        Var_e =      0.9500

Estimated power:
        power =      0.8977

. power twoway, varrowcolumn(0.01) varerror(0.99)  nrows(2) ncols(2) n(200)
Estimated power for two-way ANOVA
F test for row-by-column effect
Ho: delta = 0   versus  Ha: delta != 0

Study parameters:
        alpha =      0.0500
            N =         200
    N per cell =          50
        delta =      0.1005
          N_r =           2
          N_c =           2
       Var_rc =      0.0100
        Var_e =      0.9900

Estimated power:
        power =      0.2930
```

When specifying the effect size, `power twoway` requires that you specify the option `varcolumn()`, `varrow()`, or `varrowcolumn()`. It is arbitrary which factor is called the row or column. In this example, CBT was the row and drug was the column. Furthermore, if you only use `power twoway` by specifying effect sizes as we have done here, it actually does not matter what you call `row`, `column`, or `rowcolumn`. The answer

will be the same. To demonstrate, we repeat the analyses above, but this time we use
`varrowcolumn()` for each calculation.

```
. power twoway, varrowcolumn(0.03) varerror(0.97)  nrows(2) ncols(2) n(200)
Estimated power for two-way ANOVA
F test for row-by-column effect
Ho: delta = 0  versus  Ha: delta != 0

Study parameters:

        alpha =    0.0500
            N =       200
    N per cell =       50
        delta =    0.1759
          N_r =         2
          N_c =         2
       Var_rc =    0.0300
        Var_e =    0.9700

Estimated power:

        power =    0.6967

. power twoway, varrowcolumn(0.05) varerror(0.95)  nrows(2) ncols(2) n(200)
Estimated power for two-way ANOVA
F test for row-by-column effect
Ho: delta = 0  versus  Ha: delta != 0

Study parameters:

        alpha =    0.0500
            N =       200
    N per cell =       50
        delta =    0.2294
          N_r =         2
          N_c =         2
       Var_rc =    0.0500
        Var_e =    0.9500

Estimated power:

        power =    0.8977

. power twoway, varrowcolumn(0.01) varerror(0.99)  nrows(2) ncols(2) n(200)
Estimated power for two-way ANOVA
F test for row-by-column effect
Ho: delta = 0  versus  Ha: delta != 0

Study parameters:

        alpha =    0.0500
            N =       200
    N per cell =       50
        delta =    0.1005
          N_r =         2
          N_c =         2
       Var_rc =    0.0100
        Var_e =    0.9900

Estimated power:

        power =    0.2930
```

8.4　The central importance of power

As seen previously, `power twoway` can also compute the sample size needed to detect an effect of a particular size at a specific power level. This can be done with a single effect size:

```
. power twoway, varrowcolumn(0.03) varerror(0.97)  nrows(2) ncols(2) power(0.8)
Performing iteration ...
Estimated sample size for two-way ANOVA
F test for row-by-column effect
Ho: delta = 0  versus  Ha: delta != 0
Study parameters:
        alpha =    0.0500
        power =    0.8000
        delta =    0.1759
          N_r =         2
          N_c =         2
       Var_rc =    0.0300
        Var_e =    0.9700
Estimated sample sizes:
            N =       256
  N per cell =        64
```

This can also be done for multiple effect sizes by using a number list and the `parallel` option (see also section 8.3.4):

```
. power twoway, varrowcolumn(0.05(0.05)0.25) varerror(0.95(0.05)0.75)  nrows(2)
> ncols(2) power(0.8) parallel
Performing iteration ...
Estimated sample size for two-way ANOVA
F test for row-by-column effect
Ho: delta = 0  versus  Ha: delta != 0
```

alpha	power	N	N_per_cell	delta	N_r	N_c	Var_rc	Var_e
.05	.8	152	38	.2294	2	2	.05	.95
.05	.8	76	19	.3333	2	2	.1	.9
.05	.8	48	12	.4201	2	2	.15	.85
.05	.8	36	9	.5	2	2	.2	.8
.05	.8	28	7	.5774	2	2	.25	.75

8.4　The central importance of power

Many disciplines, including psychology, use null-hypothesis tests as a key piece of evidence in research studies. Specifically, p-values are seen as a critical arbiter of the evidential value (Greenwald 1975; Meehl 1978). If $p < 0.05$, effects are considered real and usually interpreted as important.[5] A difficulty in using p-values is that they must be interpreted in light of statistical power because power has implications above and

---

5. Other aspects of research studies are also considered, such as good measurement, random assignment to condition, and cross-sectional versus longitudinal designs. Thus, p-values are not the sole criteria. However, p-values are usually seen as a key indicator of evidential value.

beyond ensuring that you will be able to reject the null when you should. Indeed, small studies using low-powered statistical tests can wreak havoc on the scientific literature because statistically significant results from small studies are often errors (Button et al. 2013).

Two classes of errors that are influenced by statistical power are type M errors and type S errors (Gelman and Carlin 2014). Each kind of error increases when statistical power is low. Consequently, if p-values are prioritized over power when evaluating the value of study results, the research literature may become rife with unreliable findings.

8.4.1 Type M and S errors

Power plays a major role in the accuracy of parameter estimates. In the case of type I and type II errors, the crucial or typical, the decision researchers make is binary—reject or retain the null. It is also important to consider how accurate our parameter estimates are. Figure 8.15 depicts the spread of the sampling distribution (x axis) of population standardized mean difference of $\delta = 0.25$ for specific power values (y axis).[6] The darkly shaded triangles are effects that are statistically significant, whereas lightly shaded values are statistically nonsignificant. When power is small (for example, less than 0.4), significant effects typically overestimate the population value, sometimes badly. When power is low, a small number of significant effects have the wrong sign. However, when power is high, significant effects do not exactly equal the population value but tend to be close to it. Gelman and Carlin (2014) discuss errors in inference with respect to the sign (that is, positive or negative) of an estimate and the magnitude of the estimate (that is, overestimated or underestimated). They termed these errors type S and type M, respectively, and showed how the errors are influenced by power.

---

6. I got the idea for this type of funnel plot from Shravan Vasishth:
http://vasishth-statistics.blogspot.de/2015/08/some-reflections-on-teaching.html.

8.4.1 Type M and S errors   231

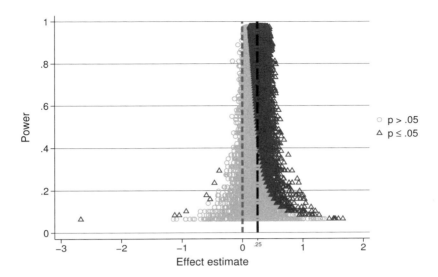

Figure 8.15. Funnel plot showing the sampling distribution of a standardized mean difference as a function of power

Type S errors

Type S errors are defined as the probability that a statistically significant sample estimate has the wrong sign. To explain this concept, Gelman and Carlin (2014) introduce the idea of a replicated effect, which they call d_{rep}. Suppose you compare two conditions, each with 10 participants, and you find a statistically significant effect. Now suppose you ran the study a second time changing nothing except for the participants. The effect from this new study is d_{rep}, the effect size in the replicated study. The type S error rate refers to the probability that d_{rep} will have the wrong sign. To compute this quantity, we have to return to (8.1), which is used to compute power for a two-tailed test:

$$\text{Power} = \Phi\left(\frac{\delta}{\sigma_D} - z_{1-\alpha/2}\right) + \Phi\left(-\frac{\delta}{\sigma_D} - z_{1-\alpha/2}\right)$$

The left-hand quantity is the probability that an estimate will be positive, and the right-hand quantity is the probability that an estimate will be negative. The sum of these two quantities is the power.

If the population effect size is positive, $\delta = 0.2$, the probability that a sample-based estimate will be negative (a type S error) is

$$\text{Type S error rate} = \frac{\text{Probability that the estimate is negative}}{\text{Power}}$$

This is equal to[7]

$$\text{Type S error rate} = \frac{\Phi\left(-\frac{\delta}{\sigma_D} - z_{1-\alpha/2}\right)}{\Phi\left(\frac{\delta}{\sigma_D} - z_{1-\alpha/2}\right) + \Phi\left(-\frac{\delta}{\sigma_D} - z_{1-\alpha/2}\right)} \quad (8.3)$$

If $\delta = 0.2$ and you use 10 patients per treatment, then the type S error rate is [using (8.3)]

```
. local delta = .2
. local s = sqrt((1/10) + (1/10))
. local probneg = normal(((-`delta'/`s') - 1.96))
. local power = normal(((`delta'/`s') - 1.96)) + normal(((-`delta'/`s') - 1.96))
. display "Type S Error Rate is " `probneg'/`power'
Type S Error Rate is .10979379
```

Figure 8.16 recreates figure 2 from Gelman and Carlin (2014) and displays type S error rates as a function of power. Type S error rates are near 0 when power is about 0.2 and become more probable when power is 0.1 or less (Gelman and Carlin 2014). Therefore, if power is very low and you rejected the null hypothesis, you will have a nontrivial probability that your significant estimate is of the wrong sign. Power can be this low when studying very small effects.

---

7. This example uses a normal distribution to compute power. If that is not appropriate for your specific scenario, you will need to substitute other power formulae.

8.4.1 Type M and S errors

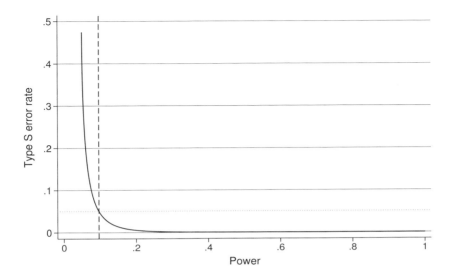

Figure 8.16. Type S error rate as a function of power: the horizontal, dotted line indicates a 5% type S error rate, while the vertical, dashed line is the power level (≈ 0.1) below which the type S error rate exceeds 5%

Type M errors

In addition to getting the sign incorrect, it is possible for an estimate to exaggerate the size of an effect. This is a type M error or magnitude error. Type M errors are indexed by an exaggeration ratio, which Gelman and Carlin (2014) define as "the expectation of the absolute value of the estimate divided by the effect size, if statistically different from zero" (p. 643). As with type S errors, this definition is conditional on the results being judged to be statistically significant. Given that most effects published are statistically significant (Fanelli 2012; Greenwald 1975), the exaggeration ratio is indexing how much the published literature on a specific effect exaggerates the effect. Ideally, the ratio would be close to 1.

The exaggeration ratio can be determined via simulation. For example, suppose you believe the true effect size for the difference between two conditions is $\delta = 0.2$ and that the standard error for the effect is $\sigma_\delta = 0.45$. We can simulate from the sampling distribution of this population effect and determine how much, on average, sample estimates exaggerate the magnitude of the population effect.

```
. clear all
. set obs 10000 // sample 10000 times
number of observations (_N) was 0, now 10,000
. set seed 29446
```

```
.  generate estimate = rnormal(.2, .45) // sample effect sizes
.  generate sig = abs(estimate)/.45 > 1.96 // is the sample effect significant?
.  by sig, sort: egen mean_est = mean(abs(estimate))
>  // what is the average of the absolute values of the estimate?
.  egen exag_tag = tag(sig) // tag a single significant value
.  generate exag = mean_est/.2 if exag_tag == 1
>  // divide the average estimated value by the population value to
>  determine exaggeration
(9,998 missing values generated)
.  list exag if exag_tag == 1 & sig == 1 // what is the exaggeration ratio?
```

	exag
9260.	5.363065

In this example, the absolute value of statistically significant effects would tend to be about five times too big.

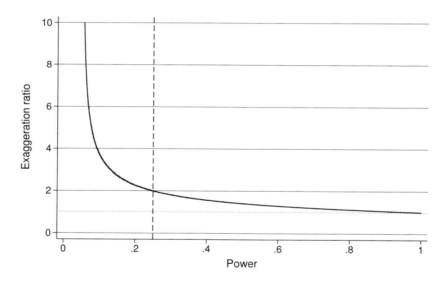

Figure 8.17. Exaggeration ratio (type M error) as a function of power: the horizontal, dotted line indicates no exaggeration, while the vertical, dashed line is the power level (≈ 0.25) below which the exaggeration ratio exceeds 2

Figure 8.17 recreates figure 2 from Gelman and Carlin (2014) and displays the exaggeration ratio as a function of power. As power exceeds 80%, the exaggeration ratio gets close to 1 (that is, no exaggeration). However, as power gets small, the exaggeration ratio grows. The vertical dashed line is drawn at a power of 25%. If power is less than 25%, the exaggeration ratio will be 2 or greater. That is, if power is 25% and you report a statistically significant effect, the absolute value of the point estimate will be about two times too big. Such exaggerations of effects will limit the quality of the research literature (Button et al. 2013).

8.5 Summary

In this chapter, we examined the concept of power. You have learned

- Why power is important for significance testing and effect estimation.
- How to use simulation to conduct power analyses.
- How to use the `power` command to estimate power, conduct sample-size analyses, and graphically display power and sample-size results for several common designs.
- How power affects the quality of the inferences we draw from our studies.

Formal power analyses as well as examination of type M and type S errors are not used as often as they should be. Fortunately, the tools in Stata help make these types of calculations simple and accessible to many researchers. Further, for more-complicated designs, Stata's simulation facilities make more-complicated power calculations possible (Muthén and Muthén 2002).

9 Multilevel models for cross-sectional data

Although substantial evidence suggests that psychotherapy is effective (Wampold and Imel 2015), a key question for patients is, "Who should I see for therapy?" Are the effective therapists the people who use a particular treatment, the people with a particular kind of degree, or the people with the best personality? Despite the fact that psychotherapy research has been around for over 50 years, not much is known about what makes a therapist effective. Furthermore, it has only been in the last 20 years that we have tried to model how much variability in outcomes is associated with therapists. One reason for this is that traditional statistical tools used in psychology, such as regression and analysis of variance (ANOVA), are not well suited for answering this question. Psychotherapy data are clustered—patients are clustered or nested within therapists—and traditional statistical methods are not well suited to modeling such data.[1] Multilevel models, which extend regression and ANOVA, are useful for psychotherapy data and other clustered data structures. This chapter introduces multilevel models for clustered data structures and illustrates how to use `mixed` to implement them.

Clustered data in psychology are common. Often, clustered data are described as hierarchical, with two or more levels to the hierarchy (Raudenbush and Bryk 2001; Singer and Willet 2003; Snijders and Bosker 2011). In a two-level hierarchy, level 2 is the upper level and lever 1 is the lower level. Level 1 units of observation are clustered or nested within level 2 units, such that each level 2 unit is connected to multiple level 1 units but each level 1 unit is associated with only one level 2 unit. Examples include the following:

- Psychotherapy where patients (level 1) see a single therapist (level 2) but each therapist sees many patients.
- Group psychotherapy where patients (level 1) participate in a single group (level 2) but each group has multiple patients.
- Family studies where each family member (level 1) is part of a single family (level 2) but each family has multiple members.
- Marriage research where each partner (level 1) is part of one dyad (level 2) but each dyad has, by definition, two partners.

---

1. A nested ANOVA can do the trick, but it can be cumbersome and inflexible as datasets get larger and more complex.

- Cluster-randomized trials where schools are randomized to conditions. Participants within a school (level 1) are part of a single school (level 2) but each school has multiple students.
- Meta-analysis where effect sizes (level 1) come from a single study (level 2).
- Longitudinal research where observations (level 1) are clustered within a person (level 2; see chapters 7 and 10).

Clustered designs can have more than two levels, such as longitudinal school-based research or longitudinal psychotherapy trials. The framework covered in this chapter can accommodate these designs (Rabe-Hesketh and Skrondal 2012b), though I focus only on designs with two levels.

In this chapter, you will learn the following:

1. Why clustered data structures matter both conceptually and statistically.
2. How multilevel models partition the variance of variables into between-clusters and within-cluster variability.
3. The difference between complete pooling, no pooling, and partial pooling of information across clusters.
4. How to estimate and interpret random effects.
5. How to examine between-clusters and within-cluster relationships.

Stata commands and functions featured in this chapter

- `mixed`: for fitting a multilevel model
- `estat icc`: for computing the intraclass correlation following a multilevel model
- `margins`: for visualizing a multilevel model
- `egen`: for computing group-level variables
- `predict`: for producing empirical Bayes estimates
- `statsby`: for exploring random slopes
- `lrtest`: for comparing the fit of random-intercept and random-slope models
- `lincom`: for testing the difference between within-cluster and between-clusters coefficients

9.1 Data used in this chapter

This chapter uses a psychotherapy dataset modeled after the data used in Baldwin, Wampold, and Imel (2007). Unfortunately, the data for that study are not available for distribution, so I have simulated similar data with the same structure. The dataset includes five variables:

- `tid`: therapist ID
- `pid`: patient ID (clustered within `tid`)
- `distress`: a measure of psychological distress taken at the end of treatment (lower values mean lower distress)
- `alliance`: a measure of the therapeutic alliance taken at the third session (higher values mean a stronger alliance)
- `motivate`: a measure of a patient's motivation to make change measured at the beginning of treatment (higher values mean more motivation)

This dataset has a hierarchical data structure with two levels, the therapist level and the patient level. Each therapist sees multiple patients, but each patient sees only a single therapist. Although `alliance`, `distress`, and `motivate` are measured at the patient level (which means that each patient has a unique value), we can partition the variance for all three variables into variability at the patient level and variability at the therapist level (see section 9.3.1).

9.2 Why clustered data structures matter

9.2.1 Statistical issues

Clustered data structures are important both statistically and conceptually. Statistically, clustered data violate the assumption of independence of observations. In a typical regression model, we assume that—once we have accounted for the covariates—observations are not correlated. That is, knowing something about one observation does not tell us something about another observation.

Clustered data nearly always introduce a correlation among observations. For example, in family research, observations between families are likely to be independent. Data from one family is not likely to be related to data from another family. However, observations within a family are typically related because the family shares the same genetics, environment, and history. Likewise, in psychotherapy studies, observations from patients of different therapists are likely to be independent. However, observations from patients of the same therapist are likely to be correlated because those patients share the experience of working with a specific therapist. For example, the therapist may be particularly skilled as compared with other therapists, and, as such, his or her patients may all have exceptionally good outcomes.

To visualize nonindependence, we can fit a regression model using the alliance data. The model predicts `distress` from `alliance`.

$$\mathtt{distress}_i = b_0 + b_1 \mathtt{alliance}_i + e_i$$

Rabe-Hesketh and Skrondal (2012a) point out that if the independence assumption is met, then residuals for patients of the same therapist should be unrelated. That is, about half of a therapist's patients should be above the regression line (a positive residual) and half below (a negative residual). Patients from the same therapist should be evenly distributed around the regression line because the observations are not correlated. However, if the observations among patients within a given therapist are correlated, then the patients of that therapist should cluster above the regression line (or below). How tightly clustered the observations are and whether all patients are above (or below) the regression line is a function of how correlated the observations within a therapist are.

In Stata, we first fit the regression and then save the residuals.

```
. use http://www.stata-press.com/data/pspus/mlm_cs_data1
. quietly regress distress alliance
. predict dist_res, residuals
```

Similar to Rabe-Hesketh and Skrondal (2012a), we create a line plot of the residuals for each patient of the first 10 therapists (more than about 10 therapists creates an unreadable plot).[2]

```
. local call
. forvalues i = 0/9 {
  2.      local call `call´ line dist_res pid if tid == `i´ + 1,
> connect(ascending) lcolor(gs`i´) ||
  3. }
. twoway `call´, legend(off) scheme(lean2)
>         xtitle("Patient ID")
```

In figure 9.1, the x axis is the patient ID (`pid`) for a patient within a therapist (that is, `pid` $= 1$ is the first patient for a given therapist). The y axis is the residual value for `distress` after accounting for `alliance`. The lines connect the patients for a therapist.

For a given therapist, patients are on one side of 0 or another. For example, most of the patients seen by the therapist represented by the solid line had residuals below 0. This consistency among the residuals reflects the correlation among the patients seen by the same therapist. Figure 9.1, which uses the alliance data for this chapter, represents a correlation of 0.16 within therapists. Given that this correlation is small (though it is still important statistically and conceptually; see section 9.2.2), most of the lines cross 0. As the correlation increases, lines will be more consistently on one side of 0. For example, figure 9.2 represents a correlation of 0.68. Some lines in figure 9.2 cross 0, as expected given that it is not a perfect correlation, but generally the lines stay on one side of 0 or the other.

---

2. To create this plot, I mimicked the code discussed in this post:
https://www.stata.com/statalist/archive/2013-01/msg01370.html.

9.2.1 Statistical issues

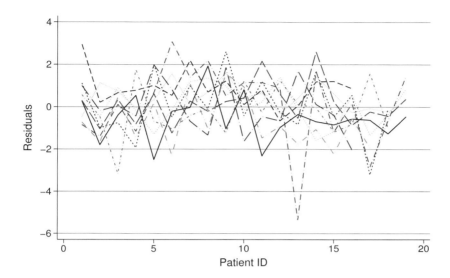

Figure 9.1. Residuals of the same therapist using the alliance data

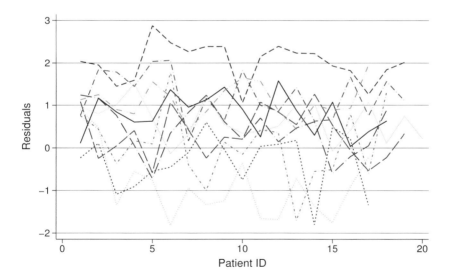

Figure 9.2. Residual of the same therapist assuming a large within-therapist correlation

Figure 9.1 suggests that the patients within therapist are not independent in the alliance data. Most commonly used statistical models, such as regression and ANOVA, assume independence of observations. Violating this assumption leads to standard errors that are too small, which in turn leads to inflated type I errors, confidence intervals (CIs) that are too narrow, and an incorrect level of power (Baldwin, Murray, and Shadish 2005; Kenny and Judd 1986; Raudenbush and Bryk 2001). In other words, assuming that observations are independent when they are not means that the analyses will look more precise than they really are and statistical inferences will be poor.

The reason for the false sense of precision is identical to the issue with the number of unique observations in repeated-measures designs (see section 7.3.1). In the alliance data, we have 720 observations on 40 therapists. Each therapist sees an average of 18 patients. If the observations are fully independent of one another, then this dataset would include 720 unique pieces of information. However, redundant information among patients of the same therapist means we have fewer than 720 unique pieces of information. If patients of the same therapists were perfectly correlated with one another, then this dataset would have 40 unique pieces of information—one for each therapist. Given that the correlation in the alliance dataset is 0.16, we have quite a bit more than 40 unique data points. Nevertheless, the analysis needs to be adjusted to account for this correlation and ensure that the standard errors are properly estimated.

9.2.2 Conceptual issues

Hierarchical data structures matter substantively too. Although within-cluster correlations, such as the correlation among patients of the same therapist, are a problem statistically, they are not simply a nuisance. The correlations may reflect important substantive processes. For example, patients of the same therapist may have similar outcomes because a given therapist is particularly skilled (or particularly poor) or the therapist may be providing a particularly useful type of treatment. In education research, students within a school may have similar achievement outcomes because their school has access to many resources and high-quality teachers. Siblings in a psychopathology study may be similar because of shared genetics or family environment. These potential sources of statistical nonindependence are critical to identifying effective therapists and teachers as well as understanding familial influence on the development of psychopathology.

Kenny et al. (2002) identify three sources of nonindependence:

1. compositional effects
2. common fate
3. mutual influence

Compositional effects mean that people within the cluster have something in common, which leads to correlated observations. This could include schools where students come from similar economic backgrounds or classrooms where all students are gifted. Another example is psychotherapy groups where all patients share the same diagnosis or therapists who consistently receive the most difficult or chronic patients.

9.2.2 Conceptual issues

Common fate indicates that all people within the same cluster share a common event or influence. Patients clustered within a therapist share the common fate of working with a specific therapist. Students in the same classroom share the common fate of working with the same teacher. Sometimes nonindependence is thought to only be relevant when there is a grouping of participants or observations where the participants interact with one another. Common fate suggests that nonindependence can arise even if the participants do not know they are clustered.

The final source of nonindependence is mutual influence. In this case, nonindependence arises because members of a cluster influence one another. Examples include jury members discussing the evidence of a case and persuading one another, and group members in group psychotherapy sharing their thoughts and feelings with one another. These interactions can create a correlation among people within a group.

In sum, ignoring nonindependence is not good statistically or conceptually. Type I error rates are inflated, and our estimates appear more precise than they are. We also miss out on unpacking the substantive meaning of nonindependence. Consequently, using the ANOVA or regression models[3] discussed in chapters 3–6 will not be useful for understanding hierarchical data. Rather, we need models that can relax the assumption of independence and a) produce correct standard errors and b) estimate parameters that contain useful substantive information about the clusters. Multilevel models meet both of these aims.

Repeated-measures data also produce nonindependence

> The repeated-measures models discussed in chapter 7 were used to model nonindependent data due to observing the same person on multiple occasions or in multiple conditions. The models used in chapter 10 are closely related to the multilevel models used in this chapter. Both the repeated-measures models and the multilevel models can be implemented using `mixed`, for example. Furthermore, some versions of the repeated-measures models are equivalent to some multilevel models when fit to the same data, such as the repeated-measures model with a compound symmetry error structure and the multilevel model with a random intercept. Multilevel models can be used to analyze repeated-measures data (see chapter 10). Finally, `mixed` can combine repeated-measures and multilevel models in the same model. The message here is that multilevel models and repeated-measures models, specifically as implemented in chapter 7, are both types of mixed-effects models.

---

3. The structural equation models and factor analysis models discussed in chapters 11, 12, and 13 also assume independence of observations.

9.3 Basics of a multilevel model

9.3.1 Partitioning sources of variance

Patients seeking treatment typically want to know which therapist they should see. Ideally, we would refer them to therapists who have good outcomes. Unfortunately, such information is not readily available, and it has only been in the last 15–20 years that researchers have begun estimating how much variability in psychotherapy outcomes is because of therapists (Crits-Christoph et al. 1991).[4] Multilevel models can separate how much variability in outcomes is associated with therapists and how much is associated with patients.

The simplest multilevel model partitions the variance in the outcome variable into two parts: a) variability between clusters and b) variability within a cluster. Between-clusters variability refers to differences among cluster means, whereas within-cluster variability refers to variability of observations around their respective cluster means. In the alliance data, the variability in distress (the outcome) is separated into variability among therapist means (that is, between-clusters variance in distress) and variability of patients around their therapist's mean (that is, within-cluster variance in distress).

A basic model that describes the total variability in distress is (Rabe-Hesketh and Skrondal 2012b)

$$\text{distress}_{ij} = b_0 + R_{ij} \tag{9.1}$$

Here b_0 is the grand mean of distress and R_{ij} is the difference (that is, residual) between the grand mean and the distress value for patient j seen by therapist i. The sum of the squared values of R_{ij} for all patients is the total variability in distress. Figure 9.3 illustrates this residual for a single patient. The horizontal line is the grand mean (b_0). The circle is the observed distress for the patient. The solid vertical line is the difference between the grand mean and the patient value (R_{ij}).

A multilevel model partitions R_{ij} into two parts:

$$R_{ij} = u_i + e_{ij}$$

u_i is the therapist residual and is the difference between the grand mean and the therapist mean. e_{ij} is the patient residual and is the difference between the therapist mean and patient distress. Thus, the basic multilevel model is as follows:

$$\text{distress}_{ij} = b_0 + u_i + e_{ij} \tag{9.2}$$

Figure 9.3 illustrates the therapist and patient residuals. The diamond is the therapist mean. The dotted line is the difference between the grand mean and the therapist

---

4. Since the beginning of psychotherapy research in the first half of the twentieth century, researchers have been interested in what makes some therapists better than others. However, most of the early research focused on comparing characteristics of therapists (for example, psychologist versus lay counselors; Beutler, Machado, and Neufeldt [1994]) or compared therapists using methods that did not allow generalizations beyond the therapists used in the study (Baldwin and Imel 2013; Wampold and Serlin 2000).

9.3.1 Partitioning sources of variance

mean (u_i), and the dashed line is the difference between the therapist mean and patient `distress` (e_{ij}).

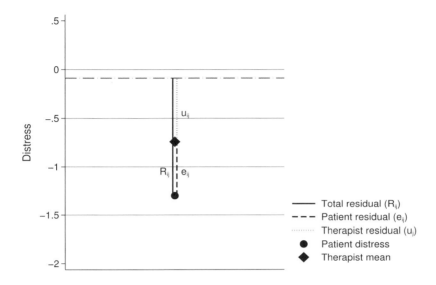

Figure 9.3. Illustration of the partitioning of the total residual

Just like the residuals in linear regression are not estimated, both the between-clusters and within-cluster residuals in a multilevel model are not estimated and thus are not model parameters. The residuals may be estimated or predicted after the fact (see section 9.3.5). In linear regression, the variance of the residuals was estimated. Likewise, in a multilevel model, the variances of the between-clusters and within-cluster residuals are estimated. Specifically, both residuals are assumed to be normally distributed with a mean of 0 and an unknown variance.

$$u_i \sim N\left(0, \sigma_u^2\right) \tag{9.3}$$
$$e_{ij} \sim N\left(0, \sigma_e^2\right) \tag{9.4}$$

The between-clusters variance, σ_u^2, is the expected variability among clusters in the population. The within-cluster variance, σ_e^2, is the expected variability among observations within clusters in the population. In the alliance data, σ_u^2 indexes how much therapists differ with respect to their patients' average `distress`, and σ_e^2 indexes how much patients of the same therapist differ from the therapist mean.

Partitioning the variance into between-clusters and within-cluster variance addresses the fact that a patient is more like other patients seen by the same therapist than patients seen by other therapists. The problem with the model in (9.1) is that it does not distinguish between therapists and thus assumes that all patients are independent from one another. In contrast, the multilevel model assumes that patients are independent from one another after accounting for the therapist each patient saw. In other words, once we account for therapist, patients seen by the same therapist are no longer correlated.

This assumption is called conditional independence. Conditional independence can be understood with respect to predicting the outcome of two patients seen by the same therapist (Rabe-Hesketh and Skrondal 2012b, 79). If we ignore therapist, then knowing the distress score of patient 1 will help us predict the distress score of patient 2. However, once we account for therapist, then we are no longer dealing with just the distress scores. Instead, we are dealing with patient 1's and patient 2's distress scores with respect to their therapist's mean—e_{ij} is the difference between the therapist mean and a patient's distress score. Therefore, knowing that patient 1 is above the therapist mean does not help us predict whether patient 2's score is above or below the therapist mean.

9.3.2 Random intercepts

The model described by (9.2)–(9.4) is called a random-intercept model. The term "random" comes from the distinction between fixed and random factors or effects (see Gelman and Hill [2007, 245] for a discussion of various ways fixed and random effects are defined). A factor is fixed if the dataset includes all levels of the factor that we are interested in making inferences about. In other words, a factor is fixed if we want to limit our conclusions to the levels of the factor included in a study. For example, we could consider therapist fixed in the alliance data, if we wanted to only draw conclusions about the 40 therapists included in the dataset. That is, if we want to learn about the alliance–distress relationship for these 40 therapists only, then we would treat therapists as fixed (Crits-Christoph and Mintz 1991; Wampold and Serlin 2000).

A factor is random if we have only included a sample of the levels we are interested in and thus want to generalize our conclusions to a broader population. In the alliance dataset, the 40 therapists could represent a sample of well-trained therapists, and we want to generalize to a population of similarly trained therapists (Wampold and Serlin 2000). There are not too many situations where researchers want to limit their conclusions to the therapists included in the study. One possibility for treating therapist as a fixed factor is an evaluation of a specific group of therapists in a performance evaluation at a clinic. Generally, however, researchers aim to have the results apply to therapists beyond those used in the study, and thus therapist will typically be a random effect.

Table 9.1 presents some other domains where factors can (and probably should) be treated as random (Judd, Westfall, and Kenny 2012). The model is called a random-intercept model because the intercept is allowed to vary as a function of a variable in the model. Indeed, the model in (9.2) has as many intercepts as there are clusters because the sum of b_0 and u_i for therapist i is a cluster-specific intercept (that is, therapist-specific).

9.3.2 Random intercepts

Table 9.1. Examples of random factors in psychology studies

Area	Possible random factor
Psychotherapy	Groups in group psychotherapy; therapists in individual and group psychotherapy; couples in couple therapy
Education	Districts; schools; classrooms; teachers
Cognitive Psychology	Words to learn in a memory experiment
Social Psychology	Juries in a legal study; faces in an emotion study
Psycholinguistics	Language; words
Psychometrics	Items on a measure; raters in an inter-rater reliability study

Allowing intercepts to vary across therapists relaxes the assumption that all therapists are equal. Consider (9.1), where therapists are not part of the model. That model assumes that all therapists have the same intercept and that the only variability in the outcome is differences among patients. A common intercept would occur in one of two situations, neither of which is particularly likely. First, therapists have no effect at all and are completely irrelevant to outcomes. Second, therapists have an effect, but it is the same for all therapists. These situations are not plausible and will not be for most random factors. Try it: substitute some of the other random factors listed in table 9.1 for therapists and see if any sound plausible.

If a parameter is fixed, then the model estimates a mean or an expectation. For example, b_0, which is a fixed parameter, is the overall mean for distress in the sample and represents the (estimated) expected value for distress in the population. If a parameter is random, then the model estimates a variance (or standard deviation), such as σ_u^2 and σ_e^2 in (9.3) and (9.4). Neither u_i nor e_{ij} is estimated, though they may be predicted following the model. For example, σ_u^2 is the estimated variance in distress among therapists in the population, and σ_e^2 is the estimated variance in distress of patients within therapists in the population. Because the target of generalization of a random factor is the population, we need to know not only the mean distress (that is, b_0) but also how much distress is expected to vary across therapists (that is, σ_u^2) and patients (that is, σ_e^2). As we will see in section 9.5, regression coefficients as well as intercepts can vary in multilevel models.

9.3.3 Estimating random intercepts

The `mixed` syntax for fitting a random-intercept model for `distress` with no predictors is

```
. mixed distress || tid:, nolog
Mixed-effects ML regression                     Number of obs     =        720
Group variable: tid                             Number of groups  =         40

                                                Obs per group:
                                                              min =         15
                                                              avg =       18.0
                                                              max =         20

                                                Wald chi2(0)      =          .
Log likelihood = -1144.6304                     Prob > chi2       =          .
```

distress	Coef.	Std. Err.	z	P>|z|	[95% Conf. Interval]
_cons	-.0841937	.1040864	-0.81	0.419	-.2881992 .1198118

Random-effects Parameters	Estimate	Std. Err.	[95% Conf. Interval]
tid: Identity			
var(_cons)	.362434	.0970346	.2144551 .6125218
var(Residual)	1.272551	.0690161	1.144224 1.415271

```
LR test vs. linear model: chibar2(01) = 107.40      Prob >= chibar2 = 0.0000
```

As with `regress`, the first argument for `mixed` is the outcome variable, in this case `distress`. The random effect for therapist is included with the || followed by `tid:`. The double pipes separate fixed from random effects, with fixed effects on the left. The double pipes also separate levels of random effects. I will not be covering those types of models in this book, but see Rabe-Hesketh and Skrondal (2012b, chap. 8) and the `mixed` documentation for examples. Immediately following the double pipes is the cluster identification variable followed by a colon, in this case `tid:`. The cluster identification variable indexes the distinct levels of the cluster that contribute to the random effect. If the code includes a cluster ID, then a random intercept is estimated by default.

The output is separated into three parts. The first part includes information about estimation and details about the sample. The number of observations is the total number of data points included in the analysis—in this case, it is the number of therapists multiplied by the number of patients per therapist. The number of groups is the number of unique values in the cluster ID variable—40 in this case. The output also provides information about the minimum and maximum number of observations within a cluster as well as the average cluster size.

The second part is the fixed effects, which looks a lot like the output from `regress`. In this instance, there is just a single fixed effect for $b_0 = -0.08$. This indicates that the expected `distress` for all patients averaging over all therapists is 0.08 (95% CI =

9.3.3 Estimating random intercepts

$[-0.29, 0.12]$). Note that the p-value and CI are constructed using a z distribution. The p-values and CIs are valid for large samples (Rabe-Hesketh and Skrondal 2012b; Faes et al. 2009). You do not have to use a z distribution but can use approximate degrees of freedom (see section 7.3.2).

The third part of the output includes the variance estimates of the random effects. The variance for clusters comes first, followed by the variance for residuals. In the section for cluster, `tid: Identity` indicates that this variance is for therapists. Identity means that the covariance matrix of the random effects at the therapist level has an identity structure, which means that there is just a single variance estimated for all random effects and no covariances among them. Given that the model has one random effect at the therapist level, identity is all that is possible. Adding more random effects allows for other possibilities (see section 9.5). The estimate labeled `var(_cons)` is the therapist variance ($\hat{\sigma}_u^2 = 0.36$), and the estimate labeled `var(Residual)` is the residual variance ($\hat{\sigma}_e^2 = 1.27$).

The output provides a standard error and CI for the variance components. The methods for computing these values assumes that the sampling distribution of variance components is normally distributed. If sample sizes are large, this is a reasonable assumption. However, in many—if not most—research contexts, the sampling distribution will not be normally distributed. A variance component has a lower bound of 0. Consequently, if the population value for the variance component is close to 0, then the sampling distribution is positively skewed. It cannot be symmetric because estimates cannot extend below 0. Therefore, interpret null hypothesis tests and CIs for variance components with caution.

Because variances are difficult to interpret, `mixed` includes the `stddeviations` option to report the dispersion in the random effects as standard deviations rather than variances. We can replay the output without needing to fit the model.

```
. mixed, stddeviations noheader nofetable
```

Random-effects Parameters	Estimate	Std. Err.	[95% Conf. Interval]
tid: Identity				
sd(_cons)	.6020249	.0805902	.4630929	.7826377
sd(Residual)	1.128074	.0305902	1.069684	1.189652

LR test vs. linear model: chibar2(01) = 107.40   Prob >= chibar2 = 0.0000

The options `noheader` and `nofetable` suppress the printing of the top portion of the output and the fixed-effects table, respectively. The therapist standard deviation is $\hat{\sigma}_u = 0.6$, and the residual standard deviation is $\hat{\sigma}_e = 1.12$. Given that the random effects are normally distributed, this model estimates that 68% of therapists will be within 0.6 `distress` units of the overall mean (that is, $b_0 = -0.08$) and 95% will be within 1.2 units. The residual standard deviation implies that 68% of patients will be within 1 unit of their therapist mean and 95% will be within 2 units.

9.3.4 Intraclass correlations

Intraclass correlations help us interpret variance components. The intraclass correlation represents the average correlation among observations within a cluster. In the alliance data, the intraclass correlation is the average correlation among patients of the same therapist. The intraclass correlation in a random-intercept model is the ratio of the cluster variance to the total variance in the data (that is, cluster plus residual variance):

$$\rho = \frac{\sigma_u^2}{\sigma_u^2 + \sigma_e^2}$$

The intraclass correlation is commonly interpreted as the proportion of variance in outcome associated with clusters. In the alliance example, this value is important substantively because it tells us how much patient outcomes are associated with therapist. If the intraclass correlation is close to 1, then most of the variability in the outcome is related to therapists. In contrast, if the intraclass correlation is close to 0, then most of the variability is at the patient level. Intraclass correlations for therapists range between 0 and 0.55, with an average of 0.05. This literature is limited by the fact that there are not many studies with large samples of therapists or patients per therapist (Baldwin and Imel 2013).

The intraclass correlation for therapists is

$$\widehat{\rho} = \frac{0.36}{0.36 + 1.28} = 0.22$$

Thus, 22% of the variability in distress is associated with therapists. The postestimation command estat icc computes the intraclass correlation following a multilevel model (not all models are supported, but it will work after a random-intercept model).

```
. estat icc
Intraclass correlation
```

Level	ICC	Std. Err.	[95% Conf. Interval]
tid	.2216742	.0474957	.1424022 .3281876

Like mixed did for variance components, the output of estat icc provides a standard error and CI for the intraclass correlation. Because the intraclass correlation is a ratio that is bounded at 0 on the lower end and 1 on the upper end, the sampling distribution will often not be normally distributed. Stata uses a logit transformation when computing the CI for the intraclass correlation to ensure that the interval is between 0 and 1 (StataCorp 2017e, 54). If inferences about variance components or intraclass correlations are central to your research question, Bayesian methods may be preferable (Baldwin and Fellingham 2013), and Stata can fit Bayesian multilevel models (see StataCorp [2017a, 682]).

9.3.4 Intraclass correlations

Figures 9.4 and 9.5 show therapist means and patient `distress` for 10 therapists. Figure 9.4 is for the alliance data used in this chapter, and the intraclass correlation is $\hat{\rho} = 0.16$. Figure 9.5 is for a dataset similar to the alliance data but that has an intraclass correlation of $\hat{\rho} = 0.68$. Comparing these two graphs allows us to see how the variability within a cluster differs when the intraclass correlation is small versus large. As expected, the patients in figure 9.5, where $\hat{\rho} = 0.68$, do not vary as much around the therapist means as the patients in figure 9.4, where $\hat{\rho} = 0.16$. Additionally, variability in the cluster means is larger when $\hat{\rho} = 0.68$ than when $\hat{\rho} = 0.16$. Thus, the larger the intraclass correlation, the more variability there is between-therapists and the less variability there is within-therapist (assuming a constant total variance).

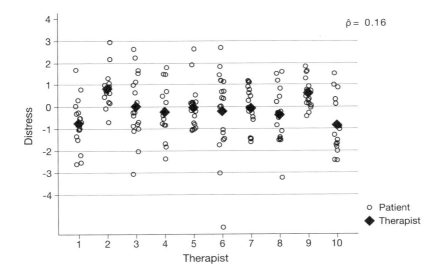

Figure 9.4. Patient and therapist data for 10 therapists and an intraclass correlation = 0.16

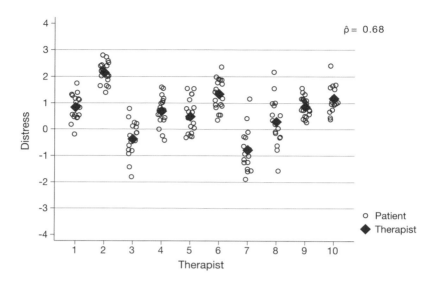

Figure 9.5. Patient and therapist data for 10 therapists and an intraclass correlation = 0.68

9.3.5 Estimating cluster means

In section 9.3.1, I noted that multilevel models estimate the variance of the clusters in the population and do not estimate the cluster means. More precisely, the multilevel models do not estimate u_i, which is how much a cluster mean differs from b_0. For example, we estimated and interpreted the variability of therapists, but we did not estimate each therapist's average distress value. The simplest way to compute means is to generate the averages for each therapist independent of the multilevel model. A characteristic of this approach is that all clusters are treated as distinct and in isolation. That is, when estimating the mean for therapist 1, we do not take into account any of the information we have about therapists 2–40.

The egen command can compute cluster means:

```
. use http://www.stata-press.com/data/pspus/mlm_cs_data1, clear
. by tid, sort: egen nopooled = mean(distress)
```

Further, the community-contributed egenmore command (Cox [2000]; if not installed, type ssc install egenmore) includes a function for egen called semean that can calculate the standard error of the mean by cluster, allowing you to compute CIs for all therapists:

```
. by tid: egen npse = semean(distress)
. generate npll = nopooled - npse*1.96
```

9.3.5 Estimating cluster means

```
. generate npul = nopooled + npse*1.96
```

Following the estimation of the standard errors, the next two lines compute the upper and lower limits of the CI.

It is common to use a "caterpillar plot" to compare clusters. In a caterpillar plot, the cluster means and CIs are ranked from best to worst and then plotted in rank order. This plot is so called because the resulting plot looks like a caterpillar, with the cluster means as the body and the CIs as the legs.

Ranking clusters requires two steps. First, we need to create a variable that identifies one observation per cluster. Note that the variable for cluster means (`nopooled`) created by `egen` changes from cluster to cluster but is constant within a cluster. When we have Stata rank the values for `nopooled`, we only want it to consider one value per cluster. We do not want Stata to see that there are a lot of "ties" in rank because `nopooled` is a constant within a cluster. The `tag` function for `egen` will create this new variable, which we will call `ther_tag`. The second step is to rank the cluster means by selecting only those observations where `ther_tag` is 1.

```
. egen ther_tag = tag(tid)
. egen nptrank = rank(nopooled) if ther_tag == 1
(680 missing values generated)
```

A caterpillar plot for therapists is created by layering a scatterplot (for the means) and a capped range plot (for the CIs). The scatterplot has the therapist means on the y axis and the therapist rank on the x axis. The capped range plot, implemented using `twoway` graph `rcap`, takes three arguments, two y-axis values and one x-axis value. The two y-axis values are the upper limit and lower limit for the cluster-specific CIs; the x-axis value is the therapist rank. Each `twoway` graph uses the tagged observations only.

The resulting figure 9.6 is a compact method for comparing each therapist as well as groups of therapists (for example, top 10 therapists versus bottom 10 therapists). Confidence intervals will also give you a quick sense of which therapists have the fewest patients (widest CIs) and which have the most (narrowest CIs).

```
. twoway scatter nopooled nptrank if ther_tag == 1 ||
>         rcap npul npll nptrank if ther_tag == 1,
>         xtitle("Therapist rank") ytitle("Distress")
>         legend(order(1 "Unpooled mean")) scheme(lean2)
```

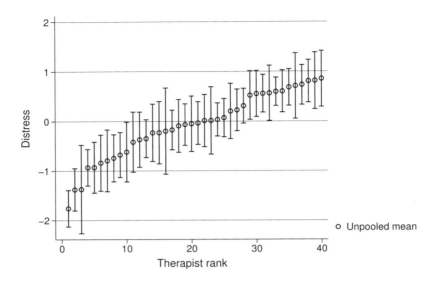

Figure 9.6. Caterpillar plot for the "no pooling" approach to computing cluster means

Computing a therapist mean without considering information from other therapists has been called the "no pooling" approach because we do not pool or combine any information across therapists (Gelman and Hill 2007).[5] The no pooling approach is reasonable in three situations. First, the no pooling approach is reasonable if we observed all patients that therapist 1 ever saw or will see. In that case, we have all the information we could ever get about therapist 1, and we do not need to draw on any information from the other therapists. This, of course, never occurs. Second, the no pooling approach is reasonable if we believe that therapists are so fundamentally different from one another that it is not reasonable to pool information across therapists. This is not common, and I have not run into this situation. Third, the no pooling approach is reasonable if we only want to draw inferences about the specific therapists used in this study. That is, if we simply want to compare the 40 therapists to one another, we may want to treat the data about each therapist as fully distinct. However, as discussed below, there are times where other methods are better, even when the focus is comparing these 40 therapists.

A second approach to estimating a therapist mean is called the "complete pooling" approach (Gelman and Hill 2007) and is the polar opposite to no pooling. With complete pooling, we estimate a single grand mean for all therapists, and there are not any specific therapist means. In this case, we make no distinctions among therapists. The notion is that the best estimate for any therapist would pool information from across all therapists. Differences among therapists are treated solely as noise or error. Con-

---

5. Although I implemented the no pooling approach with **egen mean**, it can also be done using a fixed-effects model for dealing with clustering (Gelman and Hill 2007).

9.3.5 Estimating cluster means

ceptually, this does not make sense unless we assume that all therapists are equivalent and that any differences among therapists are only due to factors such as measurement error. This approach is generally not advisable conceptually or statistically (because it ignores nonindependence of observations).

A third, middle-ground approach to estimating cluster means is to use "partial pooling" (Gelman and Hill 2007). In partial pooling, we draw strength from the other therapists when estimating the mean of a specific therapist. For example, when estimating the mean for therapist 1, we rely not only on the data for therapist 1 but also on what we learned from therapists 2–40. Drawing strength is not the same as combining completely (the complete pooling approach). The mean for a specific therapist is still based on the data for that therapist. However, the mean is informed (that is, weighted) by what we know about other therapists in the population.

The partial pooling depends upon Bayes theorem (Rabe-Hesketh and Skrondal 2012b; Raudenbush and Bryk 2001). In short, Bayes theorem says that our "conclusions" about our data are a combination of the data and our beliefs about the data before the data were collected. These beliefs are called priors, and the conclusions are the posterior because they come after the data and prior are combined. Bayes theorem can be written as

$$\text{Posterior} \propto \text{Data} \times \text{Prior}$$

In words, Bayes theorem says that the posterior is proportional to the data times the prior. A thorough discussion of Bayes theorem is beyond the scope of this chapter—we let Stata do the heavy lifting when it comes to partial pooling. Interested readers can consult McElreath (2016) for a readable treatment of Bayes theorem and Bayesian analysis.

When estimating the posterior mean for therapist 1, the data refer to the observed data for therapist 1, and the prior is what we know about the distribution of therapists in the population. The multilevel model states that the therapists (u_i) are normally distributed with a mean of 0 and a variance of σ_u^2 [see (9.3)]. Given that σ_u^2 is not known, we use the estimate of σ_u^2 from our model in its place. Therefore, we draw on what we learned from all therapists in the sample when estimating the posterior mean for therapist 1. Using the data to estimate σ_u^2 makes the posterior means "empirical Bayes" estimates, meaning that the data are used to construct the prior.

Because the posterior mean for a therapist combines the data for that therapist with what is known about the population of therapists, we say that the posterior mean draws strength from the population of therapists. The degree to which the prior influences a particular posterior mean is a function of the number of observations for a specific therapist. If the sample size for a therapist is relatively large, the prior has little impact on the posterior mean. However, if the sample size for a therapist is relatively small, the posterior mean draws heavily on the prior.

A thought experiment can illustrate what it means to draw strength. Suppose you want to establish whether a coin is biased toward heads or tails. What kind of data are required for you to conclude that a particular coin is biased (that is, has a probability of heads greater than 0.5 or less than 0.5)? Would 2 flips of the coin be good? 10? 1,000? If I flipped a coin twice and obtained heads both times, would it be reasonable to conclude the coin is biased? If you said no (or hesitated) because most coins you have run into are fair and it is not that rare to have two heads with two flips, you are implicitly using Bayes theorem. Indeed, based on prior experience, it is likely that two heads in a row is just a chance occurrence (that is, sampling error). Furthermore, it is reasonable to assume that the coin is probably like most others you have encountered. In other words, you are drawing strength from your experience (that is, your prior) regarding the behavior of coins to inform your conclusions about this particular coin. If I flipped a coin 1,000 times and obtained 950 heads, would it be reasonable to conclude the coin is biased? In that case, it likely you said yes. Regardless of your prior experience with coins, the evidence in this case is overwhelming and suggests that this coin is different. Thus, your prior experience does not influence your inference much in this instance, if at all.

Let's consider why obtaining two heads with two flips of the coin is not likely to convince you that a coin is biased. Suppose again that you strongly believe, prior to seeing the data, that your coin is not biased and thus the probability of heads is 0.5. Figure 9.7 is a probability distribution showing the probability of obtaining 0, 1, or 2 heads if you flipped a coin twice and the probability of a heads for that coin was 0.5.[6] The probability of 2 heads is 0.25 or 1 in 4. Given that obtaining 2 heads is relatively likely, you would probably not conclude that the probability of heads for your coin is 0.25. Instead, you could use what you believe from your prior—that the probability of heads is 0.5—to add to what you learned from the data to come up with an estimate. Given two data points, the prior would carry a lot of weight, and the estimate for your coin would be just below 0.5. Therefore, the estimate would draw strength from the prior and would be pulled toward the prior.

---

6. This is not a prior distribution. I'm simply trying to illustrate conceptually how prior information could inform the estimate from a specific study.

9.3.5 Estimating cluster means

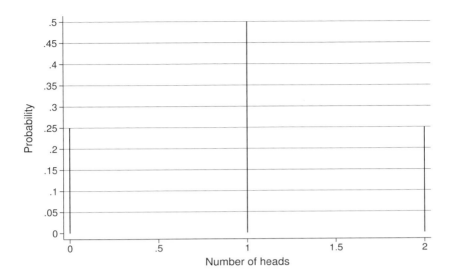

Figure 9.7. Probability of 0, 1, or 2 heads when flipping a coin twice and the probability of heads is 0.5

Figure 9.8 is a probability distribution showing the probability of obtaining 0–1,000 heads if you flipped a coin 1,000 times and the probability of heads is 0.5. The probability of obtaining 950 heads is 8.83×10^{-217}. It is so small that it does not even show up on the graph. Given that obtaining 950 heads is improbable, you would likely conclude that the probability of heads for your coin is close to 0.95. That is, you would not need to draw strength from the prior—the evidence from the data is sufficiently strong that the prior provides no additional information.

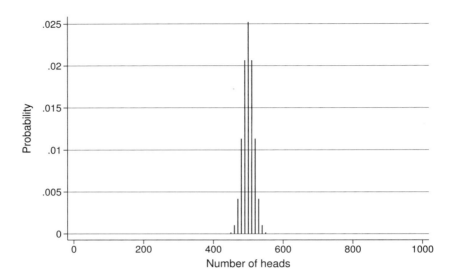

Figure 9.8. Probability of 0–1,000 heads when flipping a coin 1,000 times and the probability of heads is 0.5

For therapists means (or cluster means generally), the prior distribution is normal with a mean of 0 and a variance of σ_u^2. More precisely, we expect therapists' deviations from the intercept to follow that normal distribution. Estimates of u_i will draw strength from the prior distribution, which means that the estimates will get pulled toward the prior mean of 0.

A common objection to partial pooling is the notion that it combines data for a specific therapist with data from all other therapists. Sometimes it seems more appropriate to just keep each therapist distinct. However, the function of partial pooling is regularization, which is the process of reducing noise and chance factors by considering all therapists simultaneously (Gelman and Hill 2007). The degree to which this regularization occurs is a combination of a sample size and how extreme the data for a particular therapist are as compared with the entire sample. If the sample size is relatively small and the data extreme, the prior exerts substantial influence. If the sample size is relatively large and the data extreme, the prior exerts some influence. If the sample size is very large and the data extreme, the prior exerts little influence. Finally, regardless of sample size, if the data are not extreme but close to the average across the sample, the prior's influence is small.

Partially pooled cluster means can be computed by using `predict` after `mixed`.

```
. quietly mixed distress || tid:, nolog
. predict u_i, reffects
. predict uses, reses
. generate pooled = _b[_cons] + u_i
```

9.3.5 Estimating cluster means

Rerun the random-intercept model with the `quietly` prefix to suppress the output. The `reffects` option for `predict` produces the partially pooled values for u_i, and the `reses` option for `predict` produces the standard errors for u_i. Finally, recall that u_i is the therapist-specific residual, which is the difference between the grand mean and the therapist mean. Thus, the partially pooled therapist mean is equal to $b_0 + u_i$. The final line of code creates the pooled mean.

Creating the caterpillar plot for partially pooled estimates follows the same steps as before: a) generate upper and lower bounds for the pooled mean, b) rank the therapists based on the pooled mean, and c) use `twoway scatter` and `rcap` to plot the pooled means and CIs (see figure 9.9).

```
. generate pll = pooled - uses*1.96
. generate pul = pooled + uses*1.96
. egen ptrank = rank(pooled) if ther_tag == 1
(680 missing values generated)
. twoway scatter pooled ptrank if ther_tag == 1 ||
>       rcap pul pll ptrank if ther_tag == 1,
>       xtitle("Therapist rank") ytitle("Distress") ylabel(-2(1)2)
>       legend(order(1 "Pooled mean")) scheme(lean2)
```

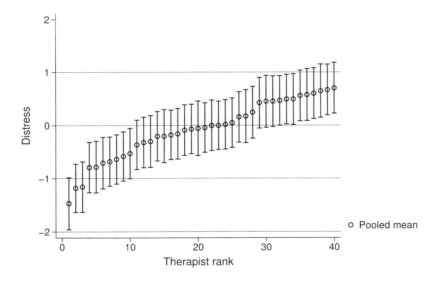

Figure 9.9. Caterpillar plot for the partial pooling approach to computing cluster means

Comparing pooled and unpooled means

In summary, the unpooled means are the average of each therapist's patients without any reference to the data from any other therapist—each therapist is treated as an island. The pooled means take into account information from other therapists to improve the quality of the estimates. Because pooled estimates are influenced by the entire distribution of therapists, the estimates are pulled toward the mean of the distribution. Consequently, a pooled estimate will be closer to the mean of the distribution than the unpooled estimate. The average of the pooled and unpooled estimates should be similar, but the variability of the pooled estimates should be smaller than the pooled estimates. Summarizing the data confirms this:

```
. summarize nopooled pooled if ther_tag == 1
    Variable |       Obs        Mean    Std. Dev.       Min        Max
-------------+--------------------------------------------------------
    nopooled |        40   -.0833593    .6671145  -1.759376   .8542917
      pooled |        40   -.0841937    .5575745  -1.472616   .7011086
```

The smaller standard deviation as well as the less extreme minimum and maximum for the pooled estimates shows the impact of the pooling. Note, however, that the pooling only affects the variability of the estimates, not the average of the estimates.

Figure 9.10 is a scatterplot of the pooled (y axis) and unpooled (x axis) means for all 40 therapists. The code to create this plot is

```
. twoway scatter pooled nopooled if ther_tag == 1
>         || function y = x, range(nopooled) xline(-0.08)
>         ytitle("Pooled mean") xtitle("Unpooled mean")
>         scheme(lean2) legend(off)
```

The command `function y = x, range(nopooled)` adds a diagonal reference line. If the pooled estimates (that is, `y`) and unpooled estimates (that is, `x`) were identical, then all the dots would fall directly on the reference line. If a dot is above the reference line, then the pooled mean is larger than the unpooled mean. If a dot is below the reference line, then the pooled mean is smaller than the unpooled mean. Figure 9.10 also includes a vertical line at the mean of all therapists (`distress` $= -0.08$), which is created by the option `xline(-0.08)`. The left-hand side of figure 9.10 shows that when a therapist is below the mean, the pooled estimate is larger than the unpooled estimate—the pooled estimates are pulled back toward the mean. Additionally, when a therapist is above the mean, the pooled estimate is smaller than the unpooled estimate.

9.3.5 Estimating cluster means

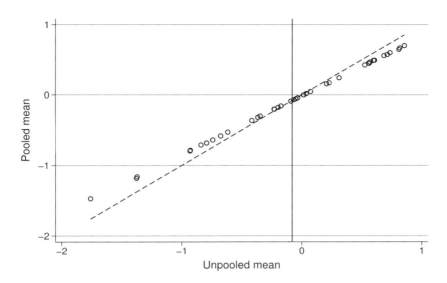

Figure 9.10. Scatterplot comparing the unpooled and pooled therapist means

A dot plot comparing the pooled and unpooled means for each therapist provides a different visualization of how the pooled means are pulled toward the overall mean. To create this kind of dot plot in Stata, the code is

```
. graph dot pooled nopooled if tid < 21 & ther_tag == 1,
>       over(tid) yline(-0.08)
>       legend(label(1 "Pooled mean") label(2 "Unpooled mean"))
>       ytitle("Distress") ylabel(-1.5(.5)1.5) scheme(lean2)
```

It does not matter whether you place pooled or nopooled first (though the reverse order will change the shading of the dots and the legend order). The code limits the plot to the first 20 therapists; otherwise, the plot is cramped. The yline(-0.08) option adds a vertical line at −0.08, the overall mean of therapists.[7]

Figure 9.11 shows the results. The pooled mean for a given therapist is closer to the overall mean than the unpooled mean. Furthermore, the more extreme means are pulled toward the overall mean more than means near the middle.

---

7. You might think it odd to use an option called yline() to add a vertical line on the x axis. The graph dot command is a categorical plot (like graph box), and as such, the categories (that is, therapists) are plotted on the x axis while outcomes (that is, distress) are plotted on the y axis. In graph dot plots, the axes are swapped.

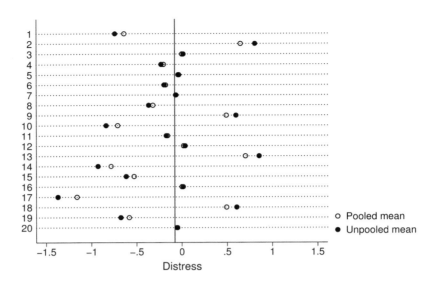

Figure 9.11. Dot plot comparing the pooled and unpooled means

9.3.6 Adding a predictor

Adding a predictor in a multilevel model is just like regression. A model with distress as the outcome and alliance as the predictor is

$$\texttt{distress}_{ij} = b_0 + b_1 \texttt{alliance}_{ij} + u_j + e_{ij}$$

The distribution of u_j and e_{ij} are the same as in (9.3) and (9.4), except now the distributions are conditional on alliance. The mixed code for this model is

```
. mixed distress alliance || tid:, nolog
Mixed-effects ML regression                     Number of obs      =        720
Group variable: tid                             Number of groups   =         40

                                                Obs per group:
                                                              min =         15
                                                              avg =       18.0
                                                              max =         20

                                                Wald chi2(1)       =       0.75
Log likelihood = -1144.2651                     Prob > chi2        =     0.3871
```

distress	Coef.	Std. Err.	z	P>|z|	[95% Conf. Interval]	
alliance	-.043794	.050636	-0.86	0.387	-.1430387	.0554507
_cons	-.0858006	.1025977	-0.84	0.403	-.2868883	.1152871

9.3.6 Adding a predictor

Random-effects Parameters	Estimate	Std. Err.	[95% Conf. Interval]
tid: Identity			
var(_cons)	.3499486	.0953096	.2051998 .5968038
var(Residual)	1.273364	.0691052	1.144875 1.416274

LR test vs. linear model: chibar2(01) = 96.54   Prob >= chibar2 = 0.0000

The slope for `alliance` is $b_1 = -0.04$, indicating that for every one-unit difference in `alliance`, `distress` is expected to differ by 0.04 units. This relationship is not significant at the $\alpha = 0.05$ level. Using `margins` can help visualize this relationship (see figure 9.12):

```
. quietly margins, at(alliance = (-3(0.5)3))
. marginsplot, addplot(scatter distress alliance, mcolor(gs5%80) below)
>         recastci(rarea) ciopts(fcolor(gs10%50))
>         xtitle("Alliance") ytitle("Expected distress")
>         title("") xlabel(-4(1)4) scheme(lean2)
>         legend(off)
  Variables that uniquely identify margins: alliance
```

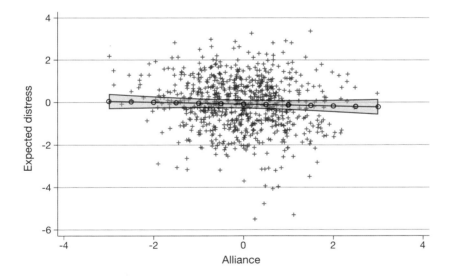

Figure 9.12. Alliance–outcome relationship from a random-intercept model

The random effects and residual variances were similar to the unconditional model, $\sigma_u^2 = 0.35$ and $\sigma_e^2 = 1.27$. Likewise, the intraclass correlation was similar:

```
. estat icc
Residual intraclass correlation
```

Level	ICC	Std. Err.	[95% Conf. Interval]
tid	.2155768	.0473734	.1369681	.3224449

This suggests that differences in `alliance` did not account for the variability among therapists in `distress`.

9.4 Between-clusters and within-cluster relationships

A construct like the therapeutic alliance is complicated because it describes the patient and therapist relationship, meaning that both the patient and the therapist contribute to the alliance. Therapist contributions to the alliance include, but are not limited to, the therapist's empathy toward the patient, interpersonal skills, multicultural awareness, technical skills, flexibility, and personality. Therapists vary with respect to these variables—for example, some therapists are more empathic than others and some therapists are more multiculturally aware than others. Consequently, therapists' alliance scores, when averaged across all patients of a therapist, will vary. Patient contributions to the alliance include the patient's personality, level of distress, diagnosis, attachment history, and interpersonal skills. Patients differ with respect to their contributions to the alliance. Thus, we expect patients to differ from one another in the strength of their alliance, even when they see the same therapist.

The alliance combines influences from both levels of a two-level design. This is in contrast to patient-level variables (such as patient diagnosis), where only the patient contributes variability, or therapist-level variables (such as therapist degree/training), where only the therapist contributes variability. When the variability of a predictor is influenced by both the upper and the lower levels of a two-level design, we get the intriguing possibility that the relationship between the predictor and the outcome is different at the upper and lower levels. In this instance, it is possible that therapist contributions to the alliance have a different relationship to the outcome than patient contributions to the alliance.

DeRubeis, Brotman, and Gibbons (2005) summarized the issue nicely:

> Perhaps a more important question concerns the meaning of the statistical relationship between outcome and the alliance, however weak or strong it may be, as it is vital that psychotherapy-process researchers come to a better understanding of the sources of variation in the alliance. Who (or what) is responsible for variability from case to case in the alliance? Aside from measurement error, there are at least four possible sources: the therapist, the

client, their interaction (in the statistical sense), and symptom improvement. The theoretical implications, or our understanding of the alliance, will be very different depending on which of these causes one focuses on. (p. 178)

Multilevel models allow us to directly examine DeRubeis, Brotman, and Gibbons's (2005) question, at least with respect to separating therapist and patient variability in the alliance and with respect to understanding how those sources of variability are related to outcome.

9.4.1 Partitioning variance in the predictor

Recall that a major purpose of the multilevel model for distress was to partition the variance of `distress` into the part that is associated with therapist and the part that is associated with patient. This was accomplished by including a random intercept for therapist in a model where `distress` was the outcome. We can fit an identical model but switch out `alliance` for `distress` as the outcome.

```
. mixed alliance || tid:, nolog
Mixed-effects ML regression                       Number of obs     =        720
Group variable: tid                               Number of groups  =         40

                                                  Obs per group:
                                                                min =         15
                                                                avg =       18.0
                                                                max =         20

                                                  Wald chi2(0)      =          .
Log likelihood = -922.78604                       Prob > chi2       =          .
```

alliance	Coef.	Std. Err.	z	P>|z|	[95% Conf. Interval]
_cons	-.0358573	.0858401	-0.42	0.676	-.2041008 .1323863

Random-effects Parameters	Estimate	Std. Err.	[95% Conf. Interval]
tid: Identity			
var(_cons)	.2569546	.0657676	.1555934 .4243474
var(Residual)	.6778808	.0367577	.6095331 .7538922

LR test vs. linear model: chibar2(01) = 152.60   Prob >= chibar2 = 0.0000

The output includes estimates of the fixed effects and the random effects. The overall intercept is equal to the grand mean for `alliance` ($b_0 = -0.04$). The random effects describe the variability at the patient and therapist level. The therapist variance was $\sigma_u^2 = 0.26$ and the patient variance was $\sigma_e^2 = 0.68$. As was the case with `distress`, interpreting these variance components directly is challenging. The intraclass correlation can help:

```
. estat icc
Intraclass correlation
```

Level	ICC	Std. Err.	[95% Conf. Interval]
tid	.2748661	.0524623	.1845288 .3883665

27% of the variability in `alliance` is at the therapist level. This is consistent with the idea that both therapists and patients contribute variance to the alliance. If only patients contributed variance, then the intraclass correlation would be 0. This would indicate that for therapists there is no relationship between the alliance scores in their caseload because the score would be fully determined by the patient. If only therapists contributed variance, then the intraclass correlation would be one. This would indicate that all patients of a specific therapists would have the exact same alliance score.[8] Given that the intraclass correlation is 0.27, patients are contributing the majority of the variance to the alliance. Nevertheless, the therapist contribution cannot be ignored given that it accounts for about a fourth of the variability in the alliance.

9.4.2 Total- versus level-specific relationships

Due to the hierarchical nature of clustered data there are three types of relationships between the X and Y variables. The first is the total relationship. This is the relationship between X and Y, without considering the specific clusters (that is, pooling across clusters). The second is the between-clusters relationship. This is the relationship between X and Y at the cluster level—the relationship between the cluster-means for X and the cluster-means for Y. The third is the within-cluster relationship—the relationship between X minus the cluster-mean for X and Y minus the cluster-mean for Y. Table 9.2 describes the three relationships in terms of X and Y and in terms of the alliance dataset.

Table 9.2. Total, between-clusters, and within-cluster relationships

Correlation		
Total	Relationship between X_{ij} and Y_{ij}	Relationship between `alliance`$_{ij}$ and `distress`$_{ij}$
Between-cluster	Relationship between $\overline{X}_j$ and $\overline{Y}_j$	Relationship between $\overline{\texttt{alliance}}_j$ and $\overline{\texttt{distress}}_j$
Within-cluster	Relationship between $X_{ij} - \overline{X}_j$ and $Y_{ij} - \overline{Y}_j$	Relationship between $\left(\texttt{alliance}_{ij} - \overline{\texttt{alliance}}_j\right)$ and $\left(\texttt{distress}_{ij} - \overline{\texttt{distress}}_j\right)$

---

8. We are ignoring measurement error in this discussion. Measurement error would introduce error at both the therapist- and patient-levels and would affect the intraclass correlation.

9.4.3 Exploring the between-clusters and within-cluster relationships

I have given a definition of the within-therapist and between-therapists relationships in another publication:

> ... we can consider relationships between variables that the therapist (that is, between-therapists correlations) and at the patient level (that is, within-therapist correlations). For example, in alliance-outcome research, the within-therapist correlation tells us how the alliance is related to outcome within a given therapist. Thus, the within-therapist correlation tests the association between patient variability in the alliance and outcome (that is, the therapist is a constant for patients nested within that particular therapist, ignoring the interaction for now). More simply, the within-therapist correlation tests whether patients who report a high alliance also report better outcomes than patients of the same therapist who report a low alliance. In contrast, the between-therapists correlation tells us how therapists' average alliance is related to their average outcome. That is, the between-therapists relationship tells us how variability among therapists in the alliance is related to outcome. (Baldwin, Wampold, and Imel 2007, 843)

Estimating only the total correlation, which is common, is a problem because it does not distinguish between the within-cluster and between-clusters relationship, making it impossible to separate therapist and patient contributions. In fact, the total relationship assumes that the between-clusters and within-cluster relationships are equal. For example, in the case of the alliance-outcome relationship, the total relationship constrains the alliance-outcome relationship at the between-therapists level to be the same as the relationship at the within-therapist level. If that constraint is not plausible, then we have a problem.

Ecological and atomistic fallacies

> Another problem that can occur with multilevel data is focusing only on the between-clusters relationships or the within-cluster relationships. Studying only between-clusters relationships and inferring that the same relationship exists at the within-cluster level is known as the ecological fallacy (Rabe-Hesketh and Skrondal 2012b). In contrast, studying only within-cluster relationships and inferring that the same relationship exists at the between-clusters level is known as the atomistic fallacy (Rabe-Hesketh and Skrondal 2012b). In either case, the problem is the same as with estimating only a total relationship—assuming that the between-clusters and within-cluster relationships are the same.

9.4.3 Exploring the between-clusters and within-cluster relationships

We can plot and estimate the between-therapists and within-therapist relationships to explore whether the relationships are similar. The between-therapists relationship is the regression of the therapists' means for distress onto the therapists' means for alliance.

$$\overline{\text{distress}}_j = b_0 + b_1 \overline{\text{alliance}}_j + e_j$$

Use egen, mean combined with bysort to create the therapist-level means. Likewise, use egen, tag to tag one observation per therapist so that we can limit between-therapists plots and analyses to one observation per therapist. An alternative would be to use collapse to create a dataset aggregated to the therapist level. However, using egen, tag makes it simpler to create plots with both therapist-level and patient-level data and means that we do not have to fiddle with preserve and restore.

```
. bysort tid: egen allmean = mean(alliance)
. bysort tid: egen dismean = mean(distress)
. egen tag = tag(tid)
```

The between-therapists alliance–distress relationship is negative and large (about half of a standard deviation):

```
. regress dismean allmean if tag == 1
```

Source	SS	df	MS
Model	2.93946212	1	2.93946212
Residual	14.4171664	38	.379399117
Total	17.3566286	39	.445041758

Number of obs	= 40
F(1, 38)	= 7.75
Prob > F	= 0.0083
R-squared	= 0.1694
Adj R-squared	= 0.1475
Root MSE	= .61595

dismean	Coef.	Std. Err.	t	P>|t|	[95% Conf. Interval]	
allmean	-.5005742	.1798385	-2.78	0.008	-.8646383	-.1365102
_cons	-.1008996	.0975945	-1.03	0.308	-.2984694	.0966701

The within-therapist relationship is equal to the association between distress and alliance after therapists' contributions to both distress and alliance are removed. To remove therapist contributions, subtract the therapist mean for distress from distress and the therapist mean for alliance from alliance. Thus, the within-therapist regression model is the patient deviations from their therapist for distress regressed on the patient deviations for alliance.

$$\left(\text{distress}_{ij} - \overline{\text{distress}}_j\right) = b_0 + b_1 \left(\text{alliance}_{ij} - \overline{\text{alliance}}_j\right) + e_{ij} - e_j$$

Note that the regression model also removes the therapist contributions to the residual, e_j, from the overall residual, e_{ij}.

Use generate to create the within-therapist deviation variables for distress and alliance, and then estimate the within-therapist regression.

```
. generate alldev = alliance - allmean
. generate disdev = distress - dismean
```

9.4.3 Exploring the between-clusters and within-cluster relationships

```
. regress disdev alldev

      Source |       SS           df       MS      Number of obs   =        720
-------------+----------------------------------   F(1, 718)       =       0.03
       Model |  .032884716         1   .032884716  Prob > F        =     0.8688
    Residual |  865.271456       718   1.20511345  R-squared       =     0.0000
-------------+----------------------------------   Adj R-squared   =    -0.0014
       Total |   865.30434       719   1.20348309  Root MSE        =     1.0978

      disdev |      Coef.   Std. Err.      t    P>|t|     [95% Conf. Interval]
-------------+----------------------------------------------------------------
      alldev |  -.0084457   .051127    -0.17   0.869    -.108822    .0919307
       _cons |  -2.18e-09   .0409117   -0.00   1.000    -.0803209   .0803209
```

The within-therapist relationship is negative but smaller than the between-therapists relationship. This implies that constraining the between-therapists and within-therapist slopes to be equal, as is done if only the total relationship is estimated, may be problematic.

Plotting the between-therapists and within-therapist data together, along with the between-therapists and within-therapist slopes, allows you to visually inspect the data and relationships. To create this plot, layer four `twoway` plots, two `scatter` plots (to plot the data), and two `lfit` plots (to plot the regression lines). I recommend adding some transparency to the within-therapist points; otherwise, it can be difficult to distinguish between the within-therapist and between-therapists data.

```
. twoway scatter disdev alldev, msymbol(plus) mcolor(gs10%40) ||
>        lfit disdev alldev ||
>        scatter dismean allmean if tag == 1, msymbol(circle_hollow)  ||
>        lfit dismean allmean if tag == 1, lpattern(solid)
>        legend(label(1 "Within-therapist deviations")
>               label(2 "Within-therapist relationship")
>               label(3 "Between-therapist means")
>               label(4 "Between-therapist relationship")
>               position(6) rows(2) cols(2))
>        xtitle("Alliance") ytitle("Distress")
>        scheme(lean2)
```

Figure 9.13 shows the stronger relationship at the between-therapists level than the within-therapist level.

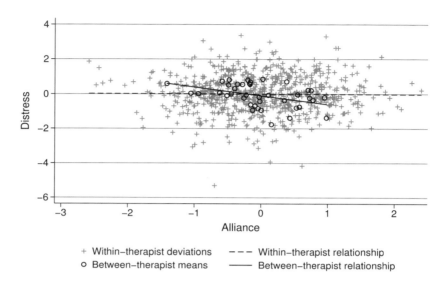

Figure 9.13. Visualizing the between-therapists and within-therapist relationships

9.4.4 Estimating the between-clusters and within-cluster effects

The regression analysis using the therapist means and the within-therapist deviations, along with figure 9.13, suggest that the between-therapists and within-therapist relationships are different. Multilevel models allow us to directly estimate the between-therapists and within-therapist relationships as well as test whether they are significantly different from one another.

The multilevel model for estimating between-clusters and within-cluster coefficients is an extension of the random-intercept model we used in section 9.3.6, where `distress` was the outcome, `alliance` was the predictor, and therapist was a random effect. The primary change is that `alliance` will be replaced by the within-therapist deviations for `alliance` and the therapist means for `alliance`. The model is

$$\texttt{distress}_{ij} = b_0 + b_1 \texttt{alldev}_{ij} + b_2 \texttt{allmean}_j + u_j + e_{ij}$$

The coefficient b_1, which is the slope for the within-therapist deviations (`alldev`), is the within-therapist relationship. The coefficient b_2, which is the slope for the between-therapists deviations (`allmean`), is the between-therapists relationship. The model is fit using `mixed` as follows:

9.4.4 Estimating the between-clusters and within-cluster effects

```
. mixed distress alldev allmean || tid:, nolog
Mixed-effects ML regression                     Number of obs    =        720
Group variable: tid                             Number of groups =         40

                                                Obs per group:
                                                            min =         15
                                                            avg =       18.0
                                                            max =         20

                                                Wald chi2(2)     =       8.29
Log likelihood = -1140.867                      Prob > chi2      =     0.0158
```

distress	Coef.	Std. Err.	z	P>|z|	[95% Conf. Interval]	
alldev	-.0084457	.0525401	-0.16	0.872	-.1114224	.0945311
allmean	-.5016727	.1744938	-2.88	0.004	-.8436742	-.1596711
_cons	-.1025786	.0948881	-1.08	0.280	-.2885559	.0833987

Random-effects Parameters	Estimate	Std. Err.	[95% Conf. Interval]	
tid: Identity				
var(_cons)	.2876249	.0804718	.1662176	.4977095
var(Residual)	1.27265	.0690294	1.144298	1.415398

LR test vs. linear model: chibar2(01) = 80.41   Prob >= chibar2 = 0.0000

The within-therapist coefficient is $b_1 = -0.01$, which indicates that for every one-unit difference in the within-therapist deviations, we expect a 0.01 difference in distress. The between-therapists coefficient is $b_2 = -0.5$, which indicates that for every one-unit difference in the between-therapists means, we expect a 0.5 difference in distress. These are similar values to what we saw when we used regression to estimate the within-therapist and between-therapists coefficients.

The within-therapist coefficient was not statistically significant ($p = 0.872$), whereas the between-therapists coefficient was significant ($p = 0.004$). This suggests there is little evidence that the relationship between alliance and distress at the within-therapist level is different from 0; the between-therapists relationship is different from 0. We know that the within-therapist coefficient is not significant and that the between-therapists coefficient is significant, but none of our information indicates whether the two coefficients are significantly different from one another (Gelman and Stern 2006; Nieuwenhuis, Forstmann, and Wagenmakers 2011).

Recall that the null hypothesis for the within-therapist coefficient is that it is equal to 0. Likewise, the null hypothesis for the between-therapists coefficient is that it is 0. Thus, rejecting the null for either coefficient does not tell us whether the coefficients are different from one another. We need to directly test that hypothesis. There are two ways to perform this test. The first method is to use the lincom postestimation command. Specifically, lincom tests whether the difference between the within-therapist and between-therapists coefficients is 0.

```
. lincom _b[allmean] - _b[alldev]
 ( 1)  - [distress]alldev + [distress]allmean = 0
```

distress	Coef.	Std. Err.	z	P>|z|	[95% Conf. Interval]	
(1)	-.493227	.1822321	-2.71	0.007	-.8503954	-.1360586

The coefficient in the `lincom` output is the difference between the within-therapist and between-therapists coefficients ($b_1 - b_2 = -0.49$, $p = 0.007$). In sum, the evidence suggests that the within-therapist relationship is $b_1 = -0.01$ and is not significantly different from 0. The between-therapists relationship is $b_2 = -0.5$ and is significantly different from 0. Further, the between-therapists and within-therapist coefficients were significantly different from one another. Taken together, this suggests that we should estimate separate within-therapist and between-therapists coefficients. Said another way, estimating only the total relationship would be misleading.

The second method for testing whether the between-therapists and within-therapist coefficients are significantly different, is to reestimate the model but use `alliance` and `allmean` as the predictors.

$$\text{distress}_{ij} = b_0 + b_1 \text{alliance}_{ij} + b_2 \text{allmean}_j + u_j + e_{ij}$$

In this model, the coefficient b_1 is the same as the within-therapist coefficient from the previous model. The coefficient b_2 is the difference between the within-therapist and the between-therapists coefficients (identical to the output from `lincom`).

```
. mixed distress alliance allmean || tid:, nolog
Mixed-effects ML regression                     Number of obs     =        720
Group variable: tid                             Number of groups  =         40

                                                Obs per group:
                                                              min =         15
                                                              avg =       18.0
                                                              max =         20

                                                Wald chi2(2)      =       8.29
Log likelihood = -1140.867                      Prob > chi2       =     0.0158
```

distress	Coef.	Std. Err.	z	P>|z|	[95% Conf. Interval]	
alliance	-.0084457	.0525401	-0.16	0.872	-.1114224	.0945311
allmean	-.493227	.1822321	-2.71	0.007	-.8503954	-.1360586
_cons	-.1025786	.0948881	-1.08	0.280	-.2885559	.0833987

Random-effects Parameters	Estimate	Std. Err.	[95% Conf. Interval]	
tid: Identity				
var(_cons)	.2876249	.0804718	.1662176	.4977095
var(Residual)	1.27265	.0690294	1.144298	1.415398

```
LR test vs. linear model: chibar2(01) = 80.41       Prob >= chibar2 = 0.0000
```

The coefficient b_2 (as well as the difference computed using `lincom`) is referred to as the contextual effect (Raudenbush and Bryk 2001) because it represents the effect of therapist contributions to `alliance` holding constant a patient's `alliance` score. In other words, the contextual effect describes how much the therapists' average `alliance` score adds above and beyond a patient's `alliance` score. Consider two patients—Katie and Darren—with the exact same `alliance` score. Katie saw a therapist who, on average, forms above average alliances with patients. Darren saw a therapist who forms below average alliances with patients. If the contextual effect is positive, Katie will have a better outcome than Daren even though they had the same `alliance` score—the therapist with whom they developed the alliance matters (Baldwin, Wampold, and Imel 2007; Baldwin and Imel 2013).

9.5 Random slopes

The models up to this point have only allowed the intercept to vary across clusters. The models have allowed clusters to have different anchor points for the regression line, but the models have constrained the slopes to be equal for all clusters. For example, in section 9.3.6, the model allowed the intercept to be specific to each therapist, but the `alliance`–`distress` relationship was the same across all therapists. This need not be the case—the slope for any patient-level variable can vary across clusters. That is, the relationship between a predictor and outcome may vary across clusters—the relationship may be large and positive in some clusters, weak and negative in others, and flat in others. Perhaps the `alliance`–`distress` relationship is particularly important for some therapists but not relevant for others. A model that allows a relationship to vary across clusters is called a random-slope or random-coefficient model.

To illustrate adding a random slope, we can explore the relationship between patients' motivation to change and outcome. In the dataset, the variable `motivate` is a patient-level variable representing patients' motivation to make changes in their life prior to beginning therapy (Prochaska, DiClemente, and Norcross 1992). Patients who are less motivated to address problematic behaviors (for example, too much drinking, a hostile relationship) may make less progress in treatment than those who are highly motivated. In this dataset, `motivate` is grand-mean centered; thus, the intercept is the expected level of `distress` for someone at the mean of `motivate`.

An initial multilevel model with `distress` predicted from `motivate` is

$$\texttt{distress}_{ij} = b_0 + b_1 \texttt{motivate}_{ij} + u_j + e_{ij} \tag{9.5}$$
$$u_j \sim N\left(0, \sigma_u^2\right)$$
$$e_{ij} \sim N\left(0, \sigma_e^2\right)$$

```
. use http://www.stata-press.com/data/pspus/mlm_cs_data1, clear
. mixed distress motivate || tid:, nolog

Mixed-effects ML regression                     Number of obs      =        720
Group variable: tid                             Number of groups   =         40

                                                Obs per group:
                                                              min =         15
                                                              avg =       18.0
                                                              max =         20

                                                Wald chi2(1)       =      55.63
Log likelihood =   -1117.84                     Prob > chi2        =     0.0000

------------------------------------------------------------------------------
    distress |      Coef.   Std. Err.      z    P>|z|     [95% Conf. Interval]
-------------+----------------------------------------------------------------
    motivate |  -.4393896   .0589129    -7.46   0.000    -.5548567   -.3239225
       _cons |  -.0851098   .0999142    -0.85   0.394    -.2809381    .1107184
------------------------------------------------------------------------------

------------------------------------------------------------------------------
  Random-effects Parameters  |   Estimate   Std. Err.     [95% Conf. Interval]
-----------------------------+------------------------------------------------
tid: Identity                |
                  var(_cons) |   .3334468   .0895332      .1970027    .5643923
-----------------------------+------------------------------------------------
               var(Residual) |   1.181803   .0640994      1.062618    1.314356
------------------------------------------------------------------------------
LR test vs. linear model: chibar2(01) = 105.17        Prob >= chibar2 = 0.0000

. estimates store randint
```

This model includes therapist-specific intercepts, allowing therapists to differ with respect to average `distress` when `motivate` is 0, but constrains the `motivate`–`distress` relationship to be equivalent across therapists. However, it seems reasonable that some therapists may be more effective in working with patients who are highly motivated (or unmotivated) than other therapists. In other words, the size of the `motivate`–`distress` relationship may vary across therapists. We can expand (9.5) to include a random effect that allows b_1, the `motivate`–`distress` relationship, to vary across therapists.

Before we look at the details of the random-slope model, let's examine how much the slope for `motivate` differs across therapists. We can do this by fitting a linear regression with `distress` as the outcome and `motivate` as the predictor, separately for each therapist. Saving the slope for `motivate` from each therapist-specific regression allows us to explore the variability in that relationship. One way to do this is with the `statsby` command.

9.5 Random slopes

```
. statsby intercept = _b[_cons] slope = _b[motivate]
>                int_se = _se[_cons] slope_se = _se[motivate]
>                df = e(df_r), by(tid): regress distress motivate
(running regress on estimation sample)
      command:  regress distress motivate
    intercept:  _b[_cons]
        slope:  _b[motivate]
       int_se:  _se[_cons]
     slope_se:  _se[motivate]
           df:  e(df_r)
           by:  tid

Statsby groups
 ─────┼─── 1 ───┼─── 2 ───┼─── 3 ───┼─── 4 ───┼─── 5
 .........................................
```

statsby is a prefix command that splits the data so that we can run analyses separately for each piece. The purpose of statsby is to save parameter estimates and other data from each analysis. Because we typed by(tid): regress distress motivate, Stata estimated and printed the output of 40 regression models. However, we will only be able to access the stored estimates from the final regression model because Stata only stores estimates for one model at a time.

The statsby syntax has three parts: i) a list of parameters to save, ii) options for statsby, including a by() option to specify how to split the data, and iii) the analysis model. In the example, the analysis model is the regression of distress on motivate, and the only option is by(tid), which splits the data by therapist. We save three things: the regression coefficients (_b), the standard errors (_se), and the residual degrees of freedom (e(df_r)). Saving the coefficients allows us to explore their variability, and the standard errors and degrees of freedom are needed to construct the CIs. The saved coefficients replace the active dataset. If you would prefer to save the dataset to your computer, statsby has a saving option.

The dataset has 40 rows, one for each therapist. Summarizing slope gives a description of the central tendency and dispersion of the slope.

```
. summarize slope
```

Variable	Obs	Mean	Std. Dev.	Min	Max
slope	40	-.3621416	1.054223	-2.396872	1.616038

The slopes range from -2.4 to 1.6, with a mean of -0.36 and standard deviation of 1.1.

We can also plot the coefficients and CIs. First, use the standard errors and degrees of freedom to create the upper and lower limits of the interval. Second, rank the therapists from low to high on slope.

```
. generate ll_b = slope - (slope_se*(-1*invt(df, .025)))
. generate ul_b = slope + (slope_se*(-1*invt(df, .025)))
. egen therrank = rank(slope)
```

Figure 9.14 combines `twoway scatter` and `twoway rcap` as we did in section 9.3.5.

```
. twoway scatter slope therrank ||
>        rcap ul_b ll_b therrank,
>        ytitle("Motivate-distress slope") xtitle("Therapist rank")
>        legend(off)
>        scheme(lean2)
```

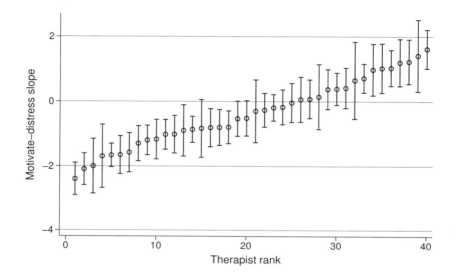

Figure 9.14. Variability in the `motivate-alliance` slope across therapists

The regression analysis by therapist also estimated a therapist-specific intercept. It is possible that the intercept and slopes are correlated within a therapist, meaning that the size of the intercept is related to the size of the slope. A positive correlation between the random intercept and slope means that larger values of the cluster-specific intercept are associated with larger values of the cluster-specific slope. Negative values mean that larger values of the cluster-specific intercept are associated with smaller values of the cluster-specific slope.

The intercept and slope are positively correlated.

```
. corr intercept slope
(obs=40)
             |  interc~t    slope
   ----------+------------------
   intercept |   1.0000
       slope |   0.4037   1.0000
```

Figure 9.15 shows that therapists with higher than average intercepts also had slopes that were higher than average. Taken together, figure 9.14 and the descriptive statistics

9.5 Random slopes

show that the motivate–distress relationship varies across therapists and is consistent with fitting a random-slope model.

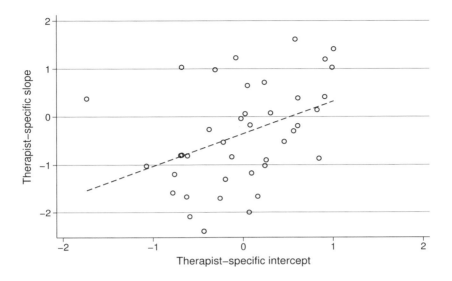

Figure 9.15. Correlation between therapist-specific intercepts and slopes from regress

The random-slope model is written as

$$\texttt{distress}_{ij} = b_0 + b_1 \texttt{motivate}_{ij} + u_{0j} + u_{1j}\texttt{motivate}_{ij} + e_{ij}$$

$$\begin{bmatrix} u_{0j} \\ u_{1j} \end{bmatrix} \sim N\left(\begin{array}{c} 0 \\ 0 \end{array}, \begin{array}{cc} \sigma^2_{u_0} & \\ \sigma_{u_0 u_1} & \sigma^2_{u_1} \end{array} \right) \quad (9.6)$$

$$e_{ij} \sim N\left(0, \sigma^2_e\right)$$

There are three changes in the random-slope model as compared with the random-intercept model. First, the regression portion of the model now includes the parameter u_{1j}, which is a random effect representing a cluster-specific deflection from the overall slope for motivate, b_1. The slope for therapist j is $b_1 + u_{1j}$. Second, the random intercept is now denoted at u_{0j}. The 0 in the subscript is to distinguish it from the random slope. Third, the random effects are drawn from a multivariate normal distribution, which means that the random intercept and the random slope are allowed to correlate. Thus, the intercept and the slope are treated as a pair rather than as two independent parameters. This is consistent with what we saw from the descriptive analysis using regress and figure 9.15. Although it is possible to constrain the correlation to 0, generally speaking it is more plausible to allow the intercept and slope to correlate. If the correlation is small, the model will estimate it as such.

The `mixed` syntax requires two additions to include random effects:

```
. use http://www.stata-press.com/data/pspus/mlm_cs_data1, clear
. mixed distress motivate || tid: motivate, covariance(unstructured) nolog
```

```
Mixed-effects ML regression                     Number of obs      =        720
Group variable: tid                             Number of groups   =         40

                                                Obs per group:
                                                              min =         15
                                                              avg =       18.0
                                                              max =         20

                                                Wald chi2(1)       =       5.10
Log likelihood = -995.66947                     Prob > chi2        =     0.0239
```

distress	Coef.	Std. Err.	z	P>|z|	[95% Conf. Interval]
motivate	-.3697524	.1636495	-2.26	0.024	-.6904994 -.0490053
_cons	-.0180389	.0980376	-0.18	0.854	-.2101891 .1741112

Random-effects Parameters	Estimate	Std. Err.	[95% Conf. Interval]
tid: Unstructured			
var(motivate)	.9753029	.240436	.6015858 1.581181
var(_cons)	.3415526	.0861327	.208354 .5599038
cov(motivate,_cons)	.2538909	.1095856	.0391071 .4686746
var(Residual)	.7226481	.0404029	.6476444 .806338

LR test vs. linear model: chi2(3) = 349.51   Prob > chi2 = 0.0000

Note: LR test is conservative and provided only for reference.

```
. estimates store randslope
```

The two additions to the syntax are the inclusion of `motivate` in the random-effects specification and the random-effects option `cov(unstructured)`. Recall that `tid:` indicates that we want therapist-specific intercepts. Adding `motivate` after `tid:` indicates that we want therapist-specific slopes for `motivate`. The option `cov(unstructured)` indicates that we want to use an unstructured covariance matrix for the random effects (see section 7.3.1), which means that we want to estimate a variance for both the random intercept and the random slope as well as the covariance between them [see (9.6)]. The random-effects output includes the two variances and the covariance. The random-intercept variance is labeled `var(_cons)` and is $\sigma^2_{u_0} = 0.34$. The random-slope variance is labeled `var(motivate)` and is $\sigma^2_{u_1} = 0.98$. Finally, the covariance between the random intercept and slope is labeled `cov(motivate,_cons)` and is $\sigma_{u_0 u_1} = 0.25$.

We stored the estimates from the random-intercept and random-slope models so that we can test whether adding the slope and covariance significantly improve fit by using a likelihood-ratio test. The null hypothesis is that the random-intercept and random-slope models fit the data equally as well. The likelihood-ratio test is the difference of the deviance from the random-slope model and random-intercept model (see section 11.9.1 for more details). The degrees of freedom is equal to the difference in the number of

9.5 Random slopes

parameters, which is two in this case—one for the random-slope variance and one for the covariance. The likelihood-ratio test is implemented with `lrtest`.

```
. lrtest randint randslope
Likelihood-ratio test                                 LR chi2(2)  =     244.34
(Assumption: randint nested in randslope)             Prob > chi2 =     0.0000
Note: The reported degrees of freedom assumes the null hypothesis is not on
      the boundary of the parameter space. If this is not true, then the
      reported test is conservative.
```

We reject the null hypothesis and conclude that the random-slope model fits the data better than the random-intercept model.

Interpreting the random-effects variances and covariances is difficult because they are like any variance or covariance. They are not standardized, and they are based on squared values (variances) or products of two variables (covariances). Consequently, it can be helpful to convert the variances to standard deviations and the covariances to correlations. We can take the square root of each of the variances to obtain the standard deviations. Additionally, the correlation is computed using the following formula (Snijders and Bosker 2011; Rabe-Hesketh and Skrondal 2012b):

$$\rho_{u_0 u_1} = \frac{\sigma_{u_0 u_1}}{\sqrt{\sigma^2_{u_0}} \sqrt{\sigma^2_{u_1}}}$$

Thus, the correlation is

$$\widehat{\rho}_{u_0 u_1} = \frac{0.25}{\sqrt{0.34}\sqrt{0.98}} = 0.43$$

Stata will print the standard deviations and the correlation if we use the `mixed` option `stddeviations`. We do not have to refit the model, but instead we can have Stata replay the results of the last `mixed` model by using the following syntax:

```
. mixed, stddeviations
Mixed-effects ML regression                     Number of obs    =        720
Group variable: tid                             Number of groups =         40
                                                Obs per group:
                                                              min =         15
                                                              avg =       18.0
                                                              max =         20
                                                Wald chi2(1)     =       5.10
Log likelihood = -995.66947                     Prob > chi2      =     0.0239
```

distress	Coef.	Std. Err.	z	P>|z|	[95% Conf. Interval]
motivate	-.3697524	.1636495	-2.26	0.024	-.6904994 -.0490053
_cons	-.0180389	.0980376	-0.18	0.854	-.2101891 .1741112

Random-effects Parameters	Estimate	Std. Err.	[95% Conf. Interval]	
tid: Unstructured				
sd(motivate)	.9875743	.1217306	.7756196	1.25745
sd(_cons)	.584425	.0736901	.4564581	.7482672
corr(motivate,_cons)	.4398945	.1453172	.1183879	.6779195
sd(Residual)	.8500871	.023764	.8047635	.8979632

LR test vs. linear model: chi2(3) = 349.51   Prob > chi2 = 0.0000

Note: LR test is conservative and provided only for reference.

Interpreting the standard deviations is challenging. Because the model includes a random slope and random intercept, we can no longer compute an intraclass correlation. It is possible to compute what is called a conditional intraclass correlation, which is the same as the intraclass correlation from the random-intercept model except that it is computed at a specific level of the predictor variable that has a random slope. Most commonly, the conditional intraclass correlation is computed by setting the predictor variable to 0 (Rabe-Hesketh and Skrondal 2012b, 191–192). This is what `estat icc` does:

```
. estat icc
Conditional intraclass correlation
```

Level	ICC	Std. Err.	[95% Conf. Interval]	
tid	.3209476	.0566366	.2211873	.4402665

Note: ICC is conditional on zero values of random-effects covariates.

Thus, the conditional intraclass correlation is 0.32, which means that 32% of the variability in `distress` is associated with therapists when `motivate` is equal to 0 (that is, when `motivate` is at its mean).

Another possibility for interpreting the random-effects standard deviations is to graphically compare them with the data. For example, we can predict the random intercept and slope following the `mixed` model. We can then generate the therapist-specific intercepts and slopes by combining the fixed effects with the random effects. Finally, we can create histograms that plot the therapist-specific estimates on top of the outcome data. We can then compare the variability of the therapist-specific intercepts, for example, to the variability in the data, which will give us a feel for how important the variability in the therapist-specific estimates is.[9]

```
. predict u1 u0, reffects
. generate est_inter = _b[_cons] + u0
```

---

9. Note that these therapist-specific estimates are the empirical Bayes estimates described in section 9.3.5. Consequently, they are pooled estimates, meaning that they will have slightly less variability than implied by the random-effects standard deviations (Rabe-Hesketh and Skrondal 2012b, chap. 4).

9.5 Random slopes

```
. generate est_slope = _b[motivate] + u1
```

These histograms can be created as follows. We use the transparency option for the Stata graph, which makes it easier to see how the variables compare given that they overlap so much.

```
. twoway histogram distress, color(gs10%30) ||
>         histogram est_inter, color(gs5%30)
>         legend(label(1 "Distress") label(2 "Intercept")) scheme(lean2)
```

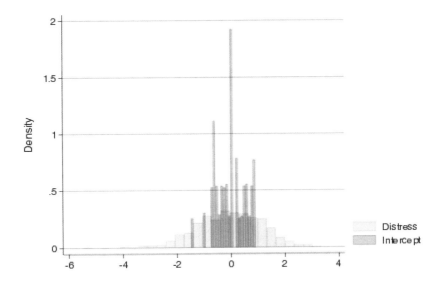

Figure 9.16. Comparison of the empirical Bayes intercept and the distribution of distress

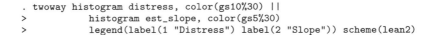

```
. twoway histogram distress, color(gs10%30) ||
>         histogram est_slope, color(gs5%30)
>         legend(label(1 "Distress") label(2 "Slope")) scheme(lean2)
```

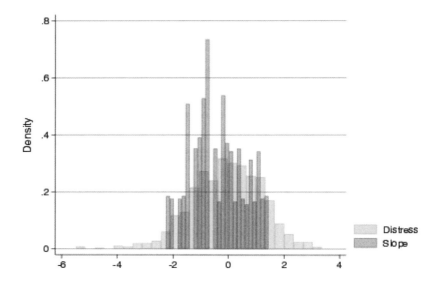

Figure 9.17. Comparison of the empirical Bayes slope and the distribution of `distress`

Figures 9.16 and 9.17 are the histograms for the random intercepts and the random slopes, respectively. In both cases, we see that the therapist variability is quite large and represents a substantial portion of the variability in the outcome variable.[10]

Another way to interpret the random effects is to consider the therapist-specific regression lines implied by the random intercept and random slope (Rabe-Hesketh and Skrondal 2012b). To compute the therapist-specific regression line, we create the predicted values of `distress` based on both the fixed and the random effects. For example, the predicted value of distress for therapist 1 is equal to $\text{distress}_{i1} = b_0 + b_1 \text{motivate}_{i1} + u_{01} + u_{11}\text{motivate}_{i1}$. Thus, we use the estimated fixed-effects intercept and slope and combine them with the predicted therapist-specific intercept and slope. This can be done following a `mixed` model by using `predict` with the option `fitted`. We can then plot the therapist-specific regression lines along with the data to visualize the variability in the `motivate–distress` relationship across therapists.

---

10. Computing values such as R^2 is not possible with a random-slope model. Consequently, these histograms are meant to only be interpreted heuristically.

9.6 Summary

```
. predict ex_distress, fitted
. sort tid motivate
. twoway scatter distress motivate, msymbol(circle_hollow) mcolor(gs10%50) ||
>          line ex_distress motivate, lpattern(solid) connect(ascending)
>          ytitle("Distress") xtitle("Motivate")
>          legend(off) scheme(lean2)
```

Figure 9.18 shows the variability in the intercepts and slopes across therapists along with the data.

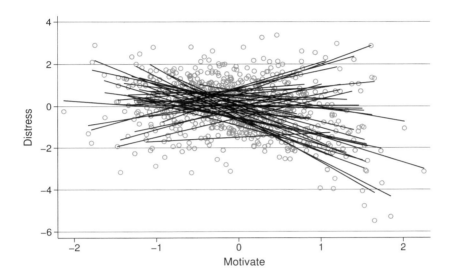

Figure 9.18. Therapist-specific regression lines (empirical Bayes) and raw data

9.6 Summary

Multilevel models are among the most commonly used statistical analyses in psychology and the social sciences. Consequently, researchers interested in enhancing their skills in analyzing data likely need to become proficient in using multilevel models. Stata's `mixed` command is able to handle many kinds of multilevel models, including many that have not been covered in this chapter. Perhaps the most important topic not covered in this chapter is the use of multilevel models for longitudinal data. Longitudinal multilevel models will be introduced in chapter 10. Below, I note a few topics for those interested in increasing their understanding of multilevel modeling as well as references to papers and texts that provide tutorials and discussions of the methods.

The predictors discussed in this chapter were all level-1 predictors, meaning they all varied within clusters. It is also possible to include level-2 predictors where the only variability is between clusters (for example, variables that describe the therapist, such as degree). Additionally, we can include cross-level interactions, which are the product of level-1 and level-2 predictors. Multilevel designs can be extended to include three or more levels, such as students (level 1) clustered within schools (level 2) clustered within states (level 3). Multilevel models can also be used to analyze cross-classified data. Cross-classified data are hierarchical data where level-1 units appear at more than one level of a level-2 variable. For example, a patient may see more than one therapist in a psychotherapy study or a student may be in more than one class in an education study. Rabe-Hesketh and Skrondal (2012b), Raudenbush and Bryk (2001), and Snijders and Bosker (2011) all provide thorough introductions to multilevel models and cover level-1 and level-2 predictors, cross-level interactions, three- or four-level designs, and cross-classified models.

Multivariate multilevel models can be used to simultaneously model two or more outcomes from a clustered design (Baldwin et al. 2014). Multilevel models can be used for nonnormal outcomes, such as binary (melogit), ordinal (meologit), count (mepoisson, menbreg), censored (mestreg, metobit), and multinomial outcomes (gsem). The me suite in Stata can handle a huge variety of models and data structures. Rabe-Hesketh and Skrondal (2012a) provide a thorough discussion of multilevel models for nonnormal data as well as how to use Stata to analyze the data. Multilevel structural equation modeling can assist with measurement and multivariate research projects, and it can be implemented in Stata using gsem (Preacher, Zyphur, and Zhang 2010).

10 Multilevel models for longitudinal data

Understanding how a person changes across time is central to many psychological theories, and longitudinal data are required to understand these change processes. With cross-sectional data, we can compare two or more people who differ with respect to their levels of a variable. For example, in a cross-sectional fMRI study, we can compare brain activity in people diagnosed with depression to activity in people without a psychiatric diagnosis. Although this cross-sectional design allows us to examine how these two groups of people differ at a given point in time, it does not allow us to examine how the brain changes as a person develops depression. To learn about how the brain changes in this regard, we need to observe a person before and after the depressive episode began. In other words, we need longitudinal data.

Examples of longitudinal designs where change is the primary focus include the following:

- Randomized trials for evaluating psychosocial treatments.
- Public health initiatives aimed at increasing physical activity within a community.
- Developmental psychology research examining change in self-identity during adolescence.
- Intensive longitudinal designs, where participants report on their mood and their activities several times a day.
- Psychopathology research that examines the course and development of psychiatric syndromes.
- Educational research that evaluates learning and skills developed after the introduction of a new curriculum.

Although we discussed repeated-measures designs in chapter 7, those models were most useful for within-subject experimental designs. In contrast, the models covered in this chapter are primarily useful for examining change in a person across time. These models are known as growth-curve models because they describe the change or growth in the outcome across time. They also allow us to include predictors of the change process, including predictors that are constant for a person across time (for example, baseline symptoms or treatment condition) and predictors that change across time (for example, stress or mood). Finally, these growth-curve models are multilevel models, meaning that they can include random intercepts and random coefficients to allow parameters, such as the rate of change over time, to vary across people. Consequently, the topics discussed in this chapter are a direct extension of the multilevel models covered in chapter 9.

In this chapter, you will learn the following:

1. The basics of a growth-curve model and why examining change over three or more time periods is useful.
2. How multilevel models can be used to fit a growth-curve model.
3. How to estimate and interpret random intercepts and random slopes in the context of a growth-curve model.
4. How to use both time-varying and time-invariant predictors.

Stata commands featured in this chapter

- `mixed`: for fitting a growth-curve model
- `margins`: for visualizing the results of the growth-curve model
- `predict`: for producing empirical Bayes estimates
- `statsby`: for exploring person-specific intercepts and slopes
- `lrtest`: for comparing the fit of random-intercept and random-slope models

10.1 Data used in this chapter

The data for this chapter are simulated data representing a longitudinal randomized trial comparing 5 weeks of cognitive behavioral therapy (CBT) to 5 weeks of treatment as usual (TAU; for example, a bona fide treatment but where the type of treatment is not controlled or dictated by the researcher). In this study, $N = 200$ participants diagnosed with depression were randomly assigned to receive either CBT or TAU. Each participant was assessed for psychological distress as well as life stressors prior to treatment and immediately after each of the five sessions, for a total of six observations per participant.

The dataset for this chapter is called `mlm_long.dta` and includes the following variables:

- `id`: patient ID
- `time`: assessment timepoint (ranging from 1 = baseline to 6 = posttreatment)
- `tx`: treatment condition (0 = TAU, 1 = CBT)
- `distress`: psychological distress (larger values equal higher psychological distress)
- `stress`: life stressors (larger values indicate higher amounts of life stressors)

10.2 Basic growth model

Box plots of `distress` and `stress` stratified by `tx` and `time` are presented in figure 10.1. Descriptive statistics for `distress` and `stress` stratified by `tx` and `time` are as follows:

```
. use http://www.stata-press.com/data/pspus/mlm_long, clear
. graph box distress stress, over(time) by(tx) ytitle("Distress") scheme(lean2)
```

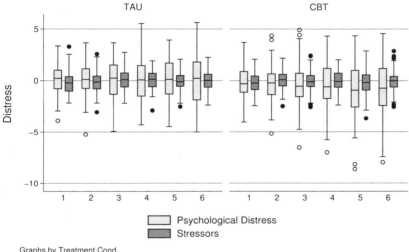

Figure 10.1. Box plot for `distress` and `stress` stratified by `tx` and `time`

```
. table tx time, c(mean distress mean stress) format(%9.2f)
```

| Treatment | Timepoint | | | | | |
Cond.	1	2	3	4	5	6
TAU	0.07	0.12	0.07	0.02	0.12	0.04
	-0.19	-0.09	0.05	0.05	-0.07	0.01
CBT	-0.17	-0.24	-0.42	-0.37	-0.95	-0.69
	-0.22	0.02	-0.01	0.00	-0.22	-0.02

10.2 Basic growth model

Researchers commonly collect longitudinal data to address the following question: "How do participants change over time?" For example, in the treatment data, we want to know whether patients receiving CBT have less psychological distress at the end of treatment than at the beginning. We also want to know whether the change that occurs in the CBT condition is more or less than the change in the TAU condition. Finally, we also want to know how a participant's life stressors are related to the change that occurs during treatment. The growth-curve models described in this chapter can address each of those questions with the treatment data.

We can start by examining how change occurs across time within the data. Figure 10.1 and the means show how `distress` varies from timepoint to timepoint. This timepoint-by-timepoint comparison is not exactly what we want because it describes changes from one timepoint to the next considering the entire sample at each timepoint. Contrast this with an analysis that considers each participant's observations and models how a given participant changes from timepoint to timepoint. The former describes between-timepoints change, whereas the latter describes within-participant change. Furthermore, the latter is often what we are interested in when studying change, especially in a treatment context. If you are a patient receiving psychotherapy, you are most likely interested in whether you will change from one session to the next.

We can examine the within-person change by describing the mean and variability in change from one session to the next. For example, compute the difference between session 1 and baseline `distress` for each participant, and then summarize and plot those changes. We can do that for any combination of sessions, though we focus only on adjacent sessions.

Generating differences between a session and its preceding session in the `mlm_long` dataset, which is in long format (see section 7.1), can be done with a single line of code:[1]

```
. use http://www.stata-press.com/data/pspus/mlm_long
. by id: generate diff = distress[_n] - distress[_n - 1]
(200 missing values generated)
```

We use by-processing (`by id:`) to separate the data into sections unique to each participant, which ensures that we only compute differences within-participant. When data are stored in long format, each participant has as many rows of data as there are timepoints, which in this case is six. The `generate` code uses the `_n` variable, which is a built-in variable that indexes observation number. If we consider the whole dataset, `_n` ranges from 1 to 1,200 because there are 1,200 observations. However, if we consider each participant individually, which is what `by id:` does, then `_n` ranges from 1 to 6. If a participant is missing an observation, then `_n` ranges from 1 to 5. The brackets following `distress` are used to index observations—`distress[1]` is the first `distress` value, `distress[2]` is the second `distress` value, and so on. What is first and what is second depends on whether by-processing is used. In other words, the `generate` code says, "After splitting the data by participant ID, create a new variable called `diff` that is equal to `distress` for observation `_n` minus `distress` for observation `_n−1`." Thus, participant 1's `distress` scores at time 2 and time 1 are 0.97 and 2.23, respectively. Thus, the `diff` at time 2 is $0.97 - 2.23 = -1.26$. The same process can be followed for comparing any two adjacent sessions.

---

1. If there are missing data, we may need to use the time-series operators (see [TS] **time series**) to make sure the differences are actually between adjacent timepoints.

10.2 Basic growth model

The values for `id`, `time`, `distress`, and `diff` for the first three participants are

```
. list id time distress diff in 1/18, clean
        id   time    distress         diff
  1.     1     1     1.476432            .
  2.     1     2     2.234554      .758122
  3.     1     3     .9713964    -1.263158
  4.     1     4     1.835834     .8644378
  5.     1     5     2.336578     .5007435
  6.     1     6     1.490801    -.8457763
  7.     2     1     2.315941            .
  8.     2     2     1.411834    -.9041065
  9.     2     3     .3069954    -1.104839
 10.     2     4    -.9715363    -1.278532
 11.     2     5     .9492834      1.92082
 12.     2     6     .5663401    -.3829433
 13.     3     1     1.560605            .
 14.     3     2     1.408401    -.1522046
 15.     3     3     1.809753     .4013528
 16.     3     4     2.576863       .76711
 17.     3     5     .1996712    -2.377192
 18.     3     6     1.425234     1.225562
```

`diff` is missing at baseline for everyone because baseline is the first observation (there is not a preceding timepoint).

The mean, standard deviation, and range of the session-to-session changes are

```
. table time, c(mean diff sd diff min diff max diff) format(%9.2f)
```

Timepoint	mean(diff)	sd(diff)	min(diff)	max(diff)
1				
2	-0.01	1.43	-4.68	4.36
3	-0.11	1.51	-3.67	5.42
4	-0.00	1.27	-3.10	3.54
5	-0.24	1.52	-4.81	3.18
6	0.09	1.45	-3.74	3.95

A histogram of the changes by session can be seen in figure 10.2.

```
. histogram diff if time > 1, by(time) xlabel(-5(1)5) xtitle("Difference")
> scheme(lean2)
```

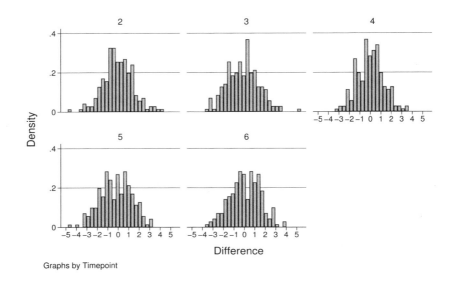

Figure 10.2. Histograms of session-to-session change

The average change from one assessment point to the next ranged from -0.24 (reduction in distress) to 0.09 (increase in distress). The minimum and maximum change ranged widely but were relatively consistent across time. This is also evident by the relatively consistent values of the standard deviation and the similar histograms over time.

Unfortunately, these descriptive analyses are not ideal for studying change because we only used two timepoints to define change. It is difficult to understand how participants changed over the course of treatment by breaking their changes into multiple parts—we lose sight of the process of change over therapy (Singer and Willet 2003). Furthermore, we do not deal with measurement error and its role in understanding change. Singer and Willet (2003) explain why this is a problem:

> If measurement error renders pretest scores too low and posttest scores too high, you might conclude erroneously that scores increase over time when a longer temporal view would suggest the opposite. In statistical terms, two-wave studies cannot describe individual trajectories of change and they confound true change with measurement error (Rogosa, Brandt, and Zimowski 1982). (p. 10)

10.2 Basic growth model

Fortunately, we can also examine change across all time points. A common graphical method is to use a spaghetti plot, which in this context is a scatterplot with `distress` on the y axis and `time` on the x axis and where each plot is for a single participant.

```
. twoway (scatter distress time, connect(L)) if id < 13,
>        by(id) xlabel(1(1)6) scheme(lean2)
```

Figure 10.3 is a spaghetti plot for the first 12 participants in the dataset. The specific participants as well as the number are arbitrary—12 leads to a readable format on the page. If you are not concerned about including the figure in an article, then you may find that more than 12 is better. We added the `connect(L)` option to add straight lines between each point. The `by(id)` option creates a separate scatterplot for each participant (sometimes called a paneled or faceted plot). Figure 10.3 illustrates that patients vary in their pattern of change. Specifically, some hardly change, some move up and down fairly dramatically but have little overall change, and others show clear improvement (or decline). Further, patients vary in their starting point (intercept) and in the type of change they make (slope).

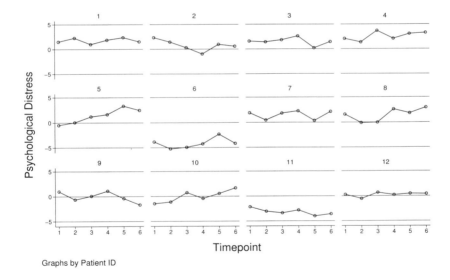

Figure 10.3. Spaghetti plot showing change for the first 12 participants

To better understand the variability in intercept and slope across participants, repeat the scatterplot. However, this time include a regression line to show the pattern of change rather than simply connecting the points (see figure 10.4).

```
. twoway (scatter distress time) (lfit distress time) if id < 13,
>        by(id, legend(off)) xlabel(1(1)6) ytitle("Psychological distress")
>        scheme(lean2)
```

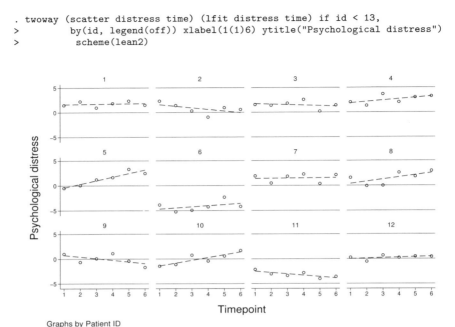

Figure 10.4. Spaghetti plot with regression line showing change for the first 12 participants

We can further explore the variability in the intercepts and slopes by estimating change for each patient. To do this, we need to a) separate the data into distinct patients, b) regress distress onto time for each patient, and c) collect the intercepts and slopes from the regression models. Although we could program a loop to accomplish this, the statsby command automates the process.

Before we estimate the regressions, we need to adjust the time variable. The time variable ranges from 1 to 6, where 1 is baseline and 6 is the posttreatment. The intercept in the model, $\text{distress}_i = b_0 + b_1 \text{time}_i + e_i$, is the expected distress when time equals 0. Given that time does not equal 0 in the dataset, it is useful to center time so that baseline (or whatever the first timepoint is) is 0. This makes b_0 the expected value of distress at baseline.

To create a new variable—in this example, called time_c—where baseline (that is, time = 1) is 0, subtract 1 from time.

```
. generate time_c = time - 1
```

10.2 Basic growth model

Use `time_c` in the `regress` portion of `statsby` (see section 9.5 for a discussion of the `statsby` command).

```
. statsby intercept = _b[_cons] slope = _b[time]
>           int_se = _se[_cons] slope_se = _se[time]
>           df = e(df_r), by(id) clear: regress distress time_c
(running regress on estimation sample)
     command:  regress distress time_c
   intercept:  _b[_cons]
       slope:  _b[time]
      int_se:  _se[_cons]
    slope_se:  _se[time]
          df:  e(df_r)
          by:  id

Statsby groups
──┼── 1 ──┼── 2 ──┼── 3 ──┼── 4 ──┼── 5
....................................................  50
....................................................  100
....................................................  150
....................................................  200
```

The dataset produced by `statsby` has 200 rows, one for each patient. Summarizing `intercept` and `slope` gives a description of the central tendency and dispersion of the intercept and rate of change in the sample of patients.

```
. summarize intercept slope
```

Variable	Obs	Mean	Std. Dev.	Min	Max
intercept	200	-.0258187	1.327343	-4.697341	3.522082
slope	200	-.069667	.4124449	-1.410108	1.011562

The intercepts, the expected `distress` value at baseline, ranged from -4.70 to 3.50, with a mean of -0.03 and standard deviation of 1.33. The slopes, the rate of change over time, ranged from -1.40 to 1.01, with a mean of -0.07 and standard deviation of 0.41. The rate of change is the expected change in `distress` for each increase in `time`. Thus, the expected change from baseline to posttreatment is $-0.07 \times 5 = -0.35$, just over one-fifth of a standard deviation for `distress`.

We can also plot the coefficients and CIs for the intercept and rate of change, though we will focus only on the rate of change. First, use the standard errors and degrees of freedom to create the upper and lower limits of the interval. Second, rank the therapists from low to high on `slope`.

```
. generate ll_b = slope - (slope_se*(-1*invt(df, .025)))
. generate ul_b = slope + (slope_se*(-1*invt(df, .025)))
. egen therrank = rank(slope)
```

Figure 10.5 shows the variability in the rate of change across patients and is created by combining `twoway scatter` and `twoway rcap` (see also sections 9.3.5 and 9.5). These analyses suggest that a full model should allow the rate of change and intercept to vary across patients (rather than fixing them to a single value).

```
. twoway scatter slope therrank ||
>          rcap ul_b ll_b therrank,
>          ytitle("Distress-time slope")
>          xtitle("Patient rank")
>          legend(off)
>          scheme(lean2)
```

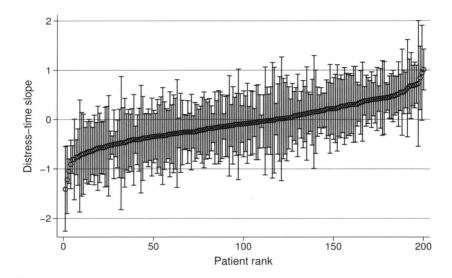

Figure 10.5. Variability in the rate of change (distress–time slope) across patients

The intercept and rate of change are also likely related to one another. We can explore this by correlating the estimates as well as plotting their relationship.

```
. corr intercept slope
(obs=200)

             |  interc~t    slope
-------------+-------------------
   intercept |   1.0000
       slope |   0.0278   1.0000

. twoway scatter slope intercept ||
>          lfit slope intercept, legend(off)
>            ytitle("Patient-specific slope")
>            xtitle("Patient-specific intercept") ylabel(-3(1)3) xlabel(-5(1)5)
>            scheme(lean2)
```

10.2.1 Multilevel model

The intercepts and rate of change have a small, positive relationship, suggesting that the higher the intercept, the more positive the rate of change.

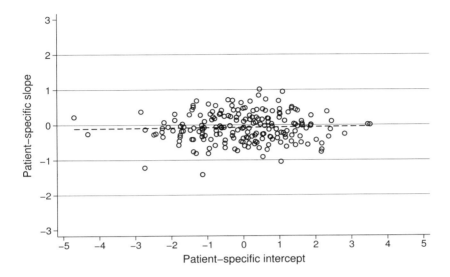

Figure 10.6. Scatterplot of the patient-specific intercepts and slopes for the `distress-time` relationship; the dotted line is the regression line of the slope predicted from the intercept

10.2.1 Multilevel model

The descriptive analyses suggest that both the intercept and the rate of change vary across patients. Consequently, the longitudinal model should allow for that variability. One option is to use a fixed-effects approach, where we include dummy variables for patient as well as the patient by time interaction. This has the same two problems that using a fixed-effects approach had for cross-sectional data: i) it requires estimating hundreds of parameters, and ii) it limits inferences to the patients used in the study. Fortunately, a second option is to use a random-effects approach implemented via a multilevel model.

The multilevel model for longitudinal data is similar to the multilevel model for cross-sectional data.[2] A model with random intercepts only is

$$\text{distress}_{ij} = b_0 + b_1 \text{time\_c}_{ij} + u_{0j} + e_{ij} \tag{10.1}$$

$$u_{0j} \sim N\left(0, \sigma_{u_0}^2\right) \tag{10.2}$$

$$e_{ij} \sim N\left(0, \sigma_e^2\right) \tag{10.3}$$

Here, distress_{ij} is distress at time i for person j, b_0 is the intercept, b_1 is the slope for time_c (also called the rate of change), u_{0j} is a person-specific residual reflecting variability in the intercept across patients, and e_{ij} is a time-specific residual reflecting variability across timepoints within a person. Both u_{0j} and e_{ij} are normally distributed with a mean of 0 and an unknown variance that will be estimated.

The multilevel model in (10.1)–(10.3) using mixed is

```
. use http://www.stata-press.com/data/pspus/mlm_long, clear
. generate time_c = time - 1
. mixed distress c.time_c || id:, nolog
```

Mixed-effects ML regression				Number of obs	=	1,200
Group variable: id				Number of groups	=	200
				Obs per group:		
				min	=	6
				avg	=	6.0
				max	=	6
				Wald chi2(1)	=	12.32
Log likelihood = -2148.6499				Prob > chi2	=	0.0004

distress	Coef.	Std. Err.	z	P>|z|	[95% Conf. Interval]	
time_c	-.069667	.0198506	-3.51	0.000	-.1085734	-.0307606
_cons	-.0258187	.1299856	-0.20	0.843	-.2805858	.2289483

Random-effects Parameters	Estimate	Std. Err.	[95% Conf. Interval]	
id: Identity				
var(_cons)	2.656834	.2888524	2.146947	3.287816
var(Residual)	1.379158	.0616778	1.263418	1.5055

LR test vs. linear model: chibar2(01) = 782.45   Prob >= chibar2 = 0.0000

```
. estimates store randint
```

The expected distress at baseline is $b_0 = -0.03$ and the expected rate of change is $b_1 = -0.07$, both of which are similar to what we saw with statsby. The variability in the intercepts is $\hat{\sigma}_u^2 = 2.66$, and the residual variability is $\hat{\sigma}_e^2 = 1.38$. We stored the estimates so that we can compare the random-intercept model with a model that has a random slope for time.

---

2. In fact, it is identical. Stata does not know if the data are hierarchical because they are longitudinal or not. Keeping track of those details is up to you as the analyst.

10.2.1 Multilevel model

A model with a random slope for time is (compare with the model described in section 9.5)

$$\text{distress}_{ij} = b_0 + b_1 \text{time\_c}_{ij} + u_{0j} + u_{1j}\text{time\_c}_{ij} + e_{ij} \qquad (10.4)$$

$$\begin{bmatrix} u_{0j} \\ u_{1j} \end{bmatrix} \sim N\left(\begin{array}{c} 0 \\ 0 \end{array}, \begin{array}{cc} \sigma^2_{u_0} & \\ \sigma_{u_0 u_1} & \sigma^2_{u_1} \end{array}\right) \qquad (10.5)$$

$$e_{ij} \sim N\left(0, \sigma^2_e\right) \qquad (10.6)$$

The fixed-effects portion of the model has not changed, but there are two random effects. The random intercept, u_{0j}, is a person-specific residual that accounts for differences in baseline distress. The random slope, u_{1j}, is a person-specific residual that accounts for differences in the rate of change in distress. The random effects are drawn from a multivariate normal distribution with a mean of 0 and with unknown variances and a single covariance. The time-specific residual, e_{ij}, is normally distributed with a mean of 0 and an unknown variance.

The multilevel model in (10.4)–(10.6) is fit using mixed:

```
. mixed distress c.time_c || id:time_c, covariance(unstructured) nolog
Mixed-effects ML regression                     Number of obs    =     1,200
Group variable: id                              Number of groups =       200

                                                Obs per group:
                                                          min =          6
                                                          avg =        6.0
                                                          max =          6

                                                Wald chi2(1)     =       5.73
Log likelihood = -2039.8658                     Prob > chi2      =     0.0166
```

distress	Coef.	Std. Err.	z	P>|z|	[95% Conf. Interval]
time_c	-.069667	.0290913	-2.39	0.017	-.1266848 -.0126492
_cons	-.0258187	.0936224	-0.28	0.783	-.2093153 .1576778

Random-effects Parameters	Estimate	Std. Err.	[95% Conf. Interval]
id: Unstructured			
var(time_c)	.1130641	.0171577	.0839756 .1522285
var(_cons)	1.237899	.1771852	.9350817 1.638781
cov(time_c,_cons)	.1556472	.0391674	.0788805 .2324139
var(Residual)	.9834339	.0491717	.8916308 1.084689

LR test vs. linear model: chi2(3) = 1000.02   Prob > chi2 = 0.0000

Note: LR test is conservative and provided only for reference.

```
. estimates store randslope
```

The fixed-effects estimates are about the same, although the standard error for b_1 is larger in the random-slope model than the random-intercept model, whereas the opposite is true for b_0. Because we asked for an unstructured covariance matrix with the `covariance(unstructured)` option, the random-effects part of the output includes three estimates at the patient level. The random-intercept variance, labeled `var(_cons)`, is $\hat{\sigma}^2_{u_0} = 1.24$. The random-slope variance, labeled `var(time_c)`, is $\hat{\sigma}^2_{u_1} = 0.11$. Finally, the covariance between the random effects, labeled `cov(time_c,_cons)`, is $\hat{\sigma}_{u_0 u_1} = 0.16$. We stored the estimates so that we can compare the fit of this model with that of the random-intercept model.

To test whether adding the random slope and the covariance between the random effects improved model fit, use a likelihood-ratio test (see section 11.9.1 for more details). The null hypothesis for this test is that the random-intercept and random-slope model fit the data equally well. The degrees of freedom is equal to the difference in the number of parameters, which is 2 in this case: one for the random-slope variance and one for the covariance. The likelihood-ratio test is implemented with `lrtest`.

```
. lrtest randint randslope
Likelihood-ratio test                              LR chi2(2)  =     217.57
(Assumption: randint nested in randslope)          Prob > chi2 =     0.0000
Note: The reported degrees of freedom assumes the null hypothesis is not on
      the boundary of the parameter space. If this is not true, then the
      reported test is conservative.
```

We reject the null hypothesis, which suggests that the random-slope model fits the data better.

The average rate of change is $b_1 = -0.06$, indicating that for every one-unit increase in time, we expect a 0.06 decrease in `distress`. `margins` and `marginsplot` can help us see what that looks like. We repeat the random-slope model to ensure that we are working with the right output, and we use `quietly` to suppress the output of both `mixed` and `margins`. Figure 10.7 shows that, on average, participants start slightly above 0 on `distress` and show a decrease in symptoms across time. The wider CIs as time gets larger show that the uncertainty in the `distress` estimates increases over time (Rabe-Hesketh and Skrondal [2012a], chapter 4 for a discussion of the impact of random slopes on the total variance).

10.2.1 Multilevel model

```
. quietly mixed distress c.time_c || id:time_c, covariance(unstructured)
. quietly margins, at(time = (0(1)5))
. marginsplot, recastci(rarea) ciopts(color(gs10%50))
>         ytitle("Expected distress") scheme(lean2)
  Variables that uniquely identify margins: time_c
```

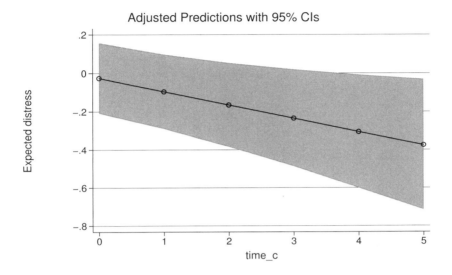

Figure 10.7. Expected rate of change for all participants

Interpreting the random-effects variability is best done via graphs because it lets us see the patient-specific intercepts and slopes and how they vary. Similar to what we did in chapter 9, we can consider the patient-specific regression lines implied by the random intercept and random slope (Rabe-Hesketh and Skrondal 2012a). To compute the patient-specific regression line, we create the predicted values of distress based on both the fixed and the random effects. For example, the predicted value of distress for patient 1 is $\text{distress}_{i1} = b_0 + b_1 \text{time\_c}_{ij} + u_{01} + u_{11}\text{time\_c}_{ij}$. Thus, we use the estimated fixed-effects intercept and slope and combine them with the predicted patient-specific intercept and slope. This can be done following a mixed model by using predict with the option fitted. We can then plot the patient-specific intercept and rate of change.

```
. predict ex_distress, fitted
. twoway scatter ex_distress time_c if time_c == 0, msymbol(circle_hollow)
> color(black) || scatter ex_distress time_c if time_c > 0,
> msymbol(triangle_hollow) color(gs5) || line ex_distress time_c,
> lpattern(solid) lcolor(gs10%50) connect(ascending)
>           ytitle("Distress") xtitle("Time")
>           legend(order(1 "Random intercepts"
>                                   2 "Expected values"
>                                   3 "Random slopes")) scheme(lean2)
```

Figure 10.8 shows the random intercepts and slopes.[3] The twoway code assigns a hollow circle to the random intercepts, a triangle to all other expected data points, and a gray line for the random slopes. This illustrates that although the average change trajectory is a small, negative change ($b_1 = -0.06$), the distribution of change trajectories in the sample varies considerably.

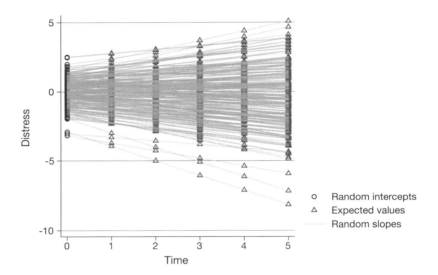

Figure 10.8. Random intercepts and random slopes for all participants

10.3 Adding a level-2 predictor

The primary aim of the treatment study was to compare whether CBT had better outcomes than TAU. To examine this question, we need to define the intervention effect. One possibility, and the one that we will use for the intervention effect, is to explore

---

3. There is a lot of data, so this plot is busy and dense. I am not too concerned about that here because the main purpose of this plot is to get a feel for how variable the intercepts and slopes are. If we wanted to reduce the number of lines in the plot, we could create a random subsample of patients by using sample.

10.3 Adding a level-2 predictor

whether patients in the CBT condition have a different rate of change than patients in the TAU condition. If patients in the CBT condition change faster, on average, than patients in the TAU condition, then they will have better outcomes. That is, given two patients who start at the same baseline level of distress and who are evaluated over the same time window, the patient who changes the fastest changes the most and will have the lowest distress at the end of treatment.

We can add condition to the model by adding the treatment condition dummy variable, tx, to the model. tx is a level-2 predictor, which means that it only varies between participants. A participant is randomized to a condition and stays in that condition for the duration of the study, meaning that there is not any within-person variability. The model including tx as a predictor is[4]

$$\text{distress}_{ij} = b_0 + b_1 \text{time}_{ij} + b_2 \text{tx}_j + u_{0j} + u_{1j}\text{time}_{ij} + e_{ij}$$

where tx_j is the treatment condition for person j. Because tx does not vary within person, it is only indexed by j.

This model will not correctly estimate the treatment effect we are interested in because it assumes that the rate of change is equal across conditions. That is, b_1 is the rate of change, holding treatment condition constant, and b_2 is the difference between CBT and TAU, holding time constant. To see what this means, fit the model and visualize the change trajectories separately for CBT and TAU.

```
. mixed distress c.time_c i.tx || id:time_c, covariance(unstructured)
> stddeviations nolog
Mixed-effects ML regression                     Number of obs     =      1,200
Group variable: id                              Number of groups  =        200
                                                Obs per group:
                                                              min =          6
                                                              avg =        6.0
                                                              max =          6
                                                Wald chi2(2)      =       7.14
Log likelihood = -2039.1802                     Prob > chi2       =     0.0281
```

distress	Coef.	Std. Err.	z	P>|z|	[95% Conf. Interval]
time_c	-.069667	.0290913	-2.39	0.017	-.1266848	-.0126492
tx						
CBT	-.2214677	.1865319	-1.19	0.235	-.5870634	.1441281
_cons	.0849151	.1319053	0.64	0.520	-.1736145	.3434447

---

[4] The distributions of the random effects do not change, so for this model and others in the rest of this chapter, I do not write those distributions out.

Random-effects Parameters	Estimate	Std. Err.	[95% Conf. Interval]	
id: Unstructured				
sd(time_c)	.33625	.0255133	.2897855	.3901647
sd(_cons)	1.10678	.0794676	.9614894	1.274026
corr(time_c,_cons)	.3992607	.11456	.1544102	.5979068
sd(Residual)	.9916823	.0247921	.944262	1.041484

LR test vs. linear model: chi2(3) = 979.00   Prob > chi2 = 0.0000

Note: LR test is conservative and provided only for reference.

```
. quietly margins tx, at(time = (0(1)5))

. marginsplot, recastci(rarea)
>       ci1opts(color(gs10%50)) ci2opts(color(gs5%50))
>       ytitle("Predicted distress") xtitle("Time")
>       title("") scheme(lean2)
  Variables that uniquely identify margins: time_c tx
```

The rate of change is $b_1 = -0.07$, indicating that when holding treatment condition constant, we expect distress to decrease by 0.07 units every session. Likewise, the treatment difference is $b_2 = -0.22$, indicating that we expect patients in the CBT condition to have, on average, 0.22 fewer units of distress than TAU when holding time constant (that is, at every assessment point).

Figure 10.9 shows what these relationships look like.

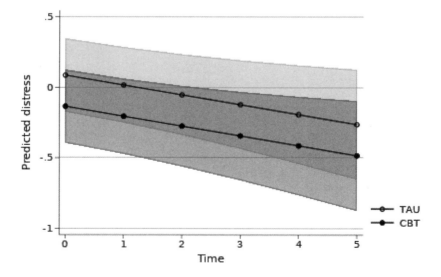

Figure 10.9. Visualization of a constant treatment difference

10.3 Adding a level-2 predictor

The lower line and the darker confidence interval is for CBT. This plot shows how the model assumes that the treatment differences are present across all timepoints. The problem is that this treatment difference may vary in size at different timepoints. For example, given random assignment to conditions, the treatment difference should be smallest at baseline (time = 0) and biggest following treatment (time = 5). Therefore, we need a model that allows the rate of change to be specific to treatment condition.

To allow for rate of change to differ across treatment conditions, add a time by tx interaction to the model.

$$\texttt{distress}_{ij} = b_0 + b_1 \texttt{time\_c}_{ij} + b_2 \texttt{tx}_j + b_3 \texttt{time\_c}_{ij} \texttt{tx}_j + u_{0j} + u_{1j} \texttt{time\_c}_{ij} + e_{ij} \quad (10.7)$$

Let's walk through the interpretation of the fixed effects (the random effects have not changed).

- We coded time so that $\texttt{time\_c}_{ij} = 0$ at baseline, and because $\texttt{tx}_j = 0$ for the TAU condition, b_0 is the expected distress for patients in the TAU condition at baseline.

- Because there is an interaction in the model, b_1 is the rate of change for patients in the TAU condition (that is, where $\texttt{tx}_j = 0$).

- b_2 is the difference in distress between the CBT and TAU conditions at baseline (that is, where $\texttt{time\_c}_{ij} = 0$).

- The interaction term, b_3, is the difference between the rate of change in the CBT condition and the rate of change in the TAU condition. Thus, b_3 is the intervention effect, and the null hypothesis for b_3 is that there is no difference in the rate of change between the conditions. If the rate of change is steeper in the CBT condition than the TAU condition, then we infer that CBT is more effective.

We estimate (10.7) in Stata:

```
. mixed distress c.time_c i.tx c.time_c#i.tx || id:time_c,
> covariance(unstructured) stddeviations nolog
```

Mixed-effects ML regression   Number of obs   =   1,200
Group variable: id   Number of groups =   200

                                                Obs per group:
                                                          min =   6
                                                          avg =   6.0
                                                          max =   6

                                                Wald chi2(3)   =   12.21
Log likelihood = -2036.7491   Prob > chi2   =   0.0067

distress	Coef.	Std. Err.	z	P>|z|	[95% Conf. Interval]	
time_c	-.0059072	.0406442	-0.15	0.884	-.0855684	.0737539
tx						
CBT	-.2275673	.1865521	-1.22	0.223	-.5932028	.1380681
tx#c.time_c						
CBT	-.1275196	.0574796	-2.22	0.027	-.2401774	-.0148617
_cons	.087965	.1319123	0.67	0.505	-.1705783	.3465082

Random-effects Parameters	Estimate	Std. Err.	[95% Conf. Interval]	
id: Unstructured				
sd(time_c)	.3301496	.0253775	.283976	.3838308
sd(_cons)	1.106776	.079467	.9614859	1.27402
corr(time_c,_cons)	.4061077	.1145155	.1607613	.6041866
sd(Residual)	.9916823	.0247921	.9442621	1.041484

LR test vs. linear model: chi2(3) = 980.26   Prob > chi2 = 0.0000
Note: LR test is conservative and provided only for reference.

The fixed effects could also be specified as c.time_c##i.tx. The expected distress at baseline in the TAU condition is $b_0 = 0.09$, the expected difference between CBT and TAU at baseline is $b_2 = -0.23$, the expected rate of change in the TAU condition is $b_1 = -0.01$, and the expected difference in the rate of change in the CBT condition and the TAU condition is $b_3 = -0.13$. The intervention effect was statistically significant, $p = 0.03$.

10.3 Adding a level-2 predictor

Repeating the call to `margins` and `marginsplot` illustrates the difference in the rate of change across conditions (see figure 10.10).

```
. quietly margins tx, at(time_c = (0(1)5))
. marginsplot, recastci(rarea)
>       ci1opts(color(gs10%50)) ci2opts(color(gs5%50))
>       ytitle("Predicted distress") xtitle("Time")
>       title("") scheme(lean2)
  Variables that uniquely identify margins: time_c tx
```

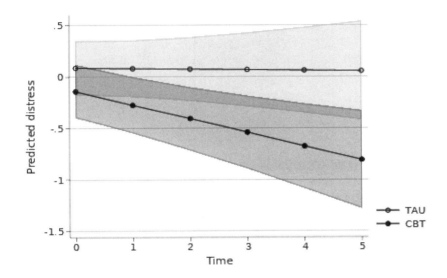

Figure 10.10. Visualization of a treatment difference

As discussed previously, (10.7) includes the intercept and slope for the TAU condition (b_0 and b_1, respectively) and the difference between CBT and TAU at baseline (b_2) and over time (b_3). We can estimate the intercept and slope for the CBT condition in two ways. First, use `lincom` to combine the appropriate elements, which will give us an estimate, standard errors, null hypothesis test, and CI for the CBT intercept and rate of change.

```
. lincom _b[_cons] + _b[1.tx]   // CBT intercept/baseline distress
 ( 1)  [distress]1.tx + [distress]_cons = 0
```

distress	Coef.	Std. Err.	z	P>|z|	[95% Conf. Interval]	
(1)	-.1396024	.1319123	-1.06	0.290	-.3981457	.1189409

```
. lincom _b[time_c] + _b[1.tx#c.time_c]  // CBT rate of change
( 1)  [distress]time_c + [distress]1.tx#c.time_c = 0
```

distress	Coef.	Std. Err.	z	P>|z|	[95% Conf. Interval]	
(1)	-.1334268	.0406442	-3.28	0.001	-.2130879	-.0537656

The expected `distress` in the CBT group at baseline is $b_0 + b_1 = -0.14$, and the expected rate of change is $b_2 + b_3 = -0.13$.

Alternatively, fit a separate intercepts, separate slopes model (section 4.5.1). This model suppresses the overall intercept and does not include a first-order coefficient for time. Consequently, we can estimate the intercept for each condition as well as the slope for each condition.

$$\texttt{distress}_{ij} = b_0 \text{TAU}_j + b_{0*} \text{CBT}_j + b_2 \texttt{time\_c}_{ij} \text{TAU}_j + b_{2*} \texttt{time\_c}_{ij} \text{CBT}_j \\ + u_{0j} + u_{1j} \texttt{time\_c}_{ij} + e_{ij} \qquad (10.8)$$

TAU_j and CBT_j are dummy variables. TAU_j is 1 if the person is in the TAU condition and is 0 otherwise. CBT_j is 1 if the person is in the CBT condition and is 0 otherwise.[5] b_0 and b_{0*} are the intercepts for the TAU and CBT conditions, and b_2 and b_{2*} are the rate of change for the TAU and CBT conditions, respectively.

Estimate the separate intercepts, separate slopes model with `mixed` as follows:

```
. mixed distress ibn.tx ibn.tx#c.time_c, noconstant || id:time_c,
> covariance(un) stddeviations nolog
Mixed-effects ML regression                     Number of obs    =     1,200
Group variable: id                              Number of groups =       200

                                                Obs per group:
                                                             min =         6
                                                             avg =       6.0
                                                             max =         6

                                                Wald chi2(4)     =     12.27
Log likelihood = -2036.7491                     Prob > chi2      =    0.0155
```

distress	Coef.	Std. Err.	z	P>|z|	[95% Conf. Interval]	
tx						
TAU	.087965	.1319123	0.67	0.505	-.1705783	.3465082
CBT	-.1396024	.1319123	-1.06	0.290	-.3981457	.1189409
tx#c.time_c						
TAU	-.0059072	.0406442	-0.15	0.884	-.0855684	.0737539
CBT	-.1334268	.0406442	-3.28	0.001	-.2130879	-.0537656

---

5. We can include both dummy variables because there is not an overall intercept.

10.4 Adding a level-1 predictor

Random-effects Parameters	Estimate	Std. Err.	[95% Conf. Interval]	
id: Unstructured				
sd(time_c)	.3301496	.0253775	.283976	.3838308
sd(_cons)	1.106776	.079467	.9614859	1.27402
corr(time_c,_cons)	.4061077	.1145155	.1607613	.6041866
sd(Residual)	.9916823	.0247921	.9442621	1.041484

LR test vs. linear model: chi2(3) = 980.26   Prob > chi2 = 0.0000
Note: LR test is conservative and provided only for reference.

The `noconstant` option in the fixed-effects portion of the model suppresses the overall intercept. Using `ibn.tx` tells Stata that we do not want a base group for `tx`, which is how we make TAU_j and CBT_j from (10.8). We must do this; otherwise, Stata will still treat the TAU condition as a comparison condition in the model, despite the fact that we have suppressed the intercept. Using `ibn.tx` in the interaction with `time` means that Stata will estimate the two interactions from (10.8). Again, we must use the `ibn` notation or the model will be wrong. Also, ensure that you do not include the first-order effect for time.

The fit of the separate intercepts, separate slopes model is the same as before, which indicates that we have simply reexpressed the parameters rather than estimated a completely different model. Furthermore, the estimates of the intercept and slope for the CBT condition are the same as the estimates from `lincom`.

10.4 Adding a level-1 predictor

Multilevel models can also accommodate variables that change over the course of the study. For example, patients in the treatment study will have different levels of life stress (for example, job stress or illness) during treatment. We may want to examine how life stress, which is called `stress` in the dataset, is related to `distress` during treatment and whether the effects of treatment hold up after controlling for stress. Variables that change over time for a given person are called within-person or level-1 variables. They are also called time-varying covariates, which is in contrast to level-2 variables that are time-invariant (Singer and Willet 2003).

Incorporating a time-varying covariate, such as `stress`, as a fixed effect means that we add `stress` as a predictor of `distress`:

$$\text{distress}_{ij} = b_0 + b_1 \text{time\_c}_{ij} + b_2 \text{tx}_j + b_3 \text{time\_c}_{ij} \text{tx}_j + b_4 \text{stress}_{ij} \\ + u_{0j} + u_{1j} \text{time\_c}_{ij} + e_{ij} \qquad (10.9)$$

b_4 is the expected difference in `distress` given a one-unit difference in `stress`, holding constant the effects of `time_c`, `tx`, and their interaction. In the language of Singer and Willet (2003), b_4 is the population average difference, over time, in `distress` for patients differing by one unit of `stress`.

The mixed syntax and output for (10.9) is

```
. mixed distress c.time_c##i.tx stress || id:time_c,
> covariance(unstructured) stddeviations nolog
Mixed-effects ML regression                     Number of obs      =      1,200
Group variable: id                              Number of groups   =        200

                                                Obs per group:
                                                              min =          6
                                                              avg =        6.0
                                                              max =          6

                                                Wald chi2(4)       =      53.70
Log likelihood = -2016.6876                     Prob > chi2        =     0.0000
```

distress	Coef.	Std. Err.	z	P>|z|	[95% Conf. Interval]	
time_c	-.0123204	.0397484	-0.31	0.757	-.0902258	.065585
tx						
CBT	-.2330082	.1834777	-1.27	0.204	-.5926178	.1266015
tx#c.time_c						
CBT	-.1224676	.0562004	-2.18	0.029	-.2326185	-.0123168
stress	.2067616	.0322309	6.42	0.000	.1435903	.269933
_cons	.1121256	.1297916	0.86	0.388	-.1422613	.3665124

Random-effects Parameters	Estimate	Std. Err.	[95% Conf. Interval]	
id: Unstructured				
sd(time_c)	.3214811	.024956	.2761075	.3743111
sd(_cons)	1.087744	.0782727	.9446592	1.252501
corr(time_c,_cons)	.4700687	.1150854	.2171058	.6638691
sd(Residual)	.9769899	.0244458	.930233	1.026097

```
LR test vs. linear model: chi2(3) = 1007.39              Prob > chi2 = 0.0000
Note: LR test is conservative and provided only for reference.
```

For every one-unit difference in stress, we expect a $b_4 = 0.21$-unit difference over time in distress, holding constant time and tx. To understand what a one-unit increase means, compute the variability in stress. The standard deviation of stress is

```
. summarize stress
```

Variable	Obs	Mean	Std. Dev.	Min	Max
stress	1,200	-.0566941	.9940125	-3.638031	3.286155

indicating that a one-unit difference in stress is a 1-standard-deviation difference. The intervention effect did not change much, nor did the first-order effects for time and tx. The intercept changes, however, because it now is the expected distress when time, tx, and stress are 0.

10.4 Adding a level-1 predictor

We can use `margins` to understand what $b_4 = 0.21$ indicates.

```
. quietly mixed distress c.time_c##i.tx stress || id:time_c,
> covariance(unstructured) stddeviations nolog
. quietly margins tx, at(time_c = (0(1)5) stress = (-1(1)1))
```

We refit the `mixed` model so that the results were available for `margins`. The call to `margins` is the same as before except that we computed the expected distress for three values of `stress` (-1, 0, and 1), which correspond to 1 standard deviation below the mean, the mean, and 1 standard deviation above the mean, respectively.

Figure 10.11 plots the results of `margins`.

```
. marginsplot, noci by(tx) byopts(title("")) ytitle("Expected distress")
>           legend(order(1 "Stress = 1" 2 "Stress = 0" 3 "Stress = 1") rows(1))
>           scheme(lean2)
   Variables that uniquely identify margins: time_c stress tx
```

Suppressing the confidence intervals with `noci` helps with the readability of the plot. Likewise, using `by(tx)` to create panels to plot the treatment conditions separately increases readability, and the option `byopts(title(""))` suppresses the default title. Finally, we have customized the labels for the legend to add spaces between the words and the equals sign, again to improve readability.

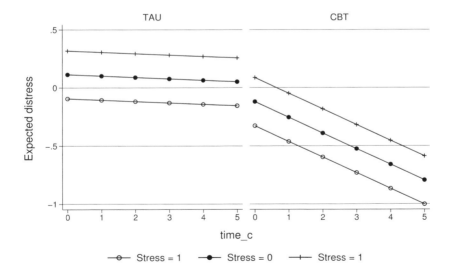

Figure 10.11. The relationship between `stress` and `distress` over time

The problem with figure 10.11 is that it implies that patients are in only one of the `stress` categories during the study—that is, the graph makes `stress` appear more like a level-2, time-invariant variable than a level-1, time-varying variable. It is best to think of patients moving between these lines, including making small deviations between lines (because the lines represent a whole 1-standard-deviation difference) throughout treatment. That is, a patient in the CBT condition might start with the average amount of `stress` at baseline, but then experience a 1-standard-deviation increase in `stress` by time 1 that remains during the rest of the study. That patient would start on the middle line and then move to the top line in the CBT panel. Of course, a patient could move between lines more than once depending on the amount of life stress that patient experienced. In short, do not forget that `stress` and other time-varying covariates change over time.

Although I have not covered it in this chapter, it is possible to allow level-1 predictors, such as `stress`, to randomly vary across people just as we did with time. Estimation can be tricky as the number of random effects increases, but it is important to consider the possibility of person-specific slopes for level-1 predictors. Likewise, level-1 variables can interact with other level-1 variables or with level-2 variables (the latter are sometimes called cross-level interactions). These interactions can be critical to unpack when understanding relationships. Interested readers wanting more instruction on random slopes and interactions can consult Singer and Willet (2003) and Rabe-Hesketh and Skrondal (2012a).

10.5 Summary

Longitudinal studies are common in psychology, and multilevel models are a flexible method for examining change across time. This chapter has given a brief introduction to some fundamental concepts of multilevel models in longitudinal designs, including how to use `mixed` to fit a basic growth model, interpret the fixed effects, interpret the random effects, interpret level-2 coefficients, interpret level-1 coefficients, and use `margins` and `marginsplot` to understand the models.

We have just scratched the surface, with respect to both multilevel models and using `mixed` to implement them for longitudinal data. For example, multilevel models can accommodate missing data (Baldwin et al. 2009; Feaster, Newman, and Rice 2003; Gottfredson, Bauer, and Baldwin 2014; Gottfredson et al. 2014) and can be combined with imputation methods, can accommodate data structures where the distance between timepoints varies across individuals, can model nonlinear and discontinuous change, and can accommodate heteroskedasticity in the random effects and the residuals. Multilevel models can be used for three-level data, such as longitudinal studies within schools—repeated observations that are clustered within students who are clustered within schools. Finally, multilevel models can accommodate nonnormal outcomes, such as binary outcomes, ordinal outcomes, and count outcomes. Readers interested in learning more about longitudinal data analysis can consult Singer and Willet (2003) and Fitzmaurice, Laird, and Ware (2011).

Stata's suite of commands and support for multilevel models and longitudinal data is large and extensive. Rabe-Hesketh and Skrondal (2012a) provides more coverage of

10.5 Summary

multilevel models using Stata and `mixed` as well as a discussion of alternative methods and Stata commands for longitudinal data. Rabe-Hesketh and Skrondal (2012b) discuss multilevel models for nonnormal outcomes using Stata. Specific commands readers may be interested in are in the `me` suite (where `me` stands for mixed effects), including `melogit`, `meologit`, `mepoisson`, `menbreg`, and `meglm` (StataCorp 2017e). The `sem` and `gsem` commands (StataCorp 2017b) are also useful for longitudinal data.

Part III

Psychometrics through the lens of factor analysis

11 Factor analysis: Reliability

For my money, the #1 neglected topic in statistics is measurement.

- Andrew Gelman[1]

Psychologists have thought a lot about measurement. A perusal of measurement texts (McDonald 1999; Nunnally and Bernstein 1994) suggests that all this thinking has led to one firm conclusion: measurement is tough, requiring careful thought and rigorous evaluation. Given that we all have experience with measurement—weighing ourselves on a bathroom scale, prepping ingredients for a recipe, getting our temperature taken during a medical exam—it may not be obvious why measurement is so tough.

Consider a measuring tape. It is simple to determine if a measuring tape is a good tool. Go to the hardware store, and use the tape to measure six feet of wood. Compare the new wood with other wood known to be six feet long; if the pieces are the same size, the tape is good. Further, if we get the same result every time we use the tape to measure a specific length, then the tape is good.

Now consider a measure of depressive symptoms. Depressive symptoms typically consist of sadness, loss of interest in pleasurable things, appetite changes, excessive guilt, sleep disturbance, memory trouble, and concentration difficulties (American Psychiatric Association 2013). How should we measure these symptoms? For example, how should we ask about sadness? People differ in how they define the word sad and what constitutes normal amounts of sadness. It is not so simple to determine whether our question about sadness is a good tool. Suppose two people differ in their response to our question. Why do they differ? Perhaps their actual levels of sadness differ, their personal definition of sadness differs, or one person is a native-English speaker and the other is not.

These concerns only scratch the surface of the challenges of measurement. These challenges are not limited to self-report measures of depressive symptoms, but include all types of measures in psychology—intelligence tests, fMRI readings, attitude measures, and academic achievement tests, just to name a few. Regardless of the measure, most of the issues in measurement involve reliability and validity. Reliability refers to the consistency of a measure, whereas validity refers to whether a measure actually assesses what it purports to measure.

This chapter provides an introduction to reliability from the perspective of confirmatory factor analysis (CFA). CFA is flexible and intuitive and can serve as a unifying

---

1. http://andrewgelman.com/2015/04/28/whats-important-thing-statistics-thats-not-textbooks/

framework for thinking about both reliability and validity. A challenge with any statistical learning is drawing connections between seemingly disparate topics. CFA will serve as the bridge between reliability and validity (see chapter 12). Consequently, this chapter begins by introducing basic concepts in CFA and showing how to fit a one-factor model. I then introduce reliability as defined by classical test theory; I show how CFA can be used to understand the assumptions of classical test theory, test those assumptions, and relax the assumptions as needed.

11.1 What you will learn in this chapter

By studying this chapter and working through the analyses, you will learn the following:

1. The definition of reliability.
2. How common and unique variances impact reliability.
3. How to use sem to estimate common and unique variances.
4. How to interpret the parameters of CFA.
5. How to evaluate model fit by using the χ^2 test.
6. How to use the output from sem to compute reliability and construct an interval estimate for reliability.
7. How reliability as computed using CFA compares with α (a commonly used measure of reliability in psychology) and how to test the assumptions of α by using CFA.
8. How to extend reliability analyses to include models with correlated residuals.

Stata commands featured in this chapter

- sem: for fitting CFA models
- nlcom: for using the output from sem to compute reliability
- bootstrap: for bootstrapping a confidence interval for the reliability estimate
- lrtest: for comparing the fit of nested CFA models
- alpha: for computing α (a measure of internal consistency reliability)

11.2 Example data

The anxiety literature distinguishes between trait anxiety and state anxiety. Trait anxiety occurs generally throughout someone's life (for example, generalized worry), whereas state anxiety is context specific (for example, worrying about an upcoming exam). The state–trait anxiety inventory (STAI) (Spielberger et al. 1983) is a self-report measure that asks about both state and trait anxiety. In this chapter, we focus on seven items from the trait subscale, which asks respondents to rate each symptom regarding how they generally feel on a scale from 1 to 4:

- 1 = Almost never
- 2 = Sometimes
- 3 = Often
- 4 = Almost always

For example, for item 2 respondents rate the statement "I feel nervous and restless" from 1 to 4.

Data for the $N = 790$ participants are in `trait.dta`.

```
. use http://www.stata-press.com/data/pspus/trait
(Trait Anxiety Items)
. describe
Contains data from trait.dta
  obs:           790                         Trait Anxiety Items
 vars:             7                         7 Nov 2018 20:49
 size:         5,530
```

variable name	storage type	display format	value label	variable label
trait2	byte	%13.0g	stai	Nervous and restless
trait4	byte	%13.0g	stai	Happy as others
trait5	byte	%13.0g	stai	Failure
trait8	byte	%13.0g	stai	Difficulties piling up
trait9	byte	%13.0g	stai	Worry too much
trait15	byte	%13.0g	stai	Inadequate
trait20	byte	%13.0g	stai	Tension

```
Sorted by:
```

11.3 Common versus unique variance

Consider item 2, "I feel nervous and restless". How can we tell if it is reliable, which is to say, how do we know if it is a consistent measure? Ideally, a person would repeatedly answer item 2. The average of the answers would provide an estimate of the person's nervousness, and the variability in the answers would provide an estimate of the imprecision of the measure (McDonald 1999). Clearly, the ideal is problematic because the repeated observations are not independent. After repeated assessments, one's understanding of the word nervous may change, one may become bored or distracted, one may become more or less nervous, one may begin to answer in a socially desirable way, or one may try to please the researcher. Measuring trait anxiety is not the same as weighing food because food does not know it is being measured and typically is not affected by the measurement process (McDonald 1999). Consequently, we need a way to obtain repeated measures on a construct that are independent of one another.

Rather than repeatedly measuring people on item 2, suppose we used the seven items from the STAI listed above. Responses to each item are believed to be indicators of trait anxiety. We can conclude the following:

1. The average (or sum) of the responses represents the best estimate of a person's trait anxiety.

2. The variability across items represents error or imprecision in measurement (McDonald 1999).

These conclusions rest on the assumption that the items share common variance. This assumption makes sense if we consider how a scale like the STAI is used and developed. Typically, researchers obtain responses to the trait items and then sum (or average) the responses to obtain a total. By summing the item scores, the researchers assume that each item provides information about trait anxiety (that is, adding construct variance from each item together).

The mean, standard deviation, and range of each trait anxiety item is

```
. summarize trait2-trait20, separator(0)
    Variable |       Obs        Mean    Std. Dev.       Min        Max
      trait2 |       790    1.887342    .7543481         1          4
      trait4 |       790    2.029114    .9764855         1          4
      trait5 |       790    1.555696    .7408499         1          4
      trait8 |       790    1.858228    .8304423         1          4
      trait9 |       790    2.070886    .8935981         1          4
     trait15 |       790    1.803797     .793341         1          4
     trait20 |       790    1.901266    .8852393         1          4
```

Why is there variability in these items (that is, why do some people score higher than others)? Why are the standard deviations different across items (that is, why do people vary more on some items than others)?

- Respondents may genuinely differ in their trait anxiety.

- Some respondents may have circled the wrong number.

- Some respondents may have felt acutely anxious and responded about how they felt at that moment rather than generally.

- Respondents may have a different understanding of what the rating scale values mean.

- Some respondents may have answered too quickly and did not give their ratings sufficient thought.

11.3 Common versus unique variance

The first possibility—genuine differences in trait anxiety—is what we hope each item measures. Indeed, given that all items were selected to measure trait anxiety, we assume that shared variance across items is due to trait anxiety.[2] The other possibilities produce noise or error in the item scores and are unique to each item.[3] Therefore, items vary for two general reasons:

1. Items vary because of what is common between them.
2. Items vary because of what is unique to each of them.

In other words, we can separate the observed variance in any item X_i into two parts:

$$\sigma^2_{X_i} = \sigma^2_C + \sigma^2_U \qquad (11.1)$$

where C stands for common and U stands for unique.

Figure 11.1 illustrates how the observed values on each trait anxiety item are a combination of common and unique variances. The value to the left of each item is the variance for the item. The reliability (ω) of the trait anxiety scale, the average (or sum) of the seven trait anxiety items, is equal to the proportion of the total variance in the scale due to the common variance (McDonald 1999, 89, (6.20a)]. The total variance in the scale is the sum of the common and unique variances.

$$\omega = \frac{\sigma^2_C}{\sigma^2_C + \sigma^2_U} \qquad (11.2)$$

If most of the variability in a given item is due to what is common among the items, then the scale will be reliable. Note that reliability is defined in terms of what is common among the items rather than by the construct that is thought to be common among the items. Reliability is about the consistency of a measure, regardless of the source of the consistency. Validity is about whether the source of consistency is the construct we are interested in.

---

2. Shared variance across items can arise for reasons other than a common construct, such as a common method of measurement (for example, all self-report or using similar words across items). These are critical issues in measurement but beyond what I am focused on at this point in the chapter.
3. This assumption is known as the conditional or local independence, which means that the items are independent of one another conditional on the latent variable. The assumption can be relaxed.

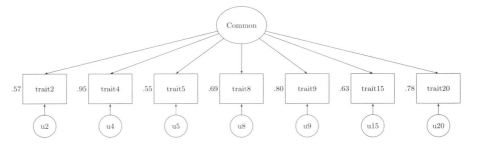

Figure 11.1. Observed variance and sources of variance for the trait anxiety items; the value to the left of each item is the variance for the item

Note that we do not directly observe the common and unique variances. We only observe the variability of the item. So how do we get an estimate of each? This is where factor analysis comes in. Specifically, we will use factor analysis to determine what proportion of the variability in the seven items is due to what is common across items and how much is unique.

11.4 One-factor model

Figure 11.2 shows a CFA model that can be used to examine reliability. As in figure 11.1, each item shares a common factor. In this model, I have named the factor TRAIT_ANX. However, remember that whether this factor is actually trait anxiety is a question of validity. I name it TRAIT_ANX here to help link the substantive question of measuring trait anxiety with the statistical question of common variance. Each item also has a unique term, which I label e2–e20, where the e stands for error.

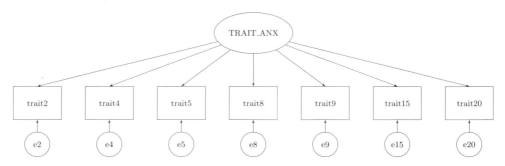

Figure 11.2. One-factor CFA model for trait anxiety

11.4.1 Parts of a path model

Figure 11.2 is a path diagram, which is a visual representation of a CFA or other structural equation model (SEM). Path diagrams have some conventions to make them consistent across researchers and publications.

- Ovals and circles represent latent (unobserved) variables.
- Rectangles and squares represent observed variables.
- Single-headed arrows represent regression paths.
- Double-headed arrows represent covariance/correlation.

Occasionally, SEM analyses are called causal models because the diagrams imply a causal structure. For example, in figure 11.2 the observed items are "caused" by two latent variables—trait anxiety and a specific error term. We say that the observed items are caused by the latent variables because the arrows move away from the latent variable to the observed items. Although these models are sometimes called causal, do not assume that we can draw valid causal inferences simply by applying the method. There is nothing special about SEM that makes causal inference automatic.

11.4.2 Where do the latent variables come from?

Given that the TRAIT_ANX and the error terms are unobserved, latent variables, where do they come from? In short, they are inferred from the data. More specifically, factor analysis is a method for estimating σ_C^2 and σ_U^2 from (11.1). Factor analysis takes the variance of each item and the covariance between the items to determine what is shared between items and what is unique.

The regression paths—the single-headed arrows—connecting TRAIT_ANX to each of the items in figure 11.2 are known as factor loadings. Factor loadings describe the strength of the relationship between the latent variable (that is, the construct) and a given item. If a lot of the variability in the items is shared across items, then factor loadings will be strong.

Figure 11.3 is a path model representing the covariance among the seven trait items. Conceptually, a factor analysis takes the covariances and turns them into a latent variable. If the covariances are strong, then the latent variable will have a strong relationship with the items, and error variances will be relatively small. Compare figure 11.2 with figure 11.3. The factor analysis depicted in figure 11.2 provides a model or structure to the relationships among the items in figure 11.3. In other words, when using factor analysis, we are taking the unstructured covariances among the items and applying the structure of a common factor model (Bollen 1989).

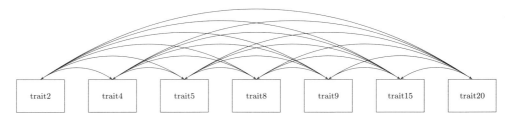

Figure 11.3. Covariance among the seven trait anxiety items

A squared correlation is interpreted as the amount of variance shared between two variables. To obtain a descriptive look at how much variance is shared between any two items, it can be useful to examine a correlation matrix of the items. The correlation matrix for the trait anxiety items is

```
. corr trait2-trait20
(obs=790)
             |   trait2    trait4    trait5    trait8    trait9   trait15   trait20
-------------+---------------------------------------------------------------------
      trait2 |   1.0000
      trait4 |   0.4553    1.0000
      trait5 |   0.4796    0.5838    1.0000
      trait8 |   0.4863    0.4959    0.5444    1.0000
      trait9 |   0.4481    0.4523    0.3808    0.4713    1.0000
     trait15 |   0.4607    0.5113    0.6537    0.4694    0.3790    1.0000
     trait20 |   0.5034    0.4505    0.4471    0.5309    0.4703    0.4470    1.0000
```

Squaring these correlations provides the shared variance. For example, the squared correlation between `trait2` and `trait4` is

```
. display  0.4553^2
.20729809
```

Thus, `trait2` and `trait4` share 21% of their variance. Rather than compute all of these by hand, we can save the results of `corr` to a matrix and square each element.

```
. corr trait2-trait20
(obs=790)
             |   trait2    trait4    trait5    trait8    trait9   trait15   trait20
-------------+---------------------------------------------------------------------
      trait2 |   1.0000
      trait4 |   0.4553    1.0000
      trait5 |   0.4796    0.5838    1.0000
      trait8 |   0.4863    0.4959    0.5444    1.0000
      trait9 |   0.4481    0.4523    0.3808    0.4713    1.0000
     trait15 |   0.4607    0.5113    0.6537    0.4694    0.3790    1.0000
     trait20 |   0.5034    0.4505    0.4471    0.5309    0.4703    0.4470    1.0000

. matrix corr = r(C)
. matrix sqcorr = J(7,7,0)
```

11.5 Prediction equation

```
. forvalues i = 1/7 {
  2.         forvalues j = 1/7 {
  3.                 matrix sqcorr[`i', `j'] = corr[`i', `j']^2
  4.         }
  5. }
. matrix list sqcorr
symmetric sqcorr[7,7]
            c1          c2          c3          c4          c5          c6          c7
r1           1
r2   .2072628           1
r3  .22997781   .3408109           1
r4  .23653116  .24588284   .2963174           1
r5  .20076974  .20454195  .14497321  .22211073           1
r6  .21225248  .26141462  .42729914  .22038251  .14363544           1
r7  .25337835  .20297188  .19988576  .28186617   .2211783  .19982213           1
```

Most of the r^2 values are in the 20% range, with a few as low as 14% and one as high as 43%. Although we cannot divine the factor loadings and error variances by looking at the r^2 values, it does give us a feel for how much the items have in common. Clearly, there is some redundancy between the items but also a fair amount of variability that is unique. The factor analysis will reveal more detail. Before we actually fit the CFA, we need to connect the ideas of factor analysis to regression.

11.5 Prediction equation

One way to think of CFA is as a series of regression models estimated simultaneously. Recall that we built prediction equations in regression [see (3.4)]. A bivariate regression is expressed as

$$y_i = \beta_0 + \beta_1 x_{1i} + \epsilon_i$$

The prediction equation for a single item in CFA is

$$x_{ij} = \tau_i + \lambda_i \xi_j + \delta_{ij} \tag{11.3}$$

where x_{ij} is the observed score for item i for person j, τ_i is the item intercept for item i, λ_i is the factor loading for item i, ξ_j is the value of the latent trait for person j, and δ_{ij} is the residual error. Although (11.3) includes an intercept, CFA often uses centered data or covariances among items as the data. In that case, the intercepts are fixed to zero and can be excluded from the equations. I will exclude them from the equations for the rest of this chapter and deal with intercepts in chapter 12.

Although (11.3) is a valid prediction equation for CFA, the more common way of expressing the prediction is not in terms of the observed score on a specific item but in terms of the model-implied covariance among the items. The prediction equation for the model-implied covariance among items is written in matrix form and is

$$\widehat{\boldsymbol{\Sigma}} = \boldsymbol{\Lambda}\boldsymbol{\Phi}\boldsymbol{\Lambda}' + \boldsymbol{\Theta} \tag{11.4}$$

where $\widehat{\boldsymbol{\Sigma}}$ is a $k \times k$ predicted (sometimes called model-implied) covariance matrix for the items (where k = the number of items), $\boldsymbol{\Lambda}$ is a $k \times p$ matrix of factor loadings ($p =$

the number of latent variables), $\boldsymbol{\Phi}$ is a $p \times p$ covariance matrix for the latent variables, and $\boldsymbol{\Theta}$ is a $k \times k$ covariance matrix for the residuals.

To understand each part of (11.4), one must understand that a key difference between ordinary least-squares regression and CFA is that in CFA the outcome variables are the observed items (that is, each of the trait anxiety items) and the predictor variable is the latent variable. We can write figure 11.2 as a series of regression equations:

$$
\begin{aligned}
\texttt{trait2} &= \text{loading}_2 \times \texttt{TRAIT\_ANX} + \text{error}_2 \\
\texttt{trait4} &= \text{loading}_4 \times \texttt{TRAIT\_ANX} + \text{error}_4 \\
\texttt{trait5} &= \text{loading}_5 \times \texttt{TRAIT\_ANX} + \text{error}_5 \\
\texttt{trait8} &= \text{loading}_8 \times \texttt{TRAIT\_ANX} + \text{error}_8 \\
\texttt{trait9} &= \text{loading}_9 \times \texttt{TRAIT\_ANX} + \text{error}_9 \\
\texttt{trait15} &= \text{loading}_{15} \times \texttt{TRAIT\_ANX} + \text{error}_{15} \\
\texttt{trait20} &= \text{loading}_{20} \times \texttt{TRAIT\_ANX} + \text{error}_{20}
\end{aligned}
$$

Thus, a participant's response to `trait2` is equal to the participant's latent trait anxiety plus measurement error. How much the latent variable contributes to the observed variables is determined by the factor loading. Specifically, the factor loading weights the contribution of trait anxiety to the participant's response to the `trait2` item.

Figure 11.4 generalizes the trait anxiety path model and introduces Greek notation so that we can connect it to (11.4). Factor loadings are symbolized with λ, and the value of the latent variable (for example, amount of trait anxiety for a given person) is denoted by ξ. The error in an item is denoted by δ. Neither ξ nor δ are observed or directly estimated because both are latent. Instead, the variability of ξ—the variability of the latent variable in the population—and the variability of δ—the error variability in the population are estimated.[4] The variance of the latent variable is denoted by ϕ and the error variance is denoted by θ.

---

4. Note the similarity between the latent values and error values in SEM and the random effects in multilevel models (see chapters 9 and 10). In both cases, we estimate the variance of those quantities but not the specific values. The relationship between SEM and multilevel models has been documented elsewhere (Curran 2003).

11.5 Prediction equation

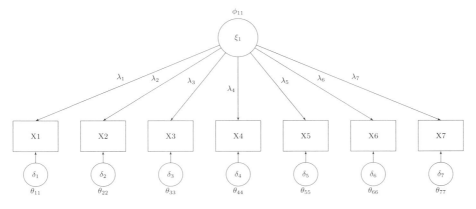

Figure 11.4. One-factor model with Greek notation

Just as we did with figure 11.2, we can write figure 11.4 as a series of regression equations.

$$
\begin{aligned}
X1 &= \lambda_1 \xi_1 + \delta_1 \\
X2 &= \lambda_2 \xi_1 + \delta_2 \\
X3 &= \lambda_3 \xi_1 + \delta_3 \\
X4 &= \lambda_4 \xi_1 + \delta_4 \\
X5 &= \lambda_5 \xi_1 + \delta_5 \\
X6 &= \lambda_6 \xi_1 + \delta_6 \\
X7 &= \lambda_7 \xi_1 + \delta_7
\end{aligned}
$$

Because these are regression equations, the interpretation of the factor loadings is just like any other regression coefficient. For example, the interpretation of λ_1 is as follows: a one-unit difference in the latent variable (for example, when two people differ by 1 on ξ_1) is associated with a λ_1 difference in X_1.

These seven equations can be rewritten in matrix form:[5]

$$
\begin{bmatrix} X1 \\ X2 \\ X3 \\ X4 \\ X5 \\ X6 \\ X7 \end{bmatrix} = \begin{bmatrix} \lambda_1 \\ \lambda_2 \\ \lambda_3 \\ \lambda_4 \\ \lambda_5 \\ \lambda_6 \\ \lambda_7 \end{bmatrix} [\xi_1] + \begin{bmatrix} \delta_1 \\ \delta_2 \\ \delta_3 \\ \delta_4 \\ \delta_5 \\ \delta_6 \\ \delta_7 \end{bmatrix}
$$

where the matrix of factor loadings is denoted by $\boldsymbol{\Lambda}$, which is the $\boldsymbol{\Lambda}$ in (11.4).

Equation (11.4) includes two other matrices on the right-hand side of the equals sign. The first is $\boldsymbol{\Phi}$, which is the covariance matrix for the latent variables. $\boldsymbol{\Phi}$ is a $p \times p$

---

5. Writing these equations in matrix form is beneficial because doing so collapses the seven equations into a single equation. That is, the seven X's are placed into a matrix, the seven loadings into a matrix, and so on.

matrix with the variance of each latent variable on the diagonal and the covariances among the latent variables on the off-diagonal. Because our example includes a single latent variable, $\mathbf{\Phi}$ contains only one element, $\mathbf{\Phi} = \phi_{11}$.

The second matrix in (11.4) is $\mathbf{\Theta}$, which is the covariance matrix of the residual errors. $\mathbf{\Theta}$ is a $k \times k$ matrix with the residual variance of each item on the diagonal and the covariances between the residuals on the off-diagonal. In our examples, we have only specified the residual variances and have not included any covariances between the residuals. Thus, the covariances are fixed to 0.

$$\mathbf{\Theta} = \begin{bmatrix} \theta_{11} & & & & & & \\ 0 & \theta_{22} & & & & & \\ 0 & 0 & \theta_{33} & & & & \\ 0 & 0 & 0 & \theta_{44} & & & \\ 0 & 0 & 0 & 0 & \theta_{55} & & \\ 0 & 0 & 0 & 0 & 0 & \theta_{66} & \\ 0 & 0 & 0 & 0 & 0 & 0 & \theta_{77} \end{bmatrix}$$

The final matrix in (11.4) is $\widehat{\mathbf{\Sigma}}$, which is the predicted covariance matrix for the items. Thus, like regression where the regression coefficients combine with the predictor variables to create a predicted value of the outcome, the factor loadings combine with the covariance matrices of the latent variables and the residuals to create a predicted covariance matrix of the items. Furthermore, this predicted value of the covariance matrix is used to assess the fit of the factor model.

11.6 Using sem to estimate CFA models

Stata provides two methods for fitting CFA models. First, one can use the SEM Model Builder, which allows the user to draw the path model. For example, one could use the SEM Model Builder to draw a path model just like figure 11.2. Second, one can use Stata syntax. I will focus on the syntax. The SEM Model Builder is a great tool and is remarkably flexible. However, I believe learning to use syntax benefits users by promoting replicability, making it simple to build multiple models quickly, and promoting a deeper understanding of what the model is actually estimating.[6]

---

6. Stata's SEM Model Builder produces the syntax for the path model when the user estimates the model. Thus, one can begin an analysis with the Model Builder, copy the syntax, and tweak the syntax in a do-file.

11.6 Using sem to estimate CFA models

The one-factor model for the trait anxiety items can be fit as follows:

```
. sem (TRAIT_ANX -> trait2 trait4 trait5 trait8
> trait9 trait15 trait20),
> var(TRAIT_ANX@1) nolog
Endogenous variables
Measurement:  trait2 trait4 trait5 trait8 trait9 trait15 trait20
Exogenous variables
Latent:       TRAIT_ANX
Structural equation model                   Number of obs     =        790
Estimation method  = ml
Log likelihood     = -5809.5247

 ( 1)  [/]var(TRAIT_ANX) = 1
```

	Coef.	OIM Std. Err.	z	P>|z|	[95% Conf. Interval]	
Measurement						
trait2						
TRAIT_ANX	.5035065	.0251842	19.99	0.000	.4541463	.5528667
_cons	1.887342	.0268215	70.37	0.000	1.834773	1.939911
trait4						
TRAIT_ANX	.6971751	.0318322	21.90	0.000	.6347852	.759565
_cons	2.029114	.0347198	58.44	0.000	1.961064	2.097163
trait5						
TRAIT_ANX	.5680217	.023666	24.00	0.000	.5216372	.6144062
_cons	1.555696	.0263416	59.06	0.000	1.504068	1.607325
trait8						
TRAIT_ANX	.5958545	.0270768	22.01	0.000	.542785	.648924
_cons	1.858228	.0295271	62.93	0.000	1.800356	1.9161
trait9						
TRAIT_ANX	.5370044	.0307768	17.45	0.000	.476683	.5973258
_cons	2.070886	.0317727	65.18	0.000	2.008613	2.133159
trait15						
TRAIT_ANX	.5705945	.0259369	22.00	0.000	.5197592	.6214299
_cons	1.803797	.0282079	63.95	0.000	1.748511	1.859084
trait20						
TRAIT_ANX	.5907191	.0296566	19.92	0.000	.5325933	.648845
_cons	1.901266	.0314755	60.40	0.000	1.839575	1.962957
var(e.trait2)	.314802	.0180901			.2812697	.3523318
var(e.trait4)	.4662638	.027836			.4147772	.5241416
var(e.trait5)	.2255151	.0148197			.1982618	.2565146
var(e.trait8)	.3337188	.0201022			.2965562	.3755385
var(e.trait9)	.5091331	.0281943			.4567665	.5675033
var(e.trait15)	.3030151	.0184572			.2689156	.3414385
var(e.trait20)	.4337075	.0250853			.3872256	.4857691
var(TRAIT_ANX)	1	(constrained)				

LR test of model vs. saturated: chi2(14) =   130.87, Prob > chi2 = 0.0000

The `sem` keyword is followed by the factor analysis model. The latent variable, `TRAIT_ANX`, is typed in caps because we need to signal to Stata that we are using a variable not contained in the dataset and that `TRAIT_ANX` should be considered latent. This requirement can be turned off with the `nocapslatent` option, but I recommend not using that option until you are comfortable specifying models. The schematic arrowhead (`->`) followed by all the trait anxiety items indicates that we want factor loadings. Indeed, we can read the material within parentheses as saying, "The latent variable `TRAIT_ANX` loads on `trait2`–`trait20`." It is critical that the head of the arrow point away from the latent variable and toward the observed variables. Variables following the arrowhead are predicted by whatever is behind the arrow—the latent variable predicts all the items just as in figure 11.2.

The final piece of the syntax is that we have fixed the variance of the latent variable to 1 by using the `variance(TRAIT_ANX@1)` option. This is a method of model identification, and it will be used in this chapter. A full discussion of model identification and the methods of model identification is in chapter 12. For now, trust me that this is a reasonable thing to do.

The `sem` routine produces a lot of output.[7] For now, we will focus on the `Coef.` (coefficient) column. Factor loadings (λ) and intercepts are reported by indicator. Ignore the intercepts (`_cons`) for now; those are discussed in chapter 12. The loading for `TRAIT_ANX` on `trait2` is $\widehat{\lambda}_2 = 0.5$, indicating that for every one-unit increase in `TRAIT_ANX`, `trait2` increases by 0.5. The other items have similar loadings, although all are a bit larger, with the loading for `trait4` being the largest.

Residual (θ) and latent variable variances (ϕ) are reported below the factor loadings. Each item name is preceded by `e.` to signal that it deals with the errors. The residual variance for `trait2` is $\theta_{22} = 0.31$. The variance of `TRAIT_ANX` is $\phi_{11} = 1$, and the output notes that this value was constrained or fixed by the user.

11.7 Model fit

How well does the CFA fit the data? In regression, we compare the predicted values from the model with the actual values, and good fit was indicated by a close relationship between the predicted and actual values (for example, casewise prediction). CFA (and SEM generally) follows the same logic. Specifically, we determine model fit by comparing the predicted covariance matrix ($\widehat{\Sigma}$) to the actual covariance among the items (S), where a close relationship between these matrices indicates good fit.

---

7. Some users find the amount of output produced by `sem`, and all SEM software, bothersome. Although the output is relatively large, the size reflects the flexibility of SEMs. In SEM, the user has control over nearly all parameters, and most parameters are free by default (unique errors for all outcomes). In regression or mixed models, users typically have less control over parameters, and parameters are often constrained by default (for example, homoskedastic errors).

11.7 Model fit

S was

```
. corr trait2-trait20, covariance
(obs=790)
```

	trait2	trait4	trait5	trait8	trait9	trait15	trait20
trait2	.569041						
trait4	.33535	.953524					
trait5	.268006	.422331	.548859				
trait8	.304667	.402105	.334902	.689634			
trait9	.302039	.394638	.252067	.349733	.798518		
trait15	.275714	.396087	.384199	.309284	.268679	.62939	
trait20	.336138	.389443	.293212	.390294	.372027	.313937	.783649

$\widehat{\Sigma}$ can be obtained with the `sem` postestimation command `estat framework, fitted`. The `fitted` option provides $\widehat{\Sigma}$, among other things.

```
. sem (TRAIT_ANX -> trait2 trait4 trait5 trait8
> trait9 trait15 trait20),
> var(TRAIT_ANX@1) nolog nocnsreport
  (output omitted)
. estat framework, fitted
  (output omitted)
```

	observed				
Sigma	trait2	trait4	trait5	trait8	trait9
observed					
trait2	.5683208				
trait4	.3510322	.9523169			
trait5	.2860026	.3960106	.5481638		
trait8	.3000166	.4154149	.3384583	.6887614	
trait9	.2703852	.3743861	.3050302	.3199765	.7975068
trait15	.2872981	.3978043	.3241101	.3399913	.3064118
trait20	.2974309	.4118347	.3355413	.3519827	.3172188
latent					
TRAIT_ANX	.5035065	.6971751	.5680217	.5958545	.5370044

	observed		latent
Sigma	trait15	trait20	TRAIT_ANX
observed			
trait15	.6285932		
trait20	.3370611	.7826566	

  (output omitted)

The interocular trauma test of the difference between the predicted and actual matrices is not significant (Savage 2009, 325). Indeed, the matrices look remarkably similar. Fortunately, we can supplement the interocular trauma test with a formal statistical test, sometimes referred to as the χ^2 test of overall fit.

The null hypothesis of the χ^2 test is that there is no difference between the actual and predicted matrices:
$$H_0 : S = \widehat{\Sigma} \tag{11.5}$$
Thus, a significant χ^2 test indicates that S and $\widehat{\Sigma}$ differ more than we expect by sampling error alone. In the example, the χ^2 test is reported at the bottom of the `sem` output and can also be obtained via the `estat gof` postestimation command.

```
. estat gof
```

Fit statistic	Value	Description
Likelihood ratio		
chi2_ms(14)	130.867	model vs. saturated
p > chi2	0.000	
chi2_bs(21)	2211.661	baseline vs. saturated
p > chi2	0.000	

We care about the first value listed.

The degrees of freedom (DF_{mod}) for the χ^2 test is equal to

$$\text{DF}_{\text{mod}} = \text{unique elements in } S + \# \text{ of observed variables} - \# \text{ of parameters} \tag{11.6}$$

The number of unique elements in S is equal to

$$\text{unique elements} = \frac{k(k+1)}{2}$$

where k is the number of observed variables (7), making 28 unique elements in our example. The total number of parameters is 21: 7 factor loadings, 7 intercepts, and 7 residual variances. Thus, the $\text{DF}_{\text{mod}} = 28 + 7 - 21 = 14$.

The test statistic is $\chi^2(14) = 130.9$, $p < 0.01$, and we reject the null hypothesis that the predicted and actual covariance matrices are identical. Ideally, the χ^2 test would not be significant, but χ^2 tests can be overly powerful as sample sizes get large (Brown 2015). As we will see in chapter 12, other tests of model fit can be used to supplement the χ^2 test. However, we use the χ^2 values to compare the fit among a set of competing models that allow us to evaluate assumptions of our reliability analysis.

11.7.1 Computing χ^2

Stata will automatically compute χ^2 when `sem` is used. However, I believe it is instructive to understand where it comes from and how χ^2 actually tests the null hypothesis that $S = \widehat{\Sigma}$. Consequently, in this section, we will learn how to manually compute χ^2 based on the one-factor model we have fit and what is known as a saturated model.

As a reminder, the null hypothesis, in words, is that the sample covariance matrix is equal to the predicted covariance matrix. To test this hypothesis, we compare two models:

11.7.1 Computing χ^2

1. The one-factor model represented in figure 11.2. This first model could be any theoretically relevant model.

2. A model that is known to perfectly reproduce the sample covariance matrix. Such a model is known as a saturated model. An example of a saturated model is figure 11.3.

Model 1 produces $\hat{\Sigma}$ and model 2 produces S. Recall that the logic of the χ^2 test is that our CFA model, if it fits the data well, ought to reproduce the covariances among the items. Thus, we want to compare the likelihood from the theoretical model and the saturated model. Likelihoods—or more precisely, log likelihoods—are produced by maximum likelihood estimation (which `sem` uses to produce parameter estimates). The details of maximum likelihood estimation are beyond the scope of this chapter. Suffice it to say that maximum likelihood estimation procedures use algorithms to determine what parameters (for example, factor loadings, error variance, means) were "most likely" to have produced the data, that is, what parameters were most likely to have produced the covariances among the items. We can think of the log likelihoods as a numerical index of how likely the parameters were (summed up across all observations). Other parameter estimates (for example, different factor loadings) would produce different log likelihoods.

We compare the log likelihoods from each model by using the following equation (StataCorp 2017i, 577):

$$\chi^2_{\text{mod}} = 2(\text{ll}_s - \text{ll}_m) \tag{11.7}$$

where ll_s is the log likelihood for the saturated model and ll_m is the log likelihood for the theoretically relevant model. Consider what the null hypothesis for the χ^2 implies about the likelihood of the theoretically relevant model and the saturated model. Specifically, the null hypothesis suggests that the likelihood is equal—both models are just as likely to have produced the data. If the null is true, then the χ^2 value from (11.7) should be small because the differences between the likelihoods would be close to 0. In contrast, if the theoretically relevant model produces parameters that are not as likely as the saturated model, then χ^2 will be relatively large.[8]

To compute χ^2 manually using `sem`, the first step is to compute the one-factor model and then save ll_m, which is automatically produced by `sem`.

```
. sem (TRAIT_ANX -> trait2 trait4 trait5 trait8
> trait9 trait15 trait20),
> variance(TRAIT_ANX@1) nolog
  (output omitted)
. scalar ll_m = e(ll)
```

---

8. You might be thinking, "What about the situation where the theoretically relevant model is more likely than the saturated model?" That cannot happen. The saturated model perfectly reproduces the data. The only possibilities are a) the theoretical model is equivalent in fit to the saturated model or b) the theoretical model has worse fit than the saturated model. Although it sounds cool, you cannot actually obtain a fit that is better than the data.

Next we compute a saturated model, which will perfectly reproduce the sample covariances. Thus, we need a model that will estimate all item variances and all possible covariances among items. The simplest way to do this is to use the following sem code:[9]

```
. sem ( <- trait2 trait4 trait5 trait8
> trait9 trait15 trait20), nolog
Exogenous variables
Observed:   trait2 trait4 trait5 trait8 trait9 trait15 trait20
Structural equation model                    Number of obs     =        790
Estimation method  = ml
Log likelihood     = -5744.0911
```

	Coef.	OIM Std. Err.	z	P>|z|	[95% Conf. Interval]	
mean(trait2)	1.887342	.0268215	70.37	0.000	1.834773	1.939911
mean(trait4)	2.029114	.0347198	58.44	0.000	1.961064	2.097163
mean(trait5)	1.555696	.0263416	59.06	0.000	1.504068	1.607325
mean(trait8)	1.858228	.0295271	62.93	0.000	1.800356	1.9161
mean(trait9)	2.070886	.0317727	65.18	0.000	2.008613	2.133159
mean(trait15)	1.803797	.0282079	63.95	0.000	1.748511	1.859084
mean(trait20)	1.901266	.0314755	60.40	0.000	1.839575	1.962957
var(trait2)	.5683208	.0285953			.5149499	.6272232
var(trait4)	.9523169	.0479163			.862885	1.051018
var(trait5)	.5481638	.0275811			.4966858	.6049771
var(trait8)	.6887614	.0346553			.6240799	.7601467
var(trait9)	.7975068	.0401269			.7226131	.8801628
var(trait15)	.6285932	.031628			.5695621	.6937424
var(trait20)	.7826566	.0393797			.7091574	.8637735

---

9. See https://stats.idre.ucla.edu/stata/faq/what-are-the-saturated-and-baseline-models-in-sem/.

11.7.1 Computing χ^2

cov(trait2, trait4)	.3349255	.0287591	11.65	0.000	.2785587	.3912922
cov(trait2, trait5)	.267667	.0220235	12.15	0.000	.2245017	.3108324
cov(trait2, trait8)	.3042814	.0247526	12.29	0.000	.2557672	.3527955
cov(trait2, trait9)	.3016568	.026247	11.49	0.000	.2502135	.3531
cov(trait2, trait15)	.2753645	.0234134	11.76	0.000	.2294751	.321254
cov(trait2, trait20)	.3357122	.026565	12.64	0.000	.2836458	.3877787
cov(trait4, trait5)	.4217962	.0297657	14.17	0.000	.3634565	.4801359
cov(trait4, trait8)	.4015959	.0321625	12.49	0.000	.3385585	.4646333
cov(trait4, trait9)	.3941388	.0340295	11.58	0.000	.3274422	.4608353
cov(trait4, trait15)	.3955856	.0309165	12.80	0.000	.3349903	.456181
cov(trait4, trait20)	.3889505	.0336892	11.55	0.000	.3229209	.4549801
cov(trait5, trait8)	.3344784	.0248904	13.44	0.000	.2856942	.3832627
cov(trait5, trait9)	.2517481	.0251713	10.00	0.000	.2024132	.3010831
cov(trait5, trait15)	.3837125	.0249508	15.38	0.000	.3348099	.4326152
cov(trait5, trait20)	.2928409	.0255268	11.47	0.000	.2428092	.3428726
cov(trait8, trait9)	.3492902	.0291503	11.98	0.000	.2921566	.4064238
cov(trait8, trait15)	.3088928	.0258615	11.94	0.000	.2582052	.3595804
cov(trait8, trait20)	.3897997	.0295752	13.18	0.000	.3318334	.4477661
cov(trait9, trait15)	.2683384	.0269391	9.96	0.000	.2155388	.321138
cov(trait9, trait20)	.3715558	.031062	11.96	0.000	.3106755	.4324362
cov(trait15, trait20)	.3135395	.0273348	11.47	0.000	.2599643	.3671147

LR test of model vs. saturated: chi2(0) =   0.00, Prob > chi2 =   .

This is a saturated model. We can verify that the model reproduces the sample covariance matrix by using the `estat framework, fitted` postestimation command.

```
. estat framework, fitted
```
(*output omitted*)

Sigma	observed trait2	trait4	trait5	trait8	trait9
observed					
trait2	.5683208				
trait4	.3349255	.9523169			
trait5	.267667	.4217962	.5481638		
trait8	.3042814	.4015959	.3344784	.6887614	
trait9	.3016568	.3941388	.2517481	.3492902	.7975068
trait15	.2753645	.3955856	.3837125	.3088928	.2683384
trait20	.3357122	.3889505	.2928409	.3897997	.3715558

Sigma	observed trait15	trait20
observed		
trait15	.6285932	
trait20	.3135395	.7826566

(*output omitted*)

Given that the saturated model reproduces the sample covariance matrix, it provides an excellent baseline against which to compare the predicted covariance matrix. We make this comparison using (11.7). We first save ll_s from the saturated model and then make the comparison.

```
. scalar ll_s = e(ll)
. display "chi2 =" 2*(ll_s - ll_m)
chi2 =130.86715
```

This is the same value produced by `sem`.

11.8 Obtaining σ_C^2 and σ_U^2

With the CFA estimated and an understanding of model fit, we can now estimate reliability. The primary task is to use the CFA model to obtain estimates of σ_C^2 and σ_U^2. As discussed above, the factor loadings are regression coefficients relating the latent variable to the indicator. In regression, we computed R^2, which told us how much variability in y was accounted for by x (or multiple x's). We can do the same thing with CFA. I will illustrate these ideas at the item level and then generalize the concepts to the scale level so that we can estimate reliability.

11.8.1 Computing R^2 for an item

R^2 for an item in CFA tells us how much variability in an observed item is associated with the common factor. To obtain R^2, we create the ratio of the variance shared between the latent variable and the item over the total variance in the item.[10] Thus, for generic item i, R^2 is [Bollen 1989, 220, (6.48)]

$$R_i^2 = \frac{(\lambda_i)^2 \phi_{11}}{(\lambda_i)^2 \phi_{11} + \theta_{ii}}$$

where λ_i is the factor loading, ϕ_{11} is the variance of the latent variable, and θ_{ii} is the error variance for item i. This formula indicates that squaring a factor loading and multiplying the square by the latent variable variance provides the shared variance between an item and the latent variable. Further, an item's total variance is equal to the squared factor loading multiplied by the latent variable variance plus the residual variance.

The factor loading for `trait2` is $\lambda_2 = 0.504$, the latent variable variance is $\phi_{11} = 1$, and the residual variance is $\theta_{22} = 0.315$. Thus, R^2 is

```
. display (.504^2)*1/((.504^2)*1 + .315)
.44641275
```

So 45% of the variance in `trait2` is associated with the latent factor. Stata will print R^2 values for all items with the `estat eqgof` postestimation command.

```
. estat eqgof
Equation-level goodness of fit
```

depvars	fitted	Variance predicted	residual	R-squared	mc	mc2
observed						
trait2	.5683208	.2535188	.314802	.446084	.6678952	.446084
trait4	.9523169	.4860531	.4662638	.5103901	.7144159	.5103901
trait5	.5481638	.3226487	.2255151	.5885991	.7672021	.5885991
trait8	.6887614	.3550426	.3337188	.5154798	.7179692	.5154798
trait9	.7975068	.2883737	.5091331	.3615941	.6013269	.3615941
trait15	.6285932	.3255781	.3030151	.5179472	.7196855	.5179472
trait20	.7826566	.3489491	.4337075	.4458521	.6677216	.4458521
overall				.8715944		

```
mc  = correlation between depvar and its prediction
mc2 = mc^2 is the Bentler-Raykov squared multiple correlation coefficient
```

R^2 values range between 36% for `trait9` ("I worry too much over something that really doesn't matter") and 59% for `trait5` ("I feel like a failure"). The slight differences

---

10. Recall that we computed R^2 in regression as SS_{Reg}—the shared variance between the outcome and predictors—divided by SS_{Total}—the total variance of the outcome. The concept is the same here—shared variance over total variance.

between the `display` output and `estat eqgof` are due to 1) rounding error and 2) the difference between how many digits past the decimal point were used in the `display` and how many digits past the decimal point were used internally by Stata in the `estat eqgof` command.

In the output of `estat eqgof`, the `Variance` section includes three columns: `fitted`, `predicted`, and `residual`. As we have seen, $\lambda_i^2 \phi_{11}$ produces the `predicted` column and θ_{ii} is the `residual` column. The `fitted` column is the actual variances of the items in the data. However, these variances differ slightly from the sample-based estimates.

```
. tabstat trait2-trait20, stat(var)
    stats |    trait2     trait4     trait5     trait8     trait9    trait15    trait20
 variance |  .5690411   .9535239   .5488585   .6896344   .7985176   .6293899   .7836486
```

The values produced by `estat eqgof` and `tabstat` are fairly close. Indeed, the differences will nearly always be unimportant. However, this difference is due to the fact that `sem` computes the asymptotic variance for each item, whereas `tabstat` or `summarize` compute the sample-based variance. The sample-based variance is

$$\sigma_X^2 = \frac{\sum (X_i - \overline{X})^2}{N - 1}$$

In contrast, the asymptotic variance for each item is

$$\sigma_X^2 = \frac{\sum (X_i - \overline{X})^2}{N}$$

We can change to sample-based values in `sem` by using the `nm1` option.

```
. sem (TRAIT_ANX -> trait2 trait4 trait5 trait8
> trait9 trait15 trait20),
> var(TRAIT_ANX@1) nolog nocnsreport nm1
Endogenous variables

Measurement:  trait2 trait4 trait5 trait8 trait9 trait15 trait20

Exogenous variables

Latent:       TRAIT_ANX
Structural equation model                       Number of obs   =        790
Estimation method  = ml
Log likelihood     = -5813.0269
```

11.8.1 Computing R^2 for an item

	Coef.	OIM Std. Err.	z	P>|z|	[95% Conf. Interval]	
Measurement						
trait2						
TRAIT_ANX	.5038255	.0252002	19.99	0.000	.4544341	.5532169
_cons	1.887342	.0268385	70.32	0.000	1.834739	1.939944
trait4						
TRAIT_ANX	.6976168	.0318523	21.90	0.000	.6351873	.7600462
_cons	2.029114	.0347418	58.41	0.000	1.961021	2.097207
trait5						
TRAIT_ANX	.5683816	.023681	24.00	0.000	.5219677	.6147955
_cons	1.555696	.0263583	59.02	0.000	1.504035	1.607357
trait8						
TRAIT_ANX	.596232	.0270939	22.01	0.000	.5431288	.6493351
_cons	1.858228	.0295458	62.89	0.000	1.800319	1.916137
trait9						
TRAIT_ANX	.5373446	.0307963	17.45	0.000	.476985	.5977042
_cons	2.070886	.0317928	65.14	0.000	2.008573	2.133199
trait15						
TRAIT_ANX	.570956	.0259533	22.00	0.000	.5200884	.6218236
_cons	1.803797	.0282258	63.91	0.000	1.748476	1.859119
trait20						
TRAIT_ANX	.5910934	.0296754	19.92	0.000	.5329307	.649256
_cons	1.901266	.0314954	60.37	0.000	1.839536	1.962996
var(e.trait2)	.315201	.0181131			.2816262	.3527784
var(e.trait4)	.4668548	.0278713			.4153029	.5248059
var(e.trait5)	.2258009	.0148384			.1985131	.2568397
var(e.trait8)	.3341418	.0201277			.2969321	.3760145
var(e.trait9)	.5097784	.0282301			.4573454	.5682226
var(e.trait15)	.3033991	.0184806			.2692564	.3418712
var(e.trait20)	.4342572	.0251171			.3877164	.4863847
var(TRAIT_ANX)	1	(constrained)				

LR test of model vs. saturated: chi2(14) = 130.87, Prob > chi2 = 0.0000

Now everything should agree. As noted above, these distinctions will not typically matter but can be a source of confusion. The Stata reference manual entry on sem estimation options (StataCorp 2017i) says it well:

> nm1 specifies that the variances and covariances used in the SEM equations be the sample variances (divided by $N-1$) and not the asymptotic variances (divided by N). This is a minor technical issue of little importance unless you are trying to match results from other software that assumes sample variances. sem assumes asymptotic variances. (p. 616)

11.8.2 Computing σ_C^2 and σ_U^2 for all items

When computing reliability, we are interested in how much variability is shared between the latent variable (what is common between the items) and all the items rather than one at a time. Fortunately, we can add up factor loadings between each item and the latent variable and square the sum to get a total. Indeed, the total shared variance between the latent variable and all items—σ_C^2—is the sum of all the factor loadings squared multiplied by the factor variance:

$$\sigma_C^2 = \left(\sum \lambda_i\right)^2 \phi_{11} \qquad (11.8)$$

For the trait anxiety items, σ_C^2 is

$$\sigma_C^2 = (\lambda_2 + \lambda_4 + \lambda_5 + \lambda_8 + \lambda_9 + \lambda_{15} + \lambda_{20})^2 \phi_{11}$$

or

$$16.51 = (0.503507 + 0.697175 + 0.568022 + 0.595855 + 0.537004 + 0.570595 + 0.590719)^2 \times 1$$

We can compute this value after `sem` using the `nlcom` postestimation command, where `nlcom` stands for nonlinear combination of parameters. Nonlinear means that the parameters are combined in a nonlinear way. In this case, we are squaring and multiplying parameters.[11]

```
. nlcom (sig2_common:(_b[trait2:TRAIT_ANX] + _b[trait4:TRAIT_ANX] +
> _b[trait5:TRAIT_ANX] + _b[trait8:TRAIT_ANX] +
> _b[trait9:TRAIT_ANX] + _b[trait15:TRAIT_ANX] +
> _b[trait20:TRAIT_ANX])^2*_b[/var(TRAIT_ANX)]), noheader
```

	Coef.	Std. Err.	z	P>|z|	[95% Conf. Interval]	
sig2_common	16.50696	.9627055	17.15	0.000	14.62009	18.39383

---

11. In contrast, linear combinations of parameters, such as adding and subtracting parameters, use `lincom` (see chapters 9 and 10 for examples).

11.8.2 Computing σ_C^2 and σ_U^2 for all items

The names for the factor loadings can be obtained using the `coeflegend` option for `sem`.

```
. sem (TRAIT_ANX -> trait2 trait4 trait5 trait8
> trait9 trait15 trait20),
> var(TRAIT_ANX@1) nolog nocnsreport coeflegend
Endogenous variables
Measurement:  trait2 trait4 trait5 trait8 trait9 trait15 trait20
Exogenous variables
Latent:       TRAIT_ANX
Structural equation model                    Number of obs    =       790
Estimation method  = ml
Log likelihood     = -5809.5247
```

	Coef.	Legend
Measurement		
trait2		
TRAIT_ANX	.5035065	_b[trait2:TRAIT_ANX]
_cons	1.887342	_b[trait2:_cons]
trait4		
TRAIT_ANX	.6971751	_b[trait4:TRAIT_ANX]
_cons	2.029114	_b[trait4:_cons]
trait5		
TRAIT_ANX	.5680217	_b[trait5:TRAIT_ANX]
_cons	1.555696	_b[trait5:_cons]
trait8		
TRAIT_ANX	.5958545	_b[trait8:TRAIT_ANX]
_cons	1.858228	_b[trait8:_cons]
trait9		
TRAIT_ANX	.5370044	_b[trait9:TRAIT_ANX]
_cons	2.070886	_b[trait9:_cons]
trait15		
TRAIT_ANX	.5705945	_b[trait15:TRAIT_ANX]
_cons	1.803797	_b[trait15:_cons]
trait20		
TRAIT_ANX	.5907191	_b[trait20:TRAIT_ANX]
_cons	1.901266	_b[trait20:_cons]
var(e.trait2)	.314802	_b[/var(e.trait2)]
var(e.trait4)	.4662638	_b[/var(e.trait4)]
var(e.trait5)	.2255151	_b[/var(e.trait5)]
var(e.trait8)	.3337188	_b[/var(e.trait8)]
var(e.trait9)	.5091331	_b[/var(e.trait9)]
var(e.trait15)	.3030151	_b[/var(e.trait15)]
var(e.trait20)	.4337075	_b[/var(e.trait20)]
var(TRAIT_ANX)	1	_b[/var(TRAIT_ANX)]

```
LR test of model vs. saturated: chi2(14) =    130.87, Prob > chi2 = 0.0000
```

The total variance for the items is

$$\sigma^2_{\text{Total}} = \sigma^2_{\text{Common}} + \sigma^2_{\text{Unique}}$$

where σ^2_C is as defined above and σ^2_U is the sum of the residual variances:

$$\sigma^2_U = \sum \theta_{ii} \qquad (11.9)$$

In the trait anxiety example, σ^2_U is

$$\sigma^2_U = \theta_2 + \theta_4 + \theta_5 + \theta_8 + \theta_9 + \theta_{15} + \theta_{20}$$

Or

$$2.59 = 0.314802 + 0.466264 + 0.225515 + 0.333719 + 0.509133 + 0.303015 + 0.433708$$

As before, compute σ^2_U by using `nlcom` and the parameter names from `coeflegend`.

```
. nlcom (sig2_unique:(_b[/var(e.trait2)] +
> _b[/var(e.trait4)] + _b[/var(e.trait5)] +
> _b[/var(e.trait8)] + _b[/var(e.trait9)] +
> _b[/var(e.trait15)] + _b[/var(e.trait20)])), noheader
```

	Coef.	Std. Err.	z	P>|z|	[95% Conf. Interval]	
sig2_unique	2.586155	.054891	47.11	0.000	2.478571	2.69374

11.8.3 Computing reliability—ω

With estimates of $\sigma^2_C = 16.51$ and $\sigma^2_U = 2.59$, we can estimate reliability (ω) using (11.2).

$$\omega = \frac{16.51}{16.51 + 2.59} = 0.86$$

This indicates that 86% of the total variance in items is shared between items. One could also say that 86% of the total variance of items is shared with the latent variable. Reliability is a matter of degree. In other words, no measure is perfectly reliable, and there is no threshold that separates reliable from unreliable measures. However, reliability of 0.86 would typically be deemed reliable.

We can generalize (11.2) by reexpressing it in terms of (11.8) and (11.9). Specifically, substituting (11.8) into the numerator of (11.2) and both (11.8) and (11.9) into the denominator of (11.2), we obtain a general form of ω expressed in the parameters of CFA (McDonald 1999; Raykov and Shrout 2002; Raykov and Grayson 2003).

$$\omega = \frac{\sum (\lambda_i)^2 \phi_{11}}{\sum (\lambda_i)^2 \phi_{11} + \sum \theta_{ii}} \qquad (11.10)$$

11.8.4 Bootstrapping the standard error and 95% confidence interval for ω

We can compute ω directly using `nlcom`.

```
. nlcom omega:(_b[trait2:TRAIT_ANX] + _b[trait4:TRAIT_ANX] +
> _b[trait5:TRAIT_ANX] + _b[trait8:TRAIT_ANX] +
> _b[trait9:TRAIT_ANX] + _b[trait15:TRAIT_ANX] +
> _b[trait20:TRAIT_ANX])^2 *_b[/var(TRAIT_ANX)] /
> ((_b[trait2:TRAIT_ANX] + _b[trait4:TRAIT_ANX] +
> _b[trait5:TRAIT_ANX] + _b[trait8:TRAIT_ANX] +
> _b[trait9:TRAIT_ANX] + _b[trait15:TRAIT_ANX] +
> _b[trait20:TRAIT_ANX])^2 *_b[/var(TRAIT_ANX)] + (_b[/var(e.trait2)] +
> _b[/var(e.trait4)] + _b[/var(e.trait5)] +
> _b[/var(e.trait8)] + _b[/var(e.trait9)] +
> _b[/var(e.trait15)] + _b[/var(e.trait20)])), noheader
```

	Coef.	Std. Err.	z	P>|z|	[95% Conf. Interval]	
omega	.8645504	.0073984	116.86	0.000	.8500499	.8790509

You might scoff at the `nlcom` code for ω, and you would be right to. It requires a fair amount of typing. However, computing ω and its standard error by hand is difficult. Fortunately, `nlcom` uses the delta method to compute the standard error of the reliability and will automatically produce the 95% confidence interval. In this case, the 95% confidence interval for ω is [0.85, 0.88].

We have a problem, however. As can be seen in (11.2) or (11.10), ω is a ratio and thus is bounded by 0 and 1. The delta method assumes the sampling distribution of ω is normally distributed (see [R] **nlcom** of the *Stata Base Reference Manual*). Given the boundaries on a ratio, some may not be willing to make a normal distribution assumption. An alternative is to use bootstrapping because bootstrapping does not make the distribution assumptions of the delta method (Raykov 1998).

11.8.4 Bootstrapping the standard error and 95% confidence interval for ω

Bootstrapping is the statistical process of resampling the data so that we can examine the precision of our estimates. Bootstrapping allows us to construct confidence intervals without relying on the normal distribution assumption. Specifically, we resample the data to mimic a sampling distribution and use the simulated sampling distribution to produce the upper and lower limits of the confidence intervals. Stata has built-in facilities for bootstrapping. We must write a program that Stata will use to conduct the bootstrap (that is, the resampling of the data). In this case, the body of the program will consist of three parts:

1. The `sem` code for estimating the CFA.
2. The `nlcom` code for computing ω.
3. A return code for saving the computed value of ω from `nlcom`.

I have written a program called `relboot` (which stands for "reliability bootstrap") that does what we need.

```
. program relboot, rclass
  1. version 15.1
  2. sem (TRAIT_ANX -> trait2* trait4 trait5 trait8
> trait9 trait15 trait20),
> var(TRAIT_ANX@1) nolog nocnsreport
  3. nlcom (_b[trait2:TRAIT_ANX] + _b[trait4:TRAIT_ANX] +
> _b[trait5:TRAIT_ANX] + _b[trait8:TRAIT_ANX] +
> _b[trait9:TRAIT_ANX] + _b[trait15:TRAIT_ANX] +
> _b[trait20:TRAIT_ANX])^2 *_b[/var(TRAIT_ANX)] /
> ((_b[trait2:TRAIT_ANX] + _b[trait4:TRAIT_ANX] +
> _b[trait5:TRAIT_ANX] + _b[trait8:TRAIT_ANX] +
> _b[trait9:TRAIT_ANX] + _b[trait15:TRAIT_ANX] +
> _b[trait20:TRAIT_ANX])^2 *_b[/var(TRAIT_ANX)] + (_b[/var(e.trait2)] +
> _b[/var(e.trait4)] + _b[/var(e.trait5)] +
> _b[/var(e.trait8)] + _b[/var(e.trait9)] +
> _b[/var(e.trait15)] + _b[/var(e.trait20)]))
  4. return scalar reliab = el(r(b), 1, 1)
  5. end
```

The first line is the `sem` code, the second line is the `nlcom` code, and the last line returns the estimated reliability.

To run `relboot`, use the `bootstrap` command.

```
. bootstrap r(reliab), reps(1000) seed(45639) nodots: relboot
Bootstrap results                               Number of obs     =        790
                                                Replications      =      1,000

      command:  relboot
        _bs_1:  r(reliab)
```

	Observed Coef.	Bootstrap Std. Err.	z	P>|z|	Normal-based [95% Conf. Interval]	
_bs_1	.8645504	.0089599	96.49	0.000	.8469893	.8821114

```
. estat bootstrap, percentile
Bootstrap results                               Number of obs     =        790
                                                Replications      =       1000

      command:  relboot
        _bs_1:  r(reliab)
```

	Observed Coef.	Bias	Bootstrap Std. Err.	[95% Conf. Interval]	
_bs_1	.86455037	-.0004437	.00895989	.8465877	.8804841 (P)

(P)   percentile confidence interval

We used `r(reliab)` to tell Stata that we want to save the reliability estimate from `relboot`. This was line 3 from the previous code. The option `reps(1000)` tells Stata to compute 1,000 bootstrapped samples. There isn't clear guidance on how many bootstrap samples are needed in a given situation. More samples will produce more precise estimates. (Mooney and Duval 1993, 37) recommend at least 1,000 samples if percentile-based confidence intervals are used.[12] As seen from our results, 1,000 replications is plenty for the trait anxiety example. Finally, after the `reps` options, we typed a colon and then the program we want to bootstrap, `relboot`.

After `bootstrap` has run, we can use the `estat bootstrap` postestimation command with the `percentile` option to obtain the bootstrapped 95% confidence interval. In this instance, the bootstrapped confidence intervals do not differ much from the delta method confidence intervals. This will be the case when sample sizes are large.

11.9 Comparing ω with α

Although it has not been explicitly stated, ω as we have used it is a measure of internal consistency reliability. A measure is internally consistent when the items contributing to the overall score (or subtest score) are similar. From the perspective of CFA, internal consistency will be reflected in consistent factor loadings across items and similar error variances. Table 11.1 illustrates the impact of including items in a measure that are not internally consistent. The table includes seven items, each with an error variance of $\theta = 0.5$. Columns 3–5 provide factor loadings. All loadings are consistent in column 3, whereas columns 4 and 5 include one or two weak loadings, respectively. The table also includes $(\sum \lambda_i)^2$ to demonstrate how the numerator of (11.10) decreases as the number of weak loadings increases. Finally, the table provides ω to illustrate the impact of lack of internal consistency on reliability: as the number of items with relatively weak loadings increases, reliability decreases. That is, a measure will not be internally consistent as the number of items with relatively weak loadings increases.

Table 11.1. The impact of weak factor loadings on internal consistency reliability

Item	Error variance	All items strong loading	One weak loading	Two weak loadings
1	0.5	0.7	0.7	0.7
2	0.5	0.6	0.6	0.6
3	0.5	0.5	0.5	0.5
4	0.5	0.7	0.1	0.1
5	0.5	0.6	0.6	0.6
6	0.5	0.7	0.7	0.1
7	0.5	0.6	0.6	0.6
	$\sum \theta_{ii} = 3.5$	$(\sum \lambda_i)^2 = 19.36$ $\omega = 0.85$	$(\sum \lambda_i)^2 = 14.44$ $\omega = 0.80$	$(\sum \lambda_i)^2 = 10.24$ $\omega = 0.75$

---

12. For a formal way of selecting the number of replications, see Poi (2004).

It turns out, ω is not the only measure of internal consistency. A commonly used measure of internal consistency is α.[13] Indeed, it would be uncommon to see a psychology study that uses multi-item measures that does not report or reference α in some way. α is a special case of ω, which is a fancy way of saying that α and ω will be equal if a certain condition is met. The condition is known as tau-equivalence. If tau-equivalence is not met, then α will be the lower bound of ω (McDonald 1999).

A set of items is tau-equivalent if all the factor loadings are equal to one another even if the error variances are unique to items. That is, if all items have the same relationship with the latent variable, then the items are tau-equivalent. Up to this point, we have fit what is called a congeneric model, which means that all factor loadings and error variances are unique (Bollen 1989). The congeneric model is written as follows. A CFA model for item j for person i is

$$X_{ij} = \lambda_i \xi_j + \delta_{ij}$$

X and δ are indexed with i and j, indicating that the X value and the residual are unique to both the person and the item. In contrast, ξ is indexed only with j, indicating that the value of the latent variable is unique to each person. Finally, and most importantly for the congeneric model, λ is indexed only with i, indicating that there is a unique factor loading for each of the items.

The tau-equivalent model involves one major change to the congeneric model. Specifically, λ no longer includes a subscript, indicating that all items share a common factor loading.

$$X_{ij} = \lambda \xi_j + \delta_{ij}$$

CFA can be used to test the assumption of tau-equivalence. Specifically, we use **sem** to fit both the congeneric and the tau-equivalent models, and then we use a likelihood-ratio (LR) or χ^2-difference test to examine whether the congeneric model fits the data better. The congeneric model is what we have been fitting all along. The tau-equivalent model can be fit with **sem**:

---

13. α is sometimes called Cronbach's α after Lee Cronbach. Although Cronbach wrote a lot about α (Cronbach 1951) and measurement generally, he did not create α. That distinction goes to Louis Guttman (Guttman 1945), leading some authors to call α the Guttman–Cronbach α (McDonald 1999).

11.9 Comparing ω with α

```
. sem (TRAIT_ANX -> trait2@l1 trait4@l1 trait5@l1 trait8@l1
> trait9@l1 trait15@l1 trait20@l1),
> var(TRAIT_ANX@1)
> nolog nocnsreport
Endogenous variables
Measurement:  trait2 trait4 trait5 trait8 trait9 trait15 trait20
Exogenous variables
Latent:       TRAIT_ANX
Structural equation model                       Number of obs    =       790
Estimation method  = ml
Log likelihood     = -5826.5778
```

	Coef.	OIM Std. Err.	z	P>|z|	[95% Conf. Interval]	
Measurement						
trait2						
TRAIT_ANX	.5740302	.0166589	34.46	0.000	.5413794	.6066809
_cons	1.887342	.0283493	66.57	0.000	1.831778	1.942905
trait4						
TRAIT_ANX	.5740302	.0166589	34.46	0.000	.5413794	.6066809
_cons	2.029114	.0325329	62.37	0.000	1.965351	2.092877
trait5						
TRAIT_ANX	.5740302	.0166589	34.46	0.000	.5413794	.6066809
_cons	1.555696	.0265574	58.58	0.000	1.503645	1.607748
trait8						
TRAIT_ANX	.5740302	.0166589	34.46	0.000	.5413794	.6066809
_cons	1.858228	.0290483	63.97	0.000	1.801294	1.915161
trait9						
TRAIT_ANX	.5740302	.0166589	34.46	0.000	.5413794	.6066809
_cons	2.070886	.0324915	63.74	0.000	2.007204	2.134568
trait15						
TRAIT_ANX	.5740302	.0166589	34.46	0.000	.5413794	.6066809
_cons	1.803797	.0282837	63.78	0.000	1.748363	1.859232
trait20						
TRAIT_ANX	.5740302	.0166589	34.46	0.000	.5413794	.6066809
_cons	1.901266	.0310713	61.19	0.000	1.840367	1.962164
var(e.trait2)	.305401	.0179937			.272094	.342785
var(e.trait4)	.5066179	.0279173			.4547523	.5643989
var(e.trait5)	.2276739	.0143271			.2012559	.2575597
var(e.trait8)	.3370943	.0195116			.3009417	.3775899
var(e.trait9)	.5044899	.0278797			.4527019	.5622023
var(e.trait15)	.3024618	.0178963			.2693432	.3396527
var(e.trait20)	.4331738	.0243152			.3880449	.4835511
var(TRAIT_ANX)	1	(constrained)				

```
LR test of model vs. saturated: chi2(20)  =    164.97, Prob > chi2 = 0.0000
```

The only difference between the code for the congeneric and tau-equivalent models is that the tau-equivalent model includes constraints for the factor loadings. We have added a @l1 after each of the item names. This tells Stata to constrain the loadings for all items to be equal to one another. This constraint does not fix the loadings to a specific value—sem will still estimate a value for the common loading. If the constraint uses letters first, then sem will treat the constraint as an unknown and estimate the value. If the constraint uses a number, then sem will treat the constraint as a known and fix the value to the number.

Computing α

The most common method for computing α is the following formula:

$$\alpha = \frac{k}{k-1}\left(1 - \frac{\sum \sigma_i^2}{\sigma_y^2}\right)$$

where k is the number of items in the scale, σ_i^2 is the variance for the ith item, and σ_y^2 is the variance of the total scale. In Stata, α is estimated using the `alpha` command.

```
. alpha trait2-trait20
Test scale = mean(unstandardized items)
Average interitem covariance:     .3378501
Number of items in the scale:            7
Scale reliability coefficient:      0.8639
```

11.9.1 Evaluating the assumption of tau-equivalence

If items do not meet the assumption of tau-equivalence, α will underestimate reliability. Often, the amount of underestimation will be small, as in our example. Raykov (1997, 347) noted that "α's slippage as a measure of scale reliability is a complex function of test length, error variances, congeneric latent variance, and deviations of individual items from essential τ-equivalence (equality among their units of measurement)". In my experience, the difference between α and ω is not typically large. However, testing the assumption of tau-equivalence and estimating both α and ω is straightforward.

When comparing the congeneric and tau-equivalent models, the question is whether constraining the factor loadings in the tau-equivalent model significantly degrades the model fit. If it does, then the assumption of tau-equivalence does not hold. We can use an LR or χ^2-difference test to compare the overall fit of the two models. These tests are valid because the congeneric and tau-equivalent models are nested models. Specifically, the tau-equivalent model is nested within the congeneric model. A model is nested within a parent model if a) the nested model includes a subset of the free parameters of the parent model and b) the exact same dataset was used to fit both models (Brown 2015). The parent model is the congeneric model and the nested model is the tau-equivalent model, and both conditions are met in our situation.

11.9.1 Evaluating the assumption of tau-equivalence

The null hypothesis for the χ^2-difference test is that the congeneric and tau-equivalent models have identical fit. In other words, the difference between the χ^2 values for each model is 0. To perform a χ^2-difference test, we compute the difference between the χ^2 values for each model.

$$\chi^2_D = \chi^2_N - \chi^2_P \tag{11.11}$$

Where χ^2_N is for the nested model (that is, tau-equivalent), χ^2_P is for the parent model (that is, congeneric), and χ^2_D is the difference. The degrees of freedom (DF) for the test is equal to the difference between the DF for the nested and parent models.

$$\text{DF}_D = \text{DF}_N - \text{DF}_P \tag{11.12}$$

Table 11.2 provides the χ^2 and DF for each model as well as their difference. The difference in fit is statistically significant, $\chi^2_D(6) = 34.1, p < 0.01$. Thus, we reject the null hypothesis and conclude that these items are not tau-equivalent.

Table 11.2. χ^2-difference test for the trait anxiety items

Model	χ^2	DF	
Tau-equivalent	164.97	20	
Congeneric	130.87	14	
Difference	34.1	6	$p < 0.01$

We can perform the χ^2-difference test directly in Stata with the following code.

```
. quietly sem (TRAIT_ANX -> trait2@l1 trait4@l1 trait5@l1 trait8@l1
> trait9@l1 trait15@l1 trait20@l1),
> var(TRAIT_ANX@1)
> nolog nocnsreport
. estimates store tauequiv
. scalar nested_chi2 = e(chi2_ms)
. scalar nested_df = e(df_ms)
. quietly sem (TRAIT_ANX -> trait2 trait4 trait5 trait8
> trait9 trait15 trait20),
> var(TRAIT_ANX@1) nolog
. estimates store congeneric
. scalar parent_chi2 = e(chi2_ms)
. scalar parent_df = e(df_ms)
. display nested_chi2
164.97338
. display nested_df
20
. display parent_chi2
130.86715
. display parent_df
14
. display 1 - chi2(nested_df - parent_df, nested_chi2 - parent_chi2)
6.417e-06
```

We fit both the tau-equivalent and congeneric models. After each model, we store the χ^2 value and associated DF for each model as the scalars `nested_chi2`, `nested_df`, `parent_chi2`, and `parent_df`. Finally, we compute the associated p-value by using the χ^2 command (as we have done in other chapters), which has two arguments: the DF and the χ^2 value. In this example, the p-value is tiny, clearly less than 0.01.

I have presented the χ^2-difference test because that is the most commonly presented method for comparing nested models in SEM textbooks (Brown 2015; Kline 2011). In Stata, the simplest way to compare nested models is to use the `lrtest` command to perform an LR test. The χ^2-difference test and the LR test are fundamentally the same thing because the χ^2 for a model is a function of the log likelihoods (see above). Thus, there is no reason to prefer one method over the other except tradition.

The LR test requires nested models just like the χ^2-difference test. The LR test involves the difference between the likelihoods for each of the models.

$$\text{LR}_D = -2(\text{ll}_N - \text{ll}_P)$$

where ll_N is the negative log likelihood for the nested model (printed near the top of the `sem` output), ll_P is the negative log likelihood for the parent model, and LR_D is the difference. LR_D is distributed as χ^2, with DF equal to the difference in the number of parameters between the nested and parent models [see (11.12)]. The null hypothesis for the LR test is the same as for the χ^2-difference test.

To perform the LR test in Stata, we store the estimates from the tau-equivalent and congeneric models with the `estimates store` command (see above). Then we can use the `lrtest` command to perform the test.

```
. lrtest tauequiv congeneric
Likelihood-ratio test                                   LR chi2(6) =      34.11
(Assumption: tauequiv nested in congeneric)             Prob > chi2 =    0.0000
```

Thus, LR_D is $\chi^2(6) = 34.11, p < 0.01$, and we reject the null hypothesis that the tau-equivalent and congeneric models have equal fit. Note that LR_D and χ^2_D are identical and produce identical results. This is simply evidence that the LR and χ^2-difference tests are equivalent.

11.9.2 Parallel items

Another concept in the reliability literature is the notion of parallel items (McDonald 1999). Parallel items, like tau-equivalent items, have a common factor loading. However, parallel items also have a common error variance. Parallel items are central to the definition of key concepts in classical test theory (McDonald 1999). Although a discussion of classical test theory is beyond the scope of the chapter, I will discuss evaluating the assumption of parallel items because it is a direct extension of the CFA models we have fit.

To fit a parallel-items CFA model, we constrain the loadings and the error variances to be equal across items.

```
. sem (TRAIT_ANX -> trait2@l1 trait4@l1 trait5@l1 trait8@l1
> trait9@l1 trait15@l1 trait20@l1),
> var(TRAIT_ANX@1)
> covstructure(e.trait2 e.trait4 e.trait5 e.trait8
> e.trait9 e.trait15 e.trait20, identity)
> nolog nocnsreport
Endogenous variables

Measurement:  trait2 trait4 trait5 trait8 trait9 trait15 trait20

Exogenous variables

Latent:       TRAIT_ANX

Structural equation model                       Number of obs    =        790
Estimation method  = ml
Log likelihood     = -5900.7372
```

	Coef.	OIM Std. Err.	z	P>|z|	[95% Conf.	Interval]
Measurement						
trait2						
TRAIT_ANX	.5808807	.0169416	34.29	0.000	.5476757	.6140857
_cons	1.887342	.0299678	62.98	0.000	1.828606	1.946078
trait4						
TRAIT_ANX	.5808807	.0169416	34.29	0.000	.5476757	.6140857
_cons	2.029114	.0299678	67.71	0.000	1.970378	2.08785
trait5						
TRAIT_ANX	.5808807	.0169416	34.29	0.000	.5476757	.6140857
_cons	1.555696	.0299678	51.91	0.000	1.49696	1.614432

trait8						
TRAIT_ANX	.5808807	.0169416	34.29	0.000	.5476757	.6140857
_cons	1.858228	.0299678	62.01	0.000	1.799492	1.916964
trait9						
TRAIT_ANX	.5808807	.0169416	34.29	0.000	.5476757	.6140857
_cons	2.070886	.0299678	69.10	0.000	2.01215	2.129622
trait15						
TRAIT_ANX	.5808807	.0169416	34.29	0.000	.5476757	.6140857
_cons	1.803797	.0299678	60.19	0.000	1.745062	1.862533
trait20						
TRAIT_ANX	.5808807	.0169416	34.29	0.000	.5476757	.6140857
_cons	1.901266	.0299678	63.44	0.000	1.84253	1.960002
var(e.trait2)	.3720518	.0076424			.3573705	.3873362
var(e.trait4)	.3720518	.0076424			.3573705	.3873362
var(e.trait5)	.3720518	.0076424			.3573705	.3873362
var(e.trait8)	.3720518	.0076424			.3573705	.3873362
var(e.trait9)	.3720518	.0076424			.3573705	.3873362
var(e.trait15)	.3720518	.0076424			.3573705	.3873362
var(e.trait20)	.3720518	.0076424			.3573705	.3873362
var(TRAIT_ANX)	1	(constrained)				

LR test of model vs. saturated: chi2(26) =   313.29, Prob > chi2 = 0.0000

```
. estimates store parallel
```

The only change from the tau-equivalent model is the addition of the `covstructure` option, which allows us to introduce constraints on the residual errors. In this case, we used the `identity` structure, which constrains all the error terms listed to be identical.

The parallel model is nested within the tau-equivalent model, and we can test whether the parallel model fits as well as the tau-equivalent model by using the LR test. The null hypothesis is that the parallel model and the tau-equivalent model have identical fit.

```
. lrtest tauequiv parallel
Likelihood-ratio test                              LR chi2(6)  =    148.32
(Assumption: parallel nested in tauequiv)          Prob > chi2 =    0.0000
```

We reject the null hypothesis and conclude that constraining the residual error significantly degrades model fit. Taken together, the results of the LR test comparing the congeneric to the tau-equivalent and the tau-equivalent to the parallel model suggest that the congeneric model fits the trait anxiety data the best.

11.10   Correlated residuals

Although we have not made it explicit, all the CFA models we have estimated at this point have assumed the residual errors are uncorrelated. Figure 11.3 shows the unstructured relationships among the seven trait anxiety items. Further, figure 11.2 shows that the CFA model assumes that all the relationships between the items can be accounted

11.10 Correlated residuals

for by the TRAIT_ANX latent variable. However, what if the items have a relationship beyond the latent variable? For example, item 5 is "I feel like a failure" and item 15 is "I feel inadequate". Given that both items require a person's judgment about their quality or success, it is possible that these two items will be correlated because of their semantic relationship. Figure 11.5 illustrates this relationship.

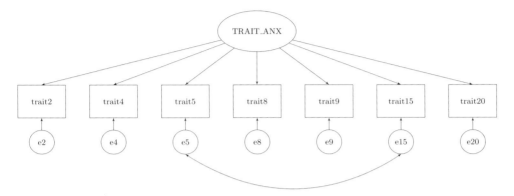

Figure 11.5. One-factor CFA model with a correlation among residuals for the trait anxiety items

Fortunately, CFA allows correlations among residuals, as does sem.[14] To include correlations among residuals, we use the covariance() option.

```
. sem (TRAIT_ANX -> trait2 trait4 trait5 trait8
> trait9 trait15 trait20),
> variance(TRAIT_ANX@1) covariance(e.trait5*e.trait15) nolog
Endogenous variables
Measurement:  trait2 trait4 trait5 trait8 trait9 trait15 trait20
Exogenous variables
Latent:       TRAIT_ANX
Structural equation model                    Number of obs     =        790
Estimation method  = ml
Log likelihood     = -5775.0544
 ( 1)  [/]var(TRAIT_ANX) = 1
```

---

14. Correlations are allowed within the conditions of identifiability of the model. Chapter 12 provides more details about identification and model revision.

	Coef.	OIM Std. Err.	z	P>|z|	[95% Conf. Interval]	
Measurement trait2						
TRAIT_ANX	.513956	.0252467	20.36	0.000	.4644733	.5634388
_cons	1.887342	.0268215	70.37	0.000	1.834773	1.939911
trait4						
TRAIT_ANX	.6930667	.0323384	21.43	0.000	.6296846	.7564488
_cons	2.029114	.0347198	58.44	0.000	1.961064	2.097163
trait5						
TRAIT_ANX	.5292642	.024624	21.49	0.000	.481002	.5775264
_cons	1.555696	.0263416	59.06	0.000	1.504068	1.607325
trait8						
TRAIT_ANX	.6091248	.0271155	22.46	0.000	.5559793	.6622702
_cons	1.858228	.0295271	62.93	0.000	1.800356	1.9161
trait9						
TRAIT_ANX	.559309	.0306771	18.23	0.000	.499183	.6194351
_cons	2.070886	.0317727	65.18	0.000	2.008613	2.133159
trait15						
TRAIT_ANX	.5239486	.0270528	19.37	0.000	.4709261	.5769712
_cons	1.803797	.0282079	63.95	0.000	1.748511	1.859084
trait20						
TRAIT_ANX	.6108859	.0295792	20.65	0.000	.5529116	.6688601
_cons	1.901266	.0314755	60.40	0.000	1.839575	1.962957
var(e.trait2)	.30417	.0180066			.2708482	.3415912
var(e.trait4)	.4719755	.0290042			.4184184	.5323878
var(e.trait5)	.2680432	.0168066			.2370465	.3030931
var(e.trait8)	.3177285	.0200341			.2807918	.3595239
var(e.trait9)	.4846802	.0275111			.4336507	.5417146
var(e.trait15)	.354071	.0209303			.3153355	.3975646
var(e.trait20)	.4094751	.0245811			.3640233	.460602
var(TRAIT_ANX)	1	(constrained)				
cov(e.trait5, e.trait15)	.1064053	.0146166	7.28	0.000	.0777572	.1350534

LR test of model vs. saturated: chi2(13) = 61.93, Prob > chi2 = 0.0000

. estimates store corr_error

The addition of covariance(e.trait5*e.trait15) is the only change from the congeneric model. This option tells Stata to estimate the covariance between the residuals for item 5 and item 15. The covariance is listed in the output below the variances.

The congeneric model with no covariances among the errors is nested within the model with a covariance between item 5 and item 15. Thus, we can use an LR test to examine whether adding the covariance significantly improved model fit.

11.11 Summary

```
. lrtest congeneric corr_error
Likelihood-ratio test                              LR chi2(1)  =     68.94
(Assumption: congeneric nested in corr_error)      Prob > chi2 =    0.0000
```

The LR test was statistically significant, suggesting an improvement in fit.

We can compute reliability for models with correlated errors with a simple extension to (11.10) (Raykov 2001).

$$\omega = \frac{(\sum \lambda_i)^2 \phi_{11}}{(\sum \lambda_i)^2 \phi_{11} + \sum \theta_{ii} + 2\sum \theta_{ij}} \qquad (11.13)$$

The only change from (11.10) is the term $2\sum \theta_{ij}$, where θ_{ij} represents the covariance between item i and item j (item 5 and item 15 in our case). The summation indicates that if there is more than one covariance, we sum over them. However, in this example we only have a single covariance, so we only need to multiply it by 2.

We can estimate this by using `nlcom`.

```
. nlcom omega_corr:(_b[trait2:TRAIT_ANX] + _b[trait4:TRAIT_ANX] +
> _b[trait5:TRAIT_ANX] + _b[trait8:TRAIT_ANX] +
> _b[trait9:TRAIT_ANX] + _b[trait15:TRAIT_ANX] +
> _b[trait20:TRAIT_ANX])^2 *_b[/var(TRAIT_ANX)] /
> ((_b[trait2:TRAIT_ANX] + _b[trait4:TRAIT_ANX] +
> _b[trait5:TRAIT_ANX] + _b[trait8:TRAIT_ANX] +
> _b[trait9:TRAIT_ANX] + _b[trait15:TRAIT_ANX] +
> _b[trait20:TRAIT_ANX])^2 *_b[/var(TRAIT_ANX)] + (_b[/var(e.trait2)] +
> _b[/var(e.trait4)] + _b[/var(e.trait5)] +
> _b[/var(e.trait8)] + _b[/var(e.trait9)] +
> _b[/var(e.trait15)] + _b[/var(e.trait20)]) + 2* _b[/cov(e.trait5,e.trait15)]),
> noheader
```

	Coef.	Std. Err.	z	P>|z|	[95% Conf. Interval]	
omega_corr	.8525176	.0082811	102.95	0.000	.836287	.8687483

Thus, including a correlated error reduced ω slightly. Based on (11.10) and (11.13), this is to be expected. Adding a covariance only affects the denominator, which will reduce reliability. Therefore, we can conclude that positively correlated residuals—relationships among items beyond the common factor—reduce the reliability of measures.

11.11 Summary

The primary goal of this chapter was to provide an introduction to CFA as a framework for understanding reliability. In doing so, this chapter has covered a lot of ground. We have

- defined reliability
- distinguished between common and unique variances

- used CFA to estimate common and unique variances
- used the output from `sem` to compute reliability (ω) and construct an interval estimate for reliability
- compared ω with α and discussed how to use CFA to test the assumptions of α
- extended CFA and ω to accommodate correlated residuals

Along the way, we have also discussed parameters in CFA models (for example, factor loadings and latent variable variance), the prediction equation in CFA, and the χ^2 test of model fit. These concepts are fundamental to understanding CFA and will be used to unpack measurement invariance and validity in the next two chapters.

12 Factor analysis: Factorial validity

Measures should be reliable and valid. Chapter 11 covered internal consistency, which is one of many forms of reliability. A measure can be reliable—consistent—but not measure what we want it to measure. A common discussion in my department is whether student ratings of instructors are valid measures of professors' teaching ability. The student ratings at my university, provided that about 20 (or more) students complete them within a semester, have an internal consistency of at least 0.70. The internal consistency can be even higher if many students complete the ratings. But do the ratings represent a professor's teaching ability or the quality of the teaching? Or do the ratings represent something else, such as the professor's likability or how easy the class is? This is a critical question because student ratings are part of hiring, promotion, and raise decisions. These questions about what the student ratings measurements capture are questions about the validity of a measure. Validity refers to whether an instrument actually provides information about the construct the instrument was designed to measure.

There are many types of validity:

- Content or face validity: Does a measure appear to tap the construct it was aimed to measure?
- Predictive or criterion validity: Does a measure adequately predict an important criterion? For example, does the Graduate Record Exam adequately predict performance in graduate school? Does a measure of employee selection predict job performance?
- Concurrent or convergent validity: Does a measure correlate with measures it should correlate with? For example, does a depression measure correlate with other depression measures?
- Divergent or discriminant validity: Does a measure not correlate with measures it should not correlate with? For example, does a depression measure not correlate with measures of dyslexia (or does it correlate more strongly with other depression measures than with dyslexia measures)?
- Factorial validity: Do items from the same scale or subscale each provide adequate information about the construct of interest? For example, does each item on a depression measure provide substantial information about a person's level of depressive symptoms? Factorial validity is the focus of this chapter.

Each of these validity measures provides information about construct validity, which is whether an instrument measures what it was designed to measure. Some types of validity are more relevant than others for a given measure. For example, the Minnesota Multiphasic Personality Inventory (Ben-Porath 2012) was not designed to have face validity but instead was designed to have strong predictive, convergent, divergent, and factorial validity.

Factor analysis is a tool to help explore the factorial validity of a measure. This chapter provides a brief introduction to exploratory factor analysis (EFA), which is an exploratory data analysis technique for examining the relationship among items within a measure (or multiple measures). Following the discussion of EFA, I discuss confirmatory factor analysis (CFA),[1] with an emphasis on examining model fit and other issues dealing with factorial validity.

I keep the description of EFA brief for two reasons. First, EFA has a long history in psychology, and while researchers should be familiar with it, I do not want to get bogged down by it; it is used far less than CFA in contemporary research, and there are many nuances, details, and controversies in the EFA literature. Second, I prefer the confirmatory nature of CFA, wherein a researcher proposes a model and uses CFA to test the adequacy of the fit.

In this chapter you will learn the following:

1. What the common factor model is.
2. How to determine the number of factors to extract in an EFA.
3. How to rotate factor loadings in EFA to obtain an interpretable model.
4. How to identify a CFA model.
5. How to interpret factor loadings and other important output in both EFA and CFA.
6. How to establish the global fit of a CFA model.

---

1. Please see chapters 11 and 13 for additional details about CFA.

12.1 Data for this chapter

Stata commands featured in this chapter

- `factor`: for estimating an EFA
- `screeplot`: for creating a screeplot, which helps the user determine the number of factors to extract in an EFA
- `paran`: for conducting a parallel analysis to help the user determine the number of factors to extract in an EFA (this command is community-contributed)
- `rotate`: for rotating an EFA solution
- `sem`: for estimating a CFA
- `estat gof` and `estat eqgof`: for producing goodness-of-fit statistics following a CFA
- `estimates store`: for saving the estimates and statistics from models
- `lrtest`: for conducting a likelihood-ratio test that compares the fit of two nested models

12.1 Data for this chapter

Like chapter 11, this chapter will use items from the state–trait anxiety inventory (STAI) (Spielberger et al. 1983). As a reminder, trait anxiety occurs generally throughout someone's life (for example, generalized worry), whereas state anxiety is context specific (for example, worrying about an upcoming exam). This chapter focuses on four items from the trait subscale of the STAI and four items from the state subscale. For the trait subscale, respondents were asked to rate items (for example, "I feel like a failure") with respect to how they feel generally. The response options are

- 1 = Almost never
- 2 = Sometimes
- 3 = Often
- 4 = Almost always

For the state subscale, respondents were asked to rate items (for example, "I feel upset") with respect to how they feel right now, at this moment. The response options are

- 1 = Not at all
- 2 = Somewhat
- 3 = Moderately so
- 4 = Very much so

Data for the $N = 790$ participants are in the dataset `state_trait.dta`.

```
. use http://www.stata-press.com/data/pspus/state_trait
. describe
Contains data from state_trait.dta
  obs:           790
 vars:             8                          25 Jul 2017 08:18
 size:         6,320
```

variable name	storage type	display format	value label	variable label
state6	byte	%13.0g	stai2	Upset
state9	byte	%13.0g	stai2	Frightened
state12	byte	%13.0g	stai2	Nervous
state17	byte	%13.0g	stai2	Worried
trait4	byte	%13.0g	stai	Happy as others
trait5	byte	%13.0g	stai	Failure
trait9	byte	%13.0g	stai	Worry too much
trait15	byte	%13.0g	stai	Inadequate

Sorted by:

12.2 Exploratory factor analysis

A key assumption of factor analysis is that items are related to one another and this relationship provides evidence of a latent (that is, unobserved) construct. That is, when people respond to items on a depression measure, their responses on each item should be correlated. This correlation is assumed to arise from their latent depression level. Of course, each item is not a perfect measure of the latent depression and could provide information about other constructs or just random error. Factor analysis provides a method for identifying common variance (what is shared between items) and unique variance (what is specific to an item).

In section 11.3, I introduced the idea of common and unique variances and how to use CFA to estimate the latent variable (what is common) and residual variance (what is unique). CFA requires that we specify a model and test the fit of that model to the data—hence, the term "confirmatory" factor analysis. EFA, on the other hand, is an exploratory technique that identifies the latent variables and how the items are related to them. Given the relationship between CFA and EFA, many of the ideas covered in chapter 11 pertain to EFA as well as to CFA, such as the definition of a latent variable and factor loadings. In fact, the fundamental equation that describes how latent variables, factor loadings, and residual errors are related to covariance matrices is identical in both EFA and CFA [see (11.4)]. An important difference between CFA and EFA is how each is implemented, which is what I turn to now.

12.2.1 Common factor model

The correlations among the four trait and four state items are

```
. corr trait4 trait5 trait9 trait15 state6 state9 state12 state17
(obs=790)

                 |   trait4    trait5    trait9   trait15    state6    state9   state12
-----------------+----------------------------------------------------------------------
          trait4 |   1.0000
          trait5 |   0.5825    1.0000
          trait9 |   0.4519    0.3788    1.0000
         trait15 |   0.5097    0.6542    0.3790    1.0000
          state6 |   0.2754    0.3122    0.1920    0.2368    1.0000
          state9 |   0.2472    0.2953    0.2354    0.2778    0.3625    1.0000
         state12 |   0.2714    0.3275    0.2630    0.2983    0.3310    0.4346    1.0000
         state17 |   0.3161    0.3968    0.3545    0.3253    0.4168    0.4153    0.4806

                 |  state17
-----------------+---------
         state17 |   1.0000
```

As noted above, EFA assumes a common factor model to explain the correlations among the items. The common factor model assumes that one or more latent variables (that is, unmeasured variables) can account for the correlation among the items. The common factor model also assumes that there is some unexplained variance in each of the items—that is, the items do not completely share variance with all other items. Consequently, the common factor model includes an error term to represent variance that is unique to a particular item.

Sometimes EFA is thought of as a data reduction technique. We introduce the idea of common factors as a way to reduce the complexity of eight anxiety items by replacing the variances and covariances of the eight items with a smaller number of parameters. For example, we can replace the covariances with just a single latent variable, if there is only one source of the covariance among items. Perhaps there are two, three, or more sources of the covariances among the items. In principle, with eight items, we can produce eight latent variables. This defeats the purpose of EFA and the common factor model because eight latent variables would be just as complex as the original data—we have not simplified the situation. Consequently, part of the task of EFA (and CFA for that matter) is determining how many common factors there are. Selecting the appropriate number of factors is covered in detail in section 12.2.7.

Assume that there are two latent variables, trait anxiety and state anxiety, influencing eight items from the STAI. We could write an equation for each item consistent with the common factor model, as follows (Brown 2015):

$$\texttt{trait4}_i = \lambda_{11}\text{State}_i + \lambda_{12}\text{Trait}_i + \epsilon_i$$
$$\texttt{trait5}_i = \lambda_{21}\text{State}_i + \lambda_{22}\text{Trait}_i + \epsilon_i$$
$$\texttt{trait9}_i = \lambda_{31}\text{State}_i + \lambda_{32}\text{Trait}_i + \epsilon_i$$
$$\texttt{trait15}_i = \lambda_{41}\text{State}_i + \lambda_{42}\text{Trait}_i + \epsilon_i$$
$$\texttt{state6}_i = \lambda_{51}\text{State}_i + \lambda_{52}\text{Trait}_i + \epsilon_i$$
$$\texttt{state9}_i = \lambda_{61}\text{State}_i + \lambda_{62}\text{Trait}_i + \epsilon_i$$
$$\texttt{state12}_i = \lambda_{71}\text{State}_i + \lambda_{72}\text{Trait}_i + \epsilon_i$$
$$\texttt{state17}_i = \lambda_{81}\text{State}_i + \lambda_{82}\text{Trait}_i + \epsilon_i$$

The λ's are factor loadings, State_i and Trait_i are the latent variable values for the ith person, and ϵ_i is the residual. Each item has a factor loading from all latent variables. Ideally, one of the loadings will be strong and all other loadings will be small. Thus, if the trait items provide more information about trait anxiety than the state items, then λ_{12}–λ_{42} should be larger than λ_{11}–λ_{41}. Just the opposite should be true for the state items. Note as discussed in section 11.4, the names of the latent factors (state and trait here) are arbitrary. Nothing about EFA or CFA tells you what to name the latent variables, so do not get fooled into thinking that because we named the first latent factor state, it means we have firmly established that it measures state anxiety. Such claims require substantial research and evidence regarding the various types of validity discussed above (Cronbach and Meehl 1955; Flake, Pek, and Hehman 2017).

I had you imagine that there are two latent variables, but it is possible that there is just one or that there are more than two. Establishing how many factors account for the relationships among the items is the primary reason for EFA. Indeed, establishing how many factors to extract and producing an interpretable result is where things get more complicated. As we will see, the trouble is that there are no hard-and-fast rules for establishing how many factors there are and for producing factor loadings that are interpretable. This can make EFA sometimes feel a bit like a "free for all".

12.2.2 Extraction methods

Researchers using EFA have a dizzying array of options for estimating loadings and other parameters. These methods include principal factor, principal-component factor, iterated principal factor, maximum likelihood, unweighted least squares, generalized least squares, alpha factoring, and image factoring; the first four are available in Stata. Each of these methods has distinct assumptions and mathematical procedures, and each can produce distinct results. Typically, the results will be similar enough to warrant the same general substantive conclusions, so I will not go into detail about the differences. Instead, for this chapter, I use the simplest extraction method—principal-component factor—to introduce you to the subject and leave you to study the details of other

12.2.2 Extraction methods

extraction methods from the numerous reference texts on factor analysis (Brown et al. 2011; Harman 1976; Nunnally and Bernstein 1994; Rencher and Christensen 2012).

Estimation of factor loadings involves decomposing a covariance or correlation matrix. In fact, you can perform a factor analysis with just a covariance or correlation matrix rather than the raw data. Decomposing a correlation matrix requires matrix algebra and the computation of quantities known as eigenvalues and eigenvectors. Readers wishing to learn about matrix algebra in this context, including how to calculate eigenvalues and eigenvectors, can study chapter 3 of Brown et al. (2011) or chapter 2 of Rencher and Christensen (2012). In this chapter, I will show you how to interpret the eigenvalues that Stata provides as part of the output and how to use the eigenvalues to help select the number of factors.

We will analyze the four state and four trait items listed above. The correlation matrix for these eight items is shown in section 12.2.1. The trait items appear to correlate more with one another, on average, than they do with the state items (r's among the trait items range from 0.38 to 0.65). The same is true for the state items (r's among the state items range from 0.33 to 0.48). This suggests there may be two factors that can account for the relationship among the items.

To explore this possibility, run a factor analysis as follows:

```
. factor trait4 trait5 trait9 trait15 state6 state9 state12 state17, pcf
(obs=790)
Factor analysis/correlation                       Number of obs    =      790
    Method: principal-component factors           Retained factors =        2
    Rotation: (unrotated)                         Number of params =       15
```

Factor	Eigenvalue	Difference	Proportion	Cumulative
Factor1	3.53128	2.33386	0.4414	0.4414
Factor2	1.19742	0.46700	0.1497	0.5911
Factor3	0.73042	0.05314	0.0913	0.6824
Factor4	0.67728	0.09155	0.0847	0.7670
Factor5	0.58573	0.08314	0.0732	0.8403
Factor6	0.50259	0.04986	0.0628	0.9031
Factor7	0.45273	0.13017	0.0566	0.9597
Factor8	0.32256	.	0.0403	1.0000

```
LR test: independent vs. saturated:  chi2(28) = 1894.52 Prob>chi2 = 0.0000
Factor loadings (pattern matrix) and unique variances
```

Variable	Factor1	Factor2	Uniqueness
trait4	0.7042	-0.4130	0.3335
trait5	0.7684	-0.3491	0.2876
trait9	0.6094	-0.2680	0.5568
trait15	0.7143	-0.4023	0.3280
state6	0.5673	0.4049	0.5142
state9	0.5975	0.4627	0.4290
state12	0.6315	0.4196	0.4252
state17	0.6969	0.3425	0.3970

The factor analysis command is `factor` followed by a variable list. The option `pcf` requests that Stata use the principal-component factor extraction method. The output is split into two parts. The first part includes the eigenvalues for all possible factors (eight in this case). The second part contains the factor loadings.

12.2.3 Interpreting loadings

Stata extracted two factors. For now, we will rely on Stata's defaults for extracting factors, but section 12.2.7 goes into detail about strategies for determining the appropriate number of factors to extract. The loadings (λ's) in the output can be interpreted as standardized regression coefficients. That is, for every one-unit change in the factor (that is, latent variable), we expect a λ-standard-deviation change in the item, holding constant the other latent variables in the model. All latent variables in EFA are scaled to have a standard deviation of 1, which is to say that they are standardized. Thus, the one-unit change in the latent variable is a one-standard-deviation change. For example, $\lambda = 0.7$ for `trait4` on factor 1, which indicates that for every one-standard-deviation change in factor 1, we expect a 0.7-standard-deviation change in `trait4`, holding constant factor 2.

At this point of the analysis, the factors are also uncorrelated, which is also known as orthogonal. Consequently, the factor loadings can also be interpreted as the correlation between the item and factor. Thus, the correlation between `trait4` and factor 1 is 0.7. In most situations where EFA is used in psychology, the factors will not remain uncorrelated, so not all factor loadings can be interpreted as correlations.

12.2.4 Eigenvalues

Eigenvalues are needed to compute factor loadings (Rencher and Christensen 2012, 444) and are part of Stata's EFA output. I do not cover how to compute eigenvalues, but I do cover how to interpret them and how they are related to other parts of the factor analysis output. You can think of eigenvalues as representing how many units of variance a given factor accounts for. For example, factor 1 accounts for 3.53 units of variance in the items, factor 2 accounts for 1.2 units, factor 3 accounts for 0.73 units, and so on. Given that factor analysis with principal-component factor extraction uses the correlation matrix as the input, the eight items each amount to 1 unit of variance for a total variance of 8. Thus, 3.53 of the 8 units (or 44%) of variance in the items is accounted for by factor 1; conditional upon the variance accounted for by factor 1, 1.2 of the 8 units (or 15%) of variance in the items is accounted for by factor 2; and so on. The `factor` output includes the proportion of variance accounted for by each factor as well as the cumulative variance across factors.

With uncorrelated principal-component (see section 12.2.7) factors in Stata, summing and squaring the factor loadings for a given factor reproduces the eigenvalues. This can be done in Stata and Mata as follows:[2]

```
. quietly factor trait4 trait5 trait9 trait15 state6 state9 state12 state17, pcf
. mata: loadings = st_matrix("e(L)") // create a Mata matrix of the loadings
. mata: loadsq = loadings :^2 // square each of the loadings
. mata: colsum(loadsq) // sum the squared loadings for a factor; equal to
> eigenvalues
                 1            2
    1    3.531279867    1.197417936

. mata: st_matrix("e(Ev)") // compare with eigenvalues computed by factor
                 1            2            3            4            5
    1    3.531279867    1.197417936    .7304200045    .6772793042    .5857300806

                 6            7            8
    1    .5025855742    .4527269731    .3225602607
```

This makes sense because the factor loadings are the correlation between the item and the factor. A squared loading, just like a squared correlation, is the shared variance

---

2. If you are not comfortable with Mata, that is fine. It is not essential to use it to do a factor analysis. For readers wanting details regarding Mata, please see the *Mata Reference Manual* (StataCorp 2017d) and Gould (2018).

between the item and the factor. Adding up the squared loadings within a factor will tell us how many variance units the factor accounts for.

Dividing the sum of the squared loadings for a factor by the total variance produces the proportion of the total variance accounted for by the factor. We did this for factors 1 and 2 above. We can also do it in Stata and Mata:

```
. mata: colsum(loadsq) :/ 8 // proportion of variance explained
                 1             2
    1   .4414099834    .149677242
```

12.2.5 Communality and uniqueness

Another quantity we care about is an item's communality, which is the proportion of the variability in an item that is accounted for by the factor. With two orthogonal factors extracted, the communality for item i is computed as

$$h_i^2 = \lambda_{i1}^2 + \lambda_{i2}^2$$

The sibling to an item's communality is its uniqueness, which is the proportion of the variability that is not accounted for by the factor—the unique variance in the common factor model. The uniqueness is computed as

$$\theta_i = 1 - h_i^2$$

The `factor` output provides uniqueness estimates for each item. The uniqueness estimates range from 29% for `trait5` to 56% for `trait9`. Thus, the two factors together do not account for nearly one-third of the variance of `trait5` and over one-half of the variance of `trait9`. Variables with low communality/high uniqueness suggest that the common factor model may not fit well for that item—some evidence against factorial validity.

We can compute the communality and uniqueness for each item via Stata and Mata by summing the squared loadings across the factors (rather than within the factors as we did with eigenvalues):

```
. mata: rowsum(loadsq) // communalities
                 1
    1   .6664563417
    2   .7123754381
    3   .4432376425
    4   .6720316542
    5   .4857913907
    6   .5710443796
    7   .5748084681
    8   .6029524878
```

```
. mata: 1 :-rowsum(loadsq) // uniqueness
                1
    1    .3335436583
    2    .2876245619
    3    .5567623575
    4    .3279683458
    5    .5142086093
    6    .4289556204
    7    .4251915319
    8    .3970475122
```

12.2.6 Factor analysis versus principal-component analysis

I would be remiss at this point if I did not mention the distinction between principal-component analysis (PCA) and factor analysis. PCA, like factor analysis, is a data-reduction technique. However, PCA does not adhere to the common factor model, in which variability is partitioned into common and unique variances, but instead aims to provide a linear combination of variables that best distinguishes among people. Rencher and Christensen (2012) describe PCA:

> In principal-component analysis, we seek to maximize the variance of a linear combination of the variables. For example, we might want to rank students on the basis of their scores on achievement tests in English, mathematics, reading, and so on. An average score would provide a single scale on which to compare the students, but with unequal weight we can spread the students out further on the scale and obtain a better ranking. (p. 405)

The weights are a product of PCA just as factor loadings are the product of EFA. In EFA, the common-factor model might assume there is a latent construct, such as academic achievement, that accounts for the relationship among the scores in the various academic domains. PCA makes no such assumption and just aims to combine the scores optimally.

PCA and the principal-component factor extraction method for factor analysis that we used in section 12.2.2 are closely related, which is why they share similar names. In fact, some might even argue that they are the same thing. They are closely related and will often produce similar results, but they are not exactly the same thing and have distinct interpretations. Rencher and Christensen (2012) explain the differences:

> The first technique [of extracting factors in factor analysis] we consider is commonly called the *principal component* method. The name is perhaps unfortunate in that it adds to the confusion between factor analysis and principal component analysis. In the principal component method for estimation of loadings, we do not actually calculate any principal components. (p. 443, emphasis in original)

> ... [T]he loadings on the jth factor [from a factor analysis using principal component extraction] are proportional to the coefficients in the jth principal component. The factors are thus related to the first m principal components. But after rotation of the loadings, the interpretation of the factors is usually different. The researcher will ordinarily prefer the rotated factors ... (p. 444)

Stata adheres to this distinction and even has a program called `pca` to perform PCA.

12.2.7 Choosing factors and rotation

How many factors should we extract?

In the preceding analysis, we relied on Stata's default for selecting the number of factors to extract (two in this case). This section introduces three methods for selecting the number of factors to extract so that we can make our choices explicit. The three methods are

1. The eigenvalue-greater-than-one rule
2. Scree plots
3. Parallel analysis

These methods require interpretation and "eyeballing", so they are not foolproof. Nevertheless, if the methods agree, we have more evidence that the number of factors we chose to extract is reasonable.

Eigenvalue-greater-than-one rule

The eigenvalue-greater-than-one rule, sometimes called the Kaiser–Guttman rule, suggests that factors with an eigenvalue greater than 1 should be retained (Guttman 1954; Kaiser 1960; Nunnally and Bernstein 1994).[3] This is the default Stata uses when the `pcf` extraction method is selected, which is why `factor` extracted two factors. The logic behind this method is that an eigenvalue greater than 1 means that the factor accounts for more variance than a single item. Likewise, factors with eigenvalues lower than 1 will have low reliability (Kaiser 1960; Nunnally and Bernstein 1994). This is simple and straightforward but can still lead to overextraction of factors. The primary problem has to do with how many items are in the analysis—as the number of items increases, a factor with an eigenvalue near 1 becomes relatively less important. In the anxiety example, an eigenvalue of 1 means that the factor accounts for 1/8th of the variance. If there were 15 items, an eigenvalue of 1 would mean that the factor accounts

---
3. The lower bound is different when using extraction methods besides principal-component factor extraction. The *Stata Multivariate Statistics Reference Manual* (StataCorp 2017f) notes that the lower bound is 0.000005 (that is, 0) when using the other extraction methods available in Stata.

12.2.7 Choosing factors and rotation

for 1/15th of the variance (Nunnally and Bernstein 1994). Thus, relying only on this rule to select factors is not recommended.

Scree plots

A scree plot is a line plot with the eigenvalue on the y axis and the factor number on the x axis. As Nunnally and Bernstein (1994, 482) note, the term "'[s]cree' denotes rubble at the bottom of a cliff, which geologists disregard in measuring a cliff's height". In a scree plot of eigenvalues, we want to see where the rubble forms with respect to factors—any factors that are part of the rubble can be disregarded (Cattell 1966; Nunnally and Bernstein 1994). Scree plots in Stata are created using the postestimation command `screeplot` after `factor`.

```
. quietly factor trait4 trait5 trait9 trait15 state6 state9 state12 state17, pcf
. screeplot, xtitle("Factor number") xlabel(0(1)8) scheme(lean2)
```

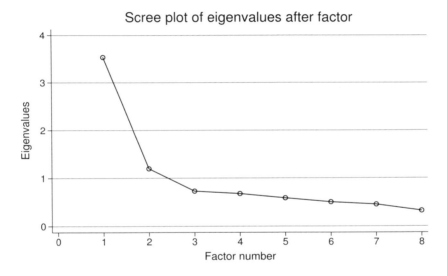

Figure 12.1. Scree plot for factors from factor analysis of anxiety items

Figure 12.1 presents the scree plot from Stata. When interpreting the scree plot, we want to focus on where it levels off, and it becomes clear that the eigenvalues are only slightly decreasing. Factors above the leveling-off point are retained. Factors at or below the leveling-off point are discarded (that is, they are the "scree"). In figure 12.1, we see a sharp decrease between factor 1 and factor 2. The decrease between factor 2 and factor 3 is more shallow but still prominent. Following factor 3, the eigenvalues do not decrease much and simply "pile up"; we would say that this plot levels off at factor 3. Based on this scree plot, we would retain two factors.

The biggest problem with scree plots is that interpreting them is notably subjective. What may look like a substantial decrease in eigenvalues to one person may be rubble to the next. There is no getting around it. Consequently, I suggest you use the scree plot in combination with the other methods to select the number of factors. Furthermore, present the scree plot in your write-up and be transparent about your interpretation.

Parallel analysis

Parallel analysis (Horn 1969) assists in selecting the number of factors by comparing the eigenvalues from the data with eigenvalues from randomly generated, uncorrelated data. We use not just one randomly generated dataset but thousands so that we can average over any random fluctuations. The logic behind the comparison is simple. Sometimes items will correlate just by chance in our data, but the extraction methods do not know the difference between chance relationships and systematic relationships. Consequently, we can compare the results from a factor analysis run on the actual data with the analysis of many datasets where there are no systematic relationships among items. Any factors extracted in the analysis of the randomly generated data have to be due to chance. We compare the results from the data with the random-data results factor by factor. If the first factor from the data has a larger eigenvalue than the average eigenvalue of the first factor from the random data, then we keep that factor. We continue making these comparisons until we reach a factor where the random-data eigenvalue is larger than the actual-data eigenvalue. If the data are not producing eigenvalues larger than what happens by chance, it is not likely the factor from the actual data is meaningful (Dinno 2009; Horn 1969).

The community-contributed program `paran` (`ssc install paran`) can perform a parallel analysis (Dinno 2009). The syntax of `paran` is nearly identical to `factor`:

```
. paran trait4 trait5 trait9 trait15 state6 state9 state12 state17,
>         factor(pcf) seed(3987) quietly
Computing: 10% 20% 30% 40% 50% 60% 70% 80% 90% 100%

Results of Horn's Parallel Analysis for principal components factors
240 iterations, using the mean estimate
--------------------------------------------------
Component    Adjusted      Unadjusted    Estimated
or Factor    Eigenvalue    Eigenvalue    Bias
--------------------------------------------------
    1        2.3104947     3.5312799     1.2207851
    2         .10061463    1.1974179     1.0968033
    3        -.30723606     .73042       1.0376561
    4        -.34602146     .6772793     1.0233008
    5        -.36814463     .58573008     .95387471
    6        -.443011       .50258557     .94559658
    7        -.44870989     .45272697     .90143687
    8        -.49798619     .32256026     .82054645
--------------------------------------------------
Criterion: retain adjusted factors > 0
```

12.2.7 Choosing factors and rotation

The `factor(pcf)` option tells `paran` to use the principal-component factor extraction method, the `seed(3987)` option sets the random-number generator seed so that your results will match what is printed here, and the `quietly` option keeps the results of the factor analysis from being printed (we have already seen that enough).

The bottom part of the output is what we care most about. There are three columns—adjusted eigenvalue, unadjusted eigenvalue, and estimated bias. The unadjusted eigenvalue column contains eigenvalues from the anxiety data; the estimated bias column contains the average eigenvalues from the randomly generated data; and the adjusted eigenvalue column contains the difference between the unadjusted and random eigenvalues. We extract factors where the unadjusted eigenvalue is larger than the random value. Said another way, we extract factors where the adjusted eigenvalue is greater than 0. Thus, we should extract factors 1 and 2, which is consistent with the eigenvalue-greater-than-one rule and the scree plot.

We can create a scree plot of both the unadjusted and the random eigenvalues by using the `graph` option for `paran`. I do not like that plot because it also includes the adjusted eigenvalues and can be challenging to customize. We can also create the plot manually by saving the posted results and matrices. Figure 12.2 shows the results and displays clearly where the eigenvalues from the data exceed the eigenvalues from the randomly generated data.

```
. matrix actual = e(UnadjustedEv)
. matrix actual = actual'
. matrix random = e(CentRandomEv)
. matrix random = random'
. svmat actual, names(actual)
. svmat random, names(random)
. generate factornum = _n in 1/8
(782 missing values generated)
. twoway scatter actual factornum, connect(L) ||
>         scatter random factornum, connect(L)
>         ytitle("Eigenvalue") xtitle("Factor number")
>         scheme(lean2)
>         legend(label(1 "Data") label(2 "Random"))
```

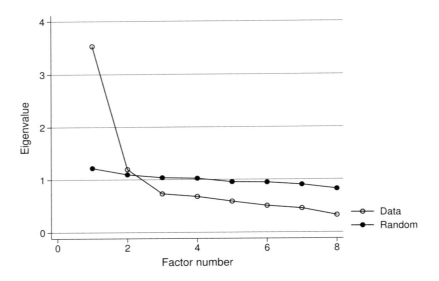

Figure 12.2. Parallel analysis plot

Orthogonal rotation—varimax

Examination of the eigenvalues suggests that two factors should be extracted. However, examination of the loadings from the factor analysis suggests that all items load most strongly on the first factor. This is a consequence of the way that the analysis was done, not necessarily a reflection of the fact that all items load strongly on the first factor. Indeed, one of the key features of the algorithms for extracting factors is that the first factor accounts for the biggest amount of variance, the second factor for the second biggest, and so on. This is useful because it means that factors that do not account for much variability can be disregarded and we can reduce the data (Brown et al. 2011). Unfortunately, this comes at an interpretative cost in the sense that it may not be (and usually is not) meaningful for the first factor to absorb the majority of the variance in this way. After all, we are examining the factor structure of the anxiety items to see if the items successfully discriminate between trait and state anxiety.

We need a way to reorganize the factor loadings so that they are more interpretable. The idea of reorganizing loadings sometimes feels a bit like cheating—after all, is not the original solution correct? The difficulty is that given two or more factors, there are an infinite number of ways to reorganize the solution that provide identical fit to the data. Indeed, there are an infinite number of solutions that provide identical communalities. Consequently, we need a criterion that helps us choose one of the solutions as the best solution. The most common criterion is called simple structure.

12.2.7 Choosing factors and rotation

Simple structure means a solution where the loadings are as large as possible for one factor and as close to 0 as possible on all other factors (Brown 2015). The primary method for reorganizing factor loadings is called rotation, and the various methods of rotation are optimized to find solutions that satisfy simple structure. Figure 12.3 is a scatterplot of the factor loadings prior to rotation. The loading plot includes two lines, a vertical line at a 0 factor loading for factor 1 and a horizontal line at a 0 factor loading for factor 2. The key observations at this point are that the loadings are largest for factor 1 and that none of the loadings is close to 0 on either factor. Thus, simple structure has not been met and rotation is warranted (as it typically is if there are two or more factors).

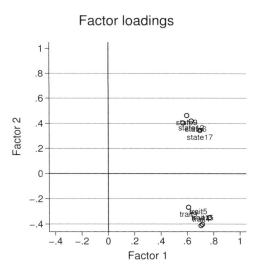

Figure 12.3. Loading plot before rotation

There are numerous rotation methods, all with fairly exotic sounding names, such as varimax, quartimax, equimax, promax, and oblimax. The rotation methods can be divided into two main categories, orthogonal and oblique. In the original factor solution produced by `factor`, the factors were orthogonal or uncorrelated with one another. Orthogonal rotations provide a simple-structure solution while keeping the factors uncorrelated with one another. Oblique rotations, on the other hand, allow the factors to be correlated. In psychology, orthogonal rotations are rarely theoretically justified because most constructs are correlated to some degree, especially constructs measured by the same instrument we are discussing in this chapter. In our example, although state and trait anxiety may be distinct, they are most likely correlated. Consequently, an oblique rotation is most justified. Nevertheless, I show the results of an orthogonal rotation to illustrate the difference from oblique rotation, and I discuss the specifics of the output.

The default rotation method in Stata is an orthogonal rotation using the varimax method. Rotation is a postestimation command, meaning that you must run the factor analysis first with `factor` and then perform the rotation. For example, the varimax rotation is done as

```
. quietly factor trait4 trait5 trait9 trait15 state6 state9 state12 state17, pcf
. matrix unrotated = e(L)
. rotate, varimax normalize
```

```
Factor analysis/correlation                    Number of obs    =      790
    Method: principal-component factors        Retained factors =        2
    Rotation: orthogonal varimax (Kaiser on)   Number of params =       15
```

Factor	Variance	Difference	Proportion	Cumulative
Factor1	2.49885	0.26900	0.3124	0.3124
Factor2	2.22985	.	0.2787	0.5911

LR test: independent vs. saturated: chi2(28) = 1894.52 Prob>chi2 = 0.0000

Rotated factor loadings (pattern matrix) and unique variances

Variable	Factor1	Factor2	Uniqueness
trait4	0.8005	0.1600	0.3335
trait5	0.8060	0.2504	0.2876
trait9	0.6333	0.2052	0.5568
trait15	0.8010	0.1746	0.3280
state6	0.1544	0.6797	0.5142
state9	0.1385	0.7429	0.4290
state12	0.1925	0.7333	0.4252
state17	0.2926	0.7193	0.3970

Factor rotation matrix

	Factor1	Factor2
Factor1	0.7467	0.6651
Factor2	-0.6651	0.7467

```
. matrix rotate_mat = e(r_T)
```

I used two options, `varimax` and `normalize`. The first option requests a varimax rotation. Although this is the default, I like to add it to be transparent. The second option, `normalize`, ensures that all rows in the loading matrix are weighted equally when performing the rotation. If you do not use the `normalize` option, then rows with high communalities will carry more weight. Additionally, other software packages automatically use normalization, so using `normalize` will ensure that Stata's results are similar.

There are three parts to the `rotate` output. The first part is similar to what `factor` produced. The second part is the rotated loadings. Now that rotation has occurred, large loadings are spread between the factors rather than limited to the first factor. The largest loadings for factor 1 are for the trait items, and the largest loadings for

12.2.7 Choosing factors and rotation

factor 2 are for the state items. This suggests that the factors may represent latent trait and latent state anxiety. Notice also that items load strongly on one factor but not the other. This is what is supposed to happen when applying the simple-structure criterion. Figure 12.4 shows that, after rotation, the loadings are pushed toward larger numbers on one axis and smaller numbers on the other. Indeed, all loadings are above 0.6 on one of the factors and below 0.30 on the other.

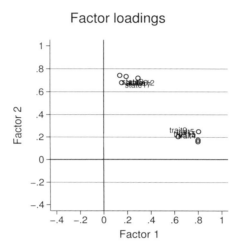

Rotation: orthogonal varimax
Method: principal–component factors

Figure 12.4. Loading plot after orthogonal rotation (varimax)

The rotated factor loadings can be interpreted as zero-order correlations between the item and the factor. Because the factors themselves are forced to be uncorrelated, we do not have to worry about redundancy in the factors. The loading of factor 1 on trait4 is $\lambda = 0.8$, which indicates that the correlation between trait4 and factor 1 is 0.8 and that the factor and the item share 64% of the variance. This indicates that trait4 is a good indicator of the latent construct. Furthermore, given that all the trait items load strongly on one factor and all the state items load strongly on the other, we have some evidence of factorial validity. Of course, we have used a rotation method that assumes the factors are uncorrelated, so we need to use an oblique rotation before we draw any conclusions.

The final part of the rotate output is the factor rotation matrix. This matrix indicates how far the axes are rotated to produce the rotated loadings. In fact, the rotated loadings are equal to the unrotated loading matrix (Λ) times the factor rotation matrix:

```
. matrix rotated = unrotated*rotate_mat
. matrix list rotated

rotated[8,2]
             Factor1     Factor2
   trait4  .80053642   .15999309
   trait5  .80602177   .25040837
   trait9  .63334168   .20522174
  trait15  .80096041   .17462552
   state6  .15438656   .67967358
   state9  .13845574   .74288249
  state12   .1925109   .73331304
  state17  .29256613   .71927571
```

Generally speaking, you will not need the factor rotation matrix.

Oblique rotation—promax

Oblique rotation, like orthogonal rotation, aims to satisfy the criterion of simple structure. However, oblique methods allow the factors to be correlated with one another, which is often far more defensible in psychology than forcing them to be orthogonal. Common choices for oblique rotation are promax and oblimin. In this chapter, we will use promax. To obtain the promax rotation, use the following:

```
. quietly factor trait4 trait5 trait9 trait15 state6 state9 state12 state17, pcf
. rotate, promax normalize
```

Factor analysis/correlation   Number of obs   =   790
    Method: principal-component factors   Retained factors =   2
    Rotation: oblique promax (Kaiser on)   Number of params =   15

Factor	Variance	Proportion	Rotated factors are correlated
Factor1	3.01454	0.3768	
Factor2	2.77715	0.3471	

LR test: independent vs. saturated: chi2(28) = 1894.52 Prob>chi2 = 0.0000

Rotated factor loadings (pattern matrix) and unique variances

Variable	Factor1	Factor2	Uniqueness
trait4	0.8327	-0.0372	0.3335
trait5	0.8147	0.0602	0.2876
trait9	0.6379	0.0566	0.5568
trait15	0.8293	-0.0213	0.3280
state6	-0.0119	0.7024	0.5142
state9	-0.0461	0.7757	0.4290
state12	0.0155	0.7509	0.4252
state17	0.1286	0.7091	0.3970

Factor rotation matrix

	Factor1	Factor2
Factor1	0.8824	0.8227
Factor2	-0.4705	0.5684

12.2.7 Choosing factors and rotation

As before, Stata prints the factor loadings and the factor rotation matrix. As shown in the loading matrix and in figure 12.5, the factor loadings are even more in line with simple structure than with the varimax rotation—the smaller loadings are even closer to 0 than before.

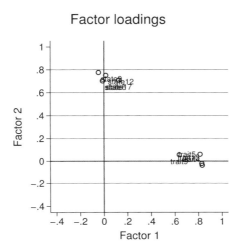

Figure 12.5. Loading plot after oblique rotation (promax)

Using an oblique rotation is not without its challenges—allowing the factors to correlate means that the factor loadings are no longer the zero-order correlation between the item and the factor. Instead, the loading is the relationship between an item and a factor after controlling for the shared variance between the factors. The loading for trait4 on factor 1 is $\lambda = 0.83$, which is interpreted like a regression coefficient. A one-unit difference in the factor variance, which is a standard deviation because the factor variance is 1, is associated with a 0.83 difference in the item, after controlling for the shared variance between factors 1 and 2.

Stata calls the loading matrix a pattern matrix when an oblique rotation is used. The term pattern matrix means that loadings are interpreted in light of the correlation among the factors. We can also request what is called the structure matrix, which provides the loadings that are interpreted as zero-order correlations between factors and items. To obtain the structure matrix, use the postestimation command `estat structure`. Do this after rotating the loadings, or else Stata will print the matrix of unrotated loadings because those are also zero-order correlations (see section 12.2.3).

```
. estat structure
```
Structure matrix: correlations between variables and promax(3) rotated common factors

Variable	Factor1	Factor2
trait4	0.8157	0.3446
trait5	0.8423	0.4338
trait9	0.6639	0.3491
trait15	0.8196	0.3589
state6	0.3101	0.6969
state9	0.3095	0.7546
state12	0.3598	0.7580
state17	0.4537	0.7680

Given that we used an oblique rotation, Stata can produce the estimate of the correlation among the factors. This correlation is important for factorial validity concerns. If the correlation is very high, then we do not have evidence that the two factors are actually separable empirically (at least with the instrument we used). If the correlation is low, but we expected the factors to be correlated, then we may not be measuring the construct we expected. To print this correlation, use `estat common` after rotation:

```
. estat common
```
Correlation matrix of the promax(3) rotated common factors

Factors	Factor1	Factor2
Factor1	1	
Factor2	.4585	1

The correlation is 0.46, suggesting a moderate relationship between the latent factors. State and trait anxiety should correlate because they are both anxiety-related constructs, but they should also be distinct and capture unique aspects of each respondent's experience. The moderate correlation is consistent with this perspective.

It is possible to use `estat common` after an orthogonal rotation or even before rotation. In both cases, Stata will show that the factors are unrelated.

```
. quietly factor trait4 trait5 trait9 trait15 state6 state9 state12 state17, pcf
. estat common
```
Correlation matrix of the common factors

Factors	Factor1	Factor2
Factor1	1	
Factor2	0	1

12.3.1 *EFA versus CFA*

```
. quietly rotate, varimax normalize
. estat common
Correlation matrix of the varimax rotated common factors
```

Factors	Factor1	Factor2
Factor1	1	
Factor2	0	1

12.3 Confirmatory factor analysis

We encountered CFA in chapter 11, but we focused on one-factor models that can help us better understand reliability. In this section, we will see how CFA can be extended to test hypotheses about the factor structure of an instrument. Along the way, we will see how CFA differs from EFA and how the CFA model can extend EFA.

12.3.1 EFA versus CFA

EFA is, as its name states, an exploratory technique. That is, we do not impose a specific pattern of factor loadings when conducting an EFA. Rather, we let the estimation algorithm determine the pattern of loadings that is most consistent with the data. This is useful, particularly while developing a measure or in a new research area where the empirical relationships among constructs is not well established. Many other times, we have a specific model that we aim to test. After all, when constructing a measure, we do not haphazardly create items but instead generate items based on theory and experience. For example, the developers of the state–trait anxiety inventory did not construct the items on that measure without any ideas about which items would go with the state subscale and which would go with the trait subscale. The items were written specifically for each subscale. Consequently, there is a model—a specific number of latent factors and pattern of factor loadings—that was used to develop the items. We can use CFA to fit a factor analysis model that corresponds to that model. Thus, CFA is a confirmatory technique because researchers dictate the details of the factor model rather than letting the data do the dictating.

In EFA, all items load on all factors. The principle of simple structure means that, following rotation, loadings will be large on one factor and small on other factors. However, even the small loadings are estimated. Figure 12.6 is a path diagram illustrating an EFA with oblique rotation. Each item has two factor loadings—both trait and state items load on the trait and state factors. Nevertheless, state items are not typically thought of as providing direct information about trait anxiety. Theoretically, state items are related to latent trait anxiety through the correlation between latent state anxiety and latent trait anxiety (that is, indirectly through correlation between state and trait anxiety). The EFA model does not match that theoretical formulation because the EFA implies a direct relationship between state items and latent trait anxiety (albeit a small relationship).

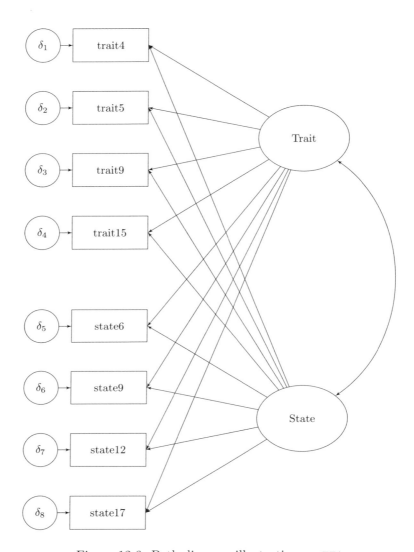

Figure 12.6. Path diagram illustrating an EFA

A CFA model can be used to fit the theoretical model. Figure 12.7 is a path diagram for such a model. The models are the same in many ways. Both the EFA and the CFA include a) two latent variables, b) a correlation among the latent variables, and c) residual errors or uniqueness. The primary difference is the pattern of the loadings. The CFA does not estimate the loadings between state items and latent trait anxiety or the loadings between trait items and latent state anxiety. The CFA model assumes that these loadings are 0 (in fact, one can explicitly constrain the loadings to 0 and obtain the same results as the model in figure 12.7). Being able to map the factor model directly to

12.3.1 EFA versus CFA

the theoretical model is a key advantage of CFA and the primary conceptual distinction between CFA and EFA.

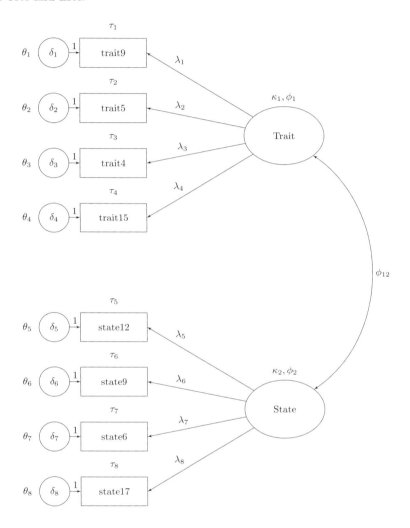

Figure 12.7. Path diagram illustrating a CFA; see sections 11.5 and 12.3.3 for a description of the Greek symbols

12.3.2 Estimating a CFA with sem

Section 11.6 provides the basics of using the sem command. To fit the CFA model in figure 12.7, the syntax is

```
. sem (TRAIT -> trait4 trait5 trait9 trait15)
>       (STATE -> state6 state9 state12 state17), nolog
Endogenous variables
Measurement:  trait4 trait5 trait9 trait15 state6 state9 state12 state17
Exogenous variables
Latent:       TRAIT STATE
Structural equation model                       Number of obs    =      790
Estimation method  = ml
Log likelihood     = -6194.105
 ( 1)  [trait4]TRAIT = 1
 ( 2)  [state6]STATE = 1
```

	Coef.	OIM Std. Err.	z	P>|z|	[95% Conf. Interval]	
Measurement						
trait4						
TRAIT	1	(constrained)				
_cons	2.03038	.0347415	58.44	0.000	1.962288	2.098472
trait5						
TRAIT	.8931898	.0466095	19.16	0.000	.8018368	.9845427
_cons	1.556962	.0263381	59.11	0.000	1.50534	1.608584
trait9						
TRAIT	.6779242	.0506463	13.39	0.000	.5786593	.7771892
_cons	2.070886	.0317727	65.18	0.000	2.008613	2.133159
trait15						
TRAIT	.8725392	.0482381	18.09	0.000	.7779942	.9670842
_cons	1.803797	.0282079	63.95	0.000	1.748511	1.859084
state6						
STATE	1	(constrained)				
_cons	1.259494	.0213997	58.86	0.000	1.217551	1.301436
state9						
STATE	.8934448	.0742939	12.03	0.000	.7478313	1.039058
_cons	1.160759	.0175664	66.08	0.000	1.12633	1.195189
state12						
STATE	1.468001	.1199652	12.24	0.000	1.232873	1.703128
_cons	1.592405	.026815	59.38	0.000	1.539849	1.644961
state17						
STATE	1.79049	.1367073	13.10	0.000	1.522549	2.058432
_cons	1.679747	.0295774	56.79	0.000	1.621776	1.737717

```
  var(e.trait4) |  .4794808   .0303288                       .4235746   .5427659
  var(e.trait5) |  .1698484   .0152018                       .1425204   .2024164
  var(e.trait9) |   .579653   .0317982                       .5205633   .6454502
 var(e.trait15) |  .267705    .0182725                       .2341838   .3060245
  var(e.state6) |  .2482167   .0144984                       .2213665   .2783236
  var(e.state9) |  .1531279   .0094571                       .1356702   .172832
 var(e.state12) |  .3233183     .0212                        .2843263   .3676577
 var(e.state17) |   .327051   .0250196                       .2815128   .3799557
     var(TRAIT) |  .4740267   .0453595                       .3929628   .5718131
     var(STATE) |  .1135603   .0151524                       .0874279   .1475037
----------------+---------------------------------------------------------------
    cov(TRAIT,
        STATE)  |  .1460129   .0154913    9.43    0.000      .1156504   .1763753

  LR test of model vs. saturated: chi2(19) =       71.01, Prob > chi2 = 0.0000
```

The code is similar to what we covered in chapter 11, except that two latent variables are included. Code for distinct latent variables is separated by parentheses, and the output is organized by latent variable. Thus, the loadings and intercepts for TRAIT come first, followed by the loadings and intercepts for STATE. The residual variances are listed next as are the latent variable variances. Finally, the covariance between the latent variables is reported. The default in sem is to estimate the covariance among the latent variables, so it does not need to be included in the syntax. To constrain the covariance to 0, similar to an orthogonal rotation, the code is covariance(TRAIT*STATE@0) (* is used to denote a covariance in sem).

12.3.3 Mean structure versus variance structure

The sem output includes the following parameters (see also figure 12.7):

1. Factor loadings (λ)
2. Latent factor variances and covariances (ϕ)
3. Residual variances (θ)
4. Item intercepts (labeled _cons; τ)
5. Factor means (not printed by default; κ)

1–3 are called the (co)variance structure of the model, and 4–5 are called the mean structure. The variance structure refers to parameters that are estimated from the covariance among the observed items and explain the variability in the data. The mean structure refers to parameters that arise from the means of the items and explain the central tendency of the data. Historically, structural equation modeling software either estimated only the variance structure or required the user to request that the mean structure be produced. Currently, most software, including sem, reports both by default.

To better understand what it means to model the variance structure of the data, remember that a goal of CFA is to reproduce the variance–covariance matrix of the items [see the CFA prediction (11.4)]. CFA does this by partitioning the variance of an item into the factor loading, the residual variance, and the factor variance. Indeed, the variance of an item can be reproduced by multiplying the squared factor loading by the factor variance and adding the error variance (Brown 2015). For example, the variance of trait5 can be reproduced as follows:

$$\text{VAR}(\texttt{trait5}) = \lambda_2^2 \phi_{11} + \theta_2$$

In Stata, we type

```
. display _b[trait5:TRAIT]^2 * _b[/var(TRAIT)] + _b[/var(e.trait5)]
.54802115
. summarize trait5, detail
```

	Failure			
	Percentiles	Smallest		
1%	1	1		
5%	1	1		
10%	1	1	Obs	790
25%	1	1	Sum of Wgt.	790
50%	1		Mean	1.556962
		Largest	Std. Dev.	.7407535
75%	2	4		
90%	2	4	Variance	.5487157
95%	3	4	Skewness	1.367008
99%	4	4	Kurtosis	4.687349

Model parameters can also be used to predict the covariance among the items. If two items load on the same factor and only load on one factor, the covariance among the items is equal to the product of the loadings for each item and the factor variance. For trait5 and trait9, the model-implied covariance is

$$\text{COV}(\texttt{trait5}, \texttt{trait9}) = \lambda_2 \phi_{11} \lambda_3$$

In Stata, we type

```
. display _b[trait5:TRAIT] * _b[/var(TRAIT)] * _b[trait9:TRAIT]
.28703025
```

This value is a predicted value based on the model and will not reproduce the covariance between trait5 and trait9 exactly unless the model is just-identified, which is uncommon with CFA (see section 12.3.4).

The predicted covariance between items that load on different, but correlated, factors is computed by multiplying the factor loading for the first item by the covariance between the factors by the loading for the second item. For trait5 and state9, the model-implied covariance is

$$\text{COV}(\texttt{trait5}, \texttt{state9}) = \lambda_2 \phi_{21} \lambda_6$$

12.3.4 Identifying models

In Stata, we type

```
. display _b[trait5:TRAIT] * _b[/cov(TRAIT,STATE)] * _b[state9:STATE]
.11652055
```

The mean structure in CFA is a little easier to understand. Commonly, and by default in Stata, the mean of latent variables is fixed to 0 (see section 12.3.4 for why). Consequently, the mean structure of the model is usually the item intercepts only (see chapter 13 for exceptions). When the latent mean is constrained to 0, the item intercepts are the means of the observed items. Compare the output of sem above with summarize:

```
. summarize trait4 trait5 trait9 trait15 state6 state9 state12 state17
```

Variable	Obs	Mean	Std. Dev.	Min	Max
trait4	790	2.03038	.9770957	1	4
trait5	790	1.556962	.7407535	1	4
trait9	790	2.070886	.8935981	1	4
trait15	790	1.803797	.793341	1	4
state6	790	1.259494	.60186	1	4
state9	790	1.160759	.4940502	1	4
state12	790	1.592405	.7541641	1	4
state17	790	1.679747	.8318562	1	4

The postestimation command `estat framework, fitted` is a simple way to obtain the model-predicted variance, covariances, and means (see section 11.7).

12.3.4 Identifying models

By default in sem, the loadings for trait4 and state6 are 1, and the output notes that these loadings are constrained. This is sem's default behavior, and these constraints on the loadings are used for identification. A CFA is identified if a unique solution—specific values of the loadings, intercepts, means, variances, and covariances—can be mathematically produced (Brown 2015; Kline 2011). Identification is possible if the number of parameters in the model is equal to or less than the number of unique pieces of information available in the data. If input data has enough information to properly estimate each parameter, then the model is identified. When the model has more parameters than available information, the model is not identified. That is, if outputs exceed the inputs, then estimation is impossible.[4]

Consider the equation

$$a + b = 15$$

What are the values of a and b? We have one piece of information as an input—the sum of a and b is 15. We have two unknowns—the values of a and b. Consequently, the outputs exceed the inputs, and with this limited information, there is no unique

---

4. This is not unique to CFA. Any statistical model with more outputs than inputs is not identified. For example, try estimating a regression with more predictors than observations. Identification is discussed in CFA and structural equation modeling because they include latent variables.

solution. In fact, there is an infinite number of solutions. The only way to solve this equation (Brown 2015; Kline 2011) is to either constrain a or b to a specific value (for example, if $a = 10$, then b must be 5) or add more information (for example, include another equation involving a and b so there is more information to work with).

Imposing constraints for identification

Regardless of how much data or how many variables we have in our dataset, we must introduce some constraints on the model for it to be identified, because CFA includes latent variables. Latent variables are by definition unobserved; all parameters involving latent variables—loadings, means, variances, and covariances—must be inferred from the observed information. Latent variables do not have a scale, meaning we do not know whether they range from -1 to 1 or -10 to 10 or anything else. Constraints are used to scale the latent variable, which when combined with sufficient information in the input data helps to identify the model.

Both the mean and the variance of the latent variable need to be scaled. Scaling the means is most commonly done in one of two ways.

1. Fix the mean of the latent variable to 0. In this case, the item intercepts (_cons in the output) are the mean of the item. This is the default in Stata and overwhelmingly the most used method to scale latent variable means.
2. Fix one of the item intercepts to 0. The item chosen is called the marker item. The latent variable mean is estimated, and the other item intercepts are interpreted as differences from the marker item mean.

Scaling the variance of latent variables is also typically done in one of two ways (a third method for scaling the mean and the variance is discussed in section 13.5.1).

1. Fix the factor loading for one item to 1. The factor variance is estimated as are the loadings for all other items. This is the default in Stata and what was used above.
2. Fix the factor variance to 1. All factor loadings are estimated.

Any combination of identifying the mean and the variance of the latent variables can be used.

So which method should you choose? Model fit will not usually be affected by how we choose to scale the latent variables. There are some situations where the scaling method can lead to differences in model fit and we must be careful (Kline 2011; Steiger 2002). Unless the research question is best answered via constraints on intercepts, I recommend fixing the mean of the latent variable to 0. Further, unless the primary research question involves the mean of the latent construct—such as wanting to compare it across groups (see section 13.4.2)—fixing the latent mean to 0 is adequate. I also recommend fixing the variance of the latent variable to 1, unless the research question is about the variance

12.3.4 Identifying models

of the latent variable. Most uses of CFA, especially when evaluating the validity of a measure, do not involve research questions about the variance of the latent variable. Consequently, for the rest of this chapter, we will constrain the mean and the variance of the latent variable.

Constraining the mean and variance of the latent variable to 0 and 1, respectively, to identify the model has specific advantages. First, all factor loadings will be estimated. Second, these constraints mean that the latent variable is a standardized normal variable (that is, has a mean of 0 and standard deviation of 1). Thus, covariances between latent variables are correlations and easy to interpret. Third, factor loadings are semi-standardized loadings. Recall that a factor loading is interpreted as the expected change in the item for every one-unit change in the latent variable. With a variance of 1, a one-unit change in the latent variable is a 1-standard-deviation change. Thus, the loadings are the expected change in the item for a 1-standard-deviation change in the latent variable. Fourth, these constraints parallel the constraints used in EFA.

The constraints that `sem` used, whether they are there by default or user-imposed constraints, are listed at the top of the output. Thus, in the original CFA where we just used the defaults, the output includes the lines

```
( 1)  [trait4]TRAIT = 1
( 2)  [state6]STATE = 1
```

This indicates that there are two constraints. The factor loading for `trait4` on the latent variable `TRAIT` and the factor loading for `state6` on the latent variable `STATE` are both fixed to 1.

How much information is needed to identify a model?

Even with constraints applied, sometimes we ask too much of CFA—we try to fit a model with too many parameters given the information available in the data. The amount of information available to contribute to estimation is defined by the model degrees of freedom (DF) [see also chapter 11, (11.6)]:

$$\text{DF}_{\text{mod}} = \text{unique elements in } S + \text{\# of observed variables} - \text{\# of parameters}$$

S is the observed covariance matrix. The unique elements in the covariance matrix identify the variance structure of the CFA model (for example, factor loadings, latent variances, residual variances, and covariances). The number of variables in the model identifies the mean structure of the CFA model (for example, intercepts and latent means). The number of unique elements is [see also chapter 11, (11.5)]

$$\text{unique elements} = \frac{k(k+1)}{2}$$

where k is the number of observed variables in the model.

With respect to identification, models fall into three categories (Kline 2011).

1. **Just-identified models**: If $\text{DF}_{\text{mod}} = 0$, a model is just-identified. This is an identified model and has the same number of parameters as unique elements in S plus the number of variables in the data. Just-identified models perfectly reproduce the data, so they are sometimes not particularly useful.

2. **Overidentified models**: If $\text{DF}_{\text{mod}} \geq 1$, a model is overidentified. This is an identified model and has fewer parameters than unique elements in S plus the number of variables in the data. All the CFA models in this book are overidentified.

3. **Underidentified models**. If $\text{DF}_{\text{mod}} < 0$, a model is underidentified. This model cannot be estimated because there is not a unique solution.

In CFA, the variance structure of the model is typically overidentified and the mean structure is just-identified. That is, the model simplifies the variance structure—it uses fewer parameters than there are unique elements in S. In contrast, models typically fit as many unique intercepts and factor means as there are means in the model. If there are specific questions about the mean structure, such as when testing measurement invariance (see chapter 13), then we introduce constraints and the mean structure will be overidentified. Stata will fail to converge when either variance or mean structures are underidentified.

12.3.5 Refitting the model with constrained latent variables

We can now refit the state–trait CFA by constraining the latent means to 0 and the latent variances to 1.

```
. sem (TRAIT -> trait4 trait5 trait9 trait15)
> (STATE -> state6 state9 state12 state17), variance(TRAIT@1) variance(STATE@1)
> nolog
Endogenous variables
Measurement:  trait4 trait5 trait9 trait15 state6 state9 state12 state17
Exogenous variables
Latent:       TRAIT STATE
Structural equation model                       Number of obs    =         790
Estimation method  = ml
Log likelihood     = -6194.105
 ( 1)  [/]var(TRAIT) = 1
 ( 2)  [/]var(STATE) = 1
```

12.3.5 Refitting the model with constrained latent variables

	Coef.	OIM Std. Err.	z	P>|z|	[95% Conf. Interval]	
Measurement						
trait4						
TRAIT	.6884959	.032941	20.90	0.000	.6239327	.7530592
_cons	2.03038	.0347415	58.44	0.000	1.962288	2.098472
trait5						
TRAIT	.6149575	.0236397	26.01	0.000	.5686245	.6612906
_cons	1.556962	.0263381	59.11	0.000	1.50534	1.608584
trait9						
TRAIT	.4667481	.0324931	14.36	0.000	.4030628	.5304334
_cons	2.070886	.0317727	65.18	0.000	2.008613	2.133159
trait15						
TRAIT	.6007397	.02594	23.16	0.000	.5498981	.6515812
_cons	1.803797	.0282079	63.95	0.000	1.748511	1.859084
state6						
STATE	.336987	.0224822	14.99	0.000	.2929227	.3810513
_cons	1.259494	.0213997	58.86	0.000	1.217551	1.301436
state9						
STATE	.3010793	.0182797	16.47	0.000	.2652517	.3369069
_cons	1.160759	.0175664	66.08	0.000	1.12633	1.195189
state12						
STATE	.4946972	.0274396	18.03	0.000	.4409165	.5484778
_cons	1.592405	.026815	59.38	0.000	1.539849	1.644961
state17						
STATE	.603372	.0298046	20.24	0.000	.544956	.661788
_cons	1.679747	.0295774	56.79	0.000	1.621776	1.737717
var(e.trait4)	.4794808	.0303288			.4235746	.5427659
var(e.trait5)	.1698484	.0152018			.1425204	.2024164
var(e.trait9)	.579653	.0317982			.5205633	.6454502
var(e.trait15)	.267705	.0182725			.2341838	.3060245
var(e.state6)	.2482167	.0144984			.2213665	.2783236
var(e.state9)	.1531279	.0094571			.1356702	.172832
var(e.state12)	.3233183	.0212			.2843263	.3676577
var(e.state17)	.327051	.0250196			.2815128	.3799557
var(TRAIT)	1	(constrained)				
var(STATE)	1	(constrained)				
cov(TRAIT, STATE)	.6293272	.0320917	19.61	0.000	.5664286	.6922259

LR test of model vs. saturated: chi2(19) =   71.01, Prob > chi2 = 0.0000

. estimates store twofactor

The output confirms the constraints at the top of the output as well as on the lines reporting the coefficients for the latent variances. The two differences between the estimates in this model and the previous model are a) all factor loadings are estimated and have different values and b) the covariance between the latent variables is now scaled as a correlation. Item intercepts and residual variances have not changed.

A general interpretation of a factor loading is the expected difference in the item for a one-unit difference in the latent variable. In the previous model, the units of the latent variable were determined by the marker item (for example, `trait4` in the case of `TRAIT`). In the current model, we have fixed the scale to a standardized normal variable. Thus, a one-unit difference in the latent variable is a 1-standard-deviation difference. For example, the loading for `trait4` is $\lambda_1 = 0.69$, indicating that a 1-standard-deviation difference in `TRAIT` anxiety is associated with a 0.69 difference in `trait4`.

The pattern and relative sizes of the loadings help us evaluate the factorial validity of the measure. There is not a specific value or threshold for factor loadings that suggests an item is a good item versus a bad item (Brown 2015). Items with small factor loadings provide less information about the latent variable than items with large factor loadings. Ultimately, the case for whether an item should be kept in a scale is informed but not determined by a CFA. Validity is a multifaceted concept, as discussed above (see Nunnally and Bernstein [1994] for a thorough discussion of validity.

For `TRAIT`, the weakest loading is `trait9` ("Worry too much"), but it is still in the direction we expect. For `STATE`, both `state6` ("Upset") and `state9` ("Frightened") had relatively weak loadings, but again they were in the expected direction. The state items were relatively weaker indicators of the latent variable than the trait items. Given that we are working with a subset of the state and trait items, it is possible that using the full set of items would change our view. It is also possible that measuring state anxiety is more difficult than measuring trait anxiety. Or something else is going on that we have not thought of.

Another aspect of factorial validity is whether the latent constructs are distinct. That is, we fit a model that included two latent variables, but it is possible that just a single latent variable (that is, anxiety in general) could account for the relationships among the items. The correlation among the latent variables was $\phi_{21} = 0.69$, indicating a strong correlation between the latent variables but not a complete overlap. We can statistically test whether the correlation is so high that a single-factor model would be better by fitting a model that constrains the correlation between the latent variables to 1 (Kline 2011, 238). We then use a likelihood-ratio test (see section 11.9.1) to compare the fit of the constrained model with the unconstrained model. The null hypothesis of the likelihood-ratio test is that a one-factor model (that is, a single anxiety factor) fits the data as well as a two-factor model (that is, separate state and trait anxiety factors). A significance test suggests a two-factor model is better.

To fit the constrained model, add the option `covariance(TRAIT*STATE@1)` to constrain the correlation between the latent factors to 1.

```
. quietly sem (TRAIT -> trait4 trait5 trait9 trait15)
>       (STATE -> state6 state9 state12 state17),
>       variance(TRAIT@1) variance(STATE@1) covariance(TRAIT*STATE@1)
. estimates store onefactor
```

12.3.6 Standardized solutions 389

Then use `lrtest` to perform the likelihood-ratio test:

```
. lrtest twofactor onefactor
Likelihood-ratio test                              LR chi2(1) =     265.90
(Assumption: onefactor nested in twofactor)        Prob > chi2 =    0.0000
```

The two-factor model fits the data better than the one-factor model, consistent with the notion of distinct trait and state anxiety constructs.

12.3.6 Standardized solutions

Constraining the latent variance to 1 standardizes the latent variables but does not provide a fully standardized solution because the loadings still represent expected changes in the scale of the original item. Stata offers a fully standardized solution, where both latent and observed metrics are standardized. Although standardized solutions are common in the literature and across structural equation modeling software, I have reservations about them. Many measures in psychology and related fields have arbitrary metrics, and standardization is a way to help researchers compare measures with disparate scaling. Standardization is just rescaling; it does not fundamentally change the relationship or fix problems that occur because of correlation among predictors or nonnormally distributed variables (Fox 2008). Be careful and thoughtful with standardized parameters. Be cautious with interpretation, and you will remain on solid ground.

To obtain the standardized solution with `sem`, add the option `standardized`:

```
. sem (TRAIT -> trait4 trait5 trait9 trait15)
>   (STATE -> state6 state9 state12 state17), var(TRAIT@1) var(STATE@1)
>   standardized nolog
Endogenous variables
Measurement:  trait4 trait5 trait9 trait15 state6 state9 state12 state17
Exogenous variables
Latent:       TRAIT STATE
Structural equation model                          Number of obs    =      790
Estimation method  = ml
Log likelihood     = -6194.105
 ( 1)  [/]var(TRAIT) = 1
 ( 2)  [/]var(STATE) = 1
```

Standardized	Coef.	OIM Std. Err.	z	P>|z|	[95% Conf. Interval]	
Measurement						
trait4						
TRAIT	.7050815	.0225151	31.32	0.000	.6609527	.7492103
_cons	2.079291	.0632628	32.87	0.000	1.955298	2.203284
trait5						
TRAIT	.8307043	.0176708	47.01	0.000	.7960702	.8653384
_cons	2.103194	.063761	32.99	0.000	1.978225	2.228163
trait9						
TRAIT	.5226553	.0299943	17.43	0.000	.4638675	.5814431
_cons	2.318937	.0683322	33.94	0.000	2.185009	2.452866
trait15						
TRAIT	.7577073	.0199398	38.00	0.000	.7186261	.7967886
_cons	2.275113	.0673933	33.76	0.000	2.143024	2.407201
state6						
STATE	.560264	.0304278	18.41	0.000	.5006267	.6199013
_cons	2.093995	.063569	32.94	0.000	1.969402	2.218588
state9						
STATE	.6097964	.0289775	21.04	0.000	.5530015	.6665912
_cons	2.350965	.0690214	34.06	0.000	2.215686	2.486245
state12						
STATE	.6563698	.0272472	24.09	0.000	.6029662	.7097734
_cons	2.112821	.0639621	33.03	0.000	1.987458	2.238184
state17						
STATE	.7257916	.0252269	28.77	0.000	.6763477	.7752355
_cons	2.020555	.0620465	32.57	0.000	1.898946	2.142164
var(e.trait4)	.50286	.03175			.4443276	.5691031
var(e.trait5)	.3099303	.0293584			.2574147	.3731598
var(e.trait9)	.7268315	.0313534			.667906	.7909555
var(e.trait15)	.4258796	.030217			.3705888	.4894195
var(e.state6)	.6861042	.0340952			.6224302	.7562921
var(e.state9)	.6281484	.0353407			.5625643	.7013784
var(e.state12)	.5691787	.0357685			.5032191	.6437839
var(e.state17)	.4732266	.036619			.4066323	.550727
var(TRAIT)	1	(constrained)				
var(STATE)	1	(constrained)				
cov(TRAIT, STATE)	.6293272	.0320917	19.61	0.000	.5664286	.6922259

LR test of model vs. saturated: chi2(19) = 71.01, Prob > chi2 = 0.0000

If you have already fit a model and want to see the standardized output, you can type sem, standardized immediately after the original model.

12.3.6 Standardized solutions

When an item only loads on a single factor, the standardized loading can be interpreted as a correlation between the item and the factor (Kline 2011, 231). Thus, the loading for trait4 is $\lambda_1 = 0.71$, which indicates that correlation between trait4 and the latent factor TRAIT is 0.7. Because this is a correlation, squaring the loading tells us how much shared variance there is between the item and the latent factor—$R^2 = \lambda^2$. In this case,

```
. display "R2 = " .7050815^2
R2 = .49713992
```

So 49.7% of the variance in trait4 is shared with TRAIT. This quantity is similar to the communality from EFA (see section 12.2.5).

The uniqueness, or proportion of variance in trait4 that is not shared with the latent variable, is uniqueness = $1 - \lambda^2$:

```
. display "The uniqueness for trait4 = " 1 - .7050815^2
The uniqueness for trait4 = .50286008
```

So 50.3% of the variance in trait4 is not shared with TRAIT.

The R^2 estimates can be obtained by using the estat eqgof (equation-level goodness-of-fit) postestimation command:

```
. estat eqgof
Equation-level goodness of fit
```

depvars	fitted	Variance predicted	residual	R-squared	mc	mc2
observed						
trait4	.9535074	.4740266	.4794808	.49714	.7050815	.49714
trait5	.5480212	.3781728	.1698484	.6900697	.8307043	.6900697
trait9	.7975068	.2178538	.579653	.2731685	.5226553	.2731685
trait15	.6285932	.3608882	.267705	.5741204	.7577073	.5741204
state6	.361777	.1135603	.2482167	.3138958	.560264	.3138958
state9	.2437766	.0906487	.1531279	.3718516	.6097964	.3718516
state12	.5680436	.2447253	.3233183	.4308213	.6563698	.4308213
state17	.6911088	.3640578	.327051	.5267734	.7257916	.5267734
overall				.9430767		

```
mc  = correlation between depvar and its prediction
mc2 = mc^2 is the Bentler-Raykov squared multiple correlation coefficient
```

In addition to R^2, estat eqgof also prints the fitted variance (observed variance of the item), variance predicted (estimated variance of the item predicted by the latent variable), and residual variance (estimated variance of the item not predicted by the latent variable). The ratio of the variance predicted to the fitted column is equal to R^2, and the ratio of the residual column to the fitted column is equal to the uniqueness. The last two columns are coefficients that are not particularly relevant to CFA models.

Note that the mean structure of the model is unaffected by the standardization. Likewise, the covariance between the latent variables is still scaled as a correlation, as it was in the model where we constrained the variance of the latent variables to 1.

12.3.7 Global fit

As we saw in section 11.7, the most common method for assessing model fit is to examine how well the model reproduces the sample covariance matrix. The χ^2 test [see (11.7)] provides a statistical test of whether the model-implied covariance matrix is different from the sample matrix. Print the χ^2 test following `sem` by using the `estat gof` postestimation command. The default is just to report χ^2, but we add the option `stats(all)` so that the output includes the other indices in this chapter:

```
. quietly sem (TRAIT -> trait4 trait5 trait9 trait15)
> (STATE -> state6 state9 state12 state17), var(TRAIT@1) var(STATE@1) nolog
. estat gof, stats(all)
```

Fit statistic	Value	Description
Likelihood ratio		
chi2_ms(19)	71.012	model vs. saturated
p > chi2	0.000	
chi2_bs(28)	1902.949	baseline vs. saturated
p > chi2	0.000	
Population error		
RMSEA	0.059	Root mean squared error of approximation
90% CI, lower bound	0.045	
upper bound	0.074	
pclose	0.146	Probability RMSEA <= 0.05
Information criteria		
AIC	12438.210	Akaike's information criterion
BIC	12555.011	Bayesian information criterion
Baseline comparison		
CFI	0.972	Comparative fit index
TLI	0.959	Tucker-Lewis index
Size of residuals		
SRMR	0.032	Standardized root mean squared residual
CD	0.943	Coefficient of determination

As we can see, $\chi^2(19) = 71.01$, $p < 0.001$. Thus, we reject the null hypothesis that the model implied and sample covariance matrices are equal.

The problem with the χ^2 test is that it is sensitive to sample size and is testing a restrictive null hypothesis that the model exactly reproduces the sample covariance. Models are representations of the data that (we hope) are useful in understanding relationships and testing theories. We do not expect them to be the truth or exact (MacCallum and Austin 2000). Consequently, researchers have developed other mea-

12.3.7 Global fit 393

sures of model fit to help users evaluate how their models perform. This chapter covers four measures of model fit that Stata provides and that are unique to structural equation modeling:

1. Root mean squared error of approximation (RMSEA)
2. Tucker–Lewis index (TLI)
3. Comparative fit index (CFI)
4. Standardized root mean squared residual (SRMR)

RMSEA

The RMSEA provides an index of model misfit. If the RMSEA is 0, then there is no model misfit, and the model-implied and sample covariances align. RMSEA values greater than 0 indicate the degree of model misfit or misspecification. An advantage of the RMSEA is that it is parsimony corrected, which means that it penalizes complex models for being complex. We can typically improve model fit by adding parameters (for example, covariances among errors or cross-loadings), but often, adding these parameters capitalizes on chance and takes of advantage of peculiarities in the dataset. Thus, the RMSEA tries to correct for this by increasing the RMSEA values as models become more complex.

As seen in section 11.7.1, the model fit [(11.7)] follows a χ^2 distribution, specifically a central χ^2 distribution (Kline 2011).[5] The central χ^2 distribution is the χ^2 distribution you know and love and learned about in your introductory statistics course. The RMSEA quantifies how much the model actually follows a noncentral χ^2 distribution. The more it follows the noncentral distribution, the poorer the fit of the model. The formula for the RMSEA is (Brown 2015)

$$\text{RMSEA} = \sqrt{\frac{\chi_m^2 - \text{DF}_m}{\text{DF}_m N}}$$

The denominator includes model DF. As models become more complex, the DF gets smaller, which in turn increases the RMSEA. This is the parsimony correction. The impact of this correction gets smaller when the sample size increases (Kline 2011).

As is common with fit indices, there are "rules of thumb" for interpreting RMSEA values. It is common to see recommendations that RMSEA should be 0.05 or less, that the confidence interval (CI) for the RMSEA should include 0.05, or that the CI should not include values that likely indicate poor fit (Kline 2015). The RMSEA, like any statistic, is subject to sampling error and lack of precision due to small sample sizes. Consequently, be careful when applying the rules of thumb when accepting or rejecting a model (see section 12.3.7).

---

5. The sample size must be sufficiently large and assumptions about multivariate normality must be met (Kline 2011).

TLI

The TLI compares the fit of the model with the fit of the baseline or null model. Thus, it quantifies how much better the proposed model fits than the baseline model. Like the RMSEA, the TLI includes a penalty for model complexity. Higher values of the TLI indicate better fit, and a common rule of thumb is that TLI values greater than 0.9 indicate good fit and greater than 0.95 indicate excellent fit. The formula for the TLI is (Brown 2015)

$$\text{TLI} = \frac{\frac{\chi_B^2}{\text{DF}_B} - \frac{\chi_M^2}{\text{DF}_M}}{\frac{\chi_B^2}{\text{DF}_B} - 1}$$

where χ_B^2 and DF_B are the χ^2 and DF from the baseline model, and χ_M^2 and DF_M are the χ^2 and DF from the model.

In Stata, and most structural equation modeling programs, the baseline model estimates the means and variances for all observed variables (for example, all items in a CFA) but fixes all covariances among items to 0. That is, the baseline model assumes that the items are all unrelated. The baseline model is not plausible and is simply used as a reference point, hence, the suggestion that values close to 1 indicate good fit. Unfortunately, for some models, the default baseline model is inappropriate, and users will need to fit their own baseline model when computing the TLI, as discussed in section 13.3.5 (Widaman and Thompson 2003).

CFI

The CFI, like the TLI, compares the fit of the model to the fit of the baseline model. The baseline model is the same as in the TLI, so the cautions discussed then apply here. The rules of thumb for interpreting the CFI are identical to those for the TLI. The CFI does not include a correction for model complexity. However, the CFI and TLI are often close in value and rarely, in my experience, lead to any meaningful differences in conclusions. The formula for the CFI is (Brown 2015)

$$\text{CFI} = 1 - \frac{\chi_M^2 - \text{DF}_M}{\max\left(\chi_B^2 - \text{DF}_B, \chi_M^2 - \text{DF}_M\right)} \tag{12.1}$$

The quantities are the same as with the TLI. The denominator includes the term max, which means that we use the biggest value of either $\chi_B^2 - \text{DF}_B$ or $\chi_M^2 - \text{DF}_M$.

SRMR

The SRMR is sometimes called a "badness-of-fit" statistic (Kline 2015, 277) because it describes how far, on average, the model-implied correlation among the observed variables differs from the actual correlations among the observed variables. Correlations are used because covariances are unstandardized and thus difficult to interpret and compare across models. If the SRMR is large, then the model does not do a good job reproducing the observed correlations. A general rule of thumb is that the SRMR should be 0.10 or lower.

A summary and a caution

Although the χ^2 was significant for the STAI analysis, the supplementary fit indices were generally consistent with good fit. RMSEA = 0.059 (90% CI = 0.045, 0.074), which is slightly over the 0.05 threshold, but the upper bound of the CI was below 0.10. CFI = 0.97, TLI = 0.96, and SRMR = 0.03, all of which are consistent with good fit. Taken together, the statistical evidence suggests that the two-factor model adequately fits the data and supports the factorial validity of the measure.

Do not be seduced into thinking that any or all of these various measures of model fit prove that a model is correct or even good (MacCallum and Austin 2000). The effects, such as the loadings, might be weak in a model with good fit. MacCallum and Austin (2000) note the following:

> In addition, a finding of good fit does not imply that effects hypothesized in the model are strong. In fact, it is entirely possible for relationships among variables to be weak, or even zero, and for a hypothesized model to fit extremely well. The resulting parameter estimates would reflect weak relationships among variables and large residual variances for endogenous variables. Thus, it is critically important to pay attention to parameter estimates, even when fit is very good. (p. 218)

Bear in mind what each index measures, know their limits, and remember that there are many models that will fit "well" for a given dataset (Kline 2011). The binary decision that a model fits well or not is a holdover from the way null hypothesis significance testing is typically used—we decide whether an effect is important via a binary decision about statistical significance. The utility of models requires far more than an RMSEA less than 0.05 or a CFI greater than 0.95. Models should be useful in prediction, consistent with theory, and replicate across datasets (Kline 2011, chap. 13).

12.3.8 Refining models further

In the case that a model does not fit well, additional exploratory analyses can be performed to examine how a model can be adjusted to improve fit. The most common method for exploration is to use modification indices. Modification indices quantify how much model fit will improve by adding parameters to models, such as allowing a variable to load on two factors or introducing a correlation among residual variances. Stata can produce modification indices with the `estat mindices` postestimation command. The details of how to use modification indices is beyond the scope of this chapter (for more details, see Acock [2013], Brown [2015], and Kline [2015]). Suffice it to say that modification indices can open a can of worms when it comes to "hacking" your model to make it look better than it is and can produce errors or models that are simply a reflection of the peculiarities of your specific dataset (Kline 2015). I regularly receive papers to review that use CFA, and the models the authors present include lots of correlations among residuals or nonintuitive cross-loadings. Most of the time, these parameters come from modification indices and do not have a good theoretical basis. This, at a minimum, makes me skeptical and often can lead me not to trust the model at all. Thus, be careful and be confident that you know what you are doing.

12.4 Summary

This chapter discussed how to use EFA and CFA to examine the factor structure, and thus factorial validity, of a measure. Generally, we saw that a two-factor solution—one for state anxiety and one for trait anxiety—fit the data well. Thus, there is some evidence for the factorial validity of the STAI. We fit both an EFA and a CFA to the data. This is not common. Typically, we would use one method or the other in a paper, or if we had a large dataset, we would split the data in two. We would perform an exploratory analysis on the initial dataset and then confirm the exploration in the second dataset.

We covered a lot of ground in this chapter. Specifically, we did the following:

- Distinguished between EFA and CFA
- Defined the common factor model
- Discussed extraction methods in EFA
- Discussed how to interpret factor loadings, eigenvalues, communalities, and uniquenesses in EFA
- Discussed how to decide how many factors to extract in EFA
- Defined and discussed some methods for rotating EFA solutions
- Distinguished between the mean and the variance structure in a CFA
- Discussed the issue of identification in CFA and discussed strategies for identifying models
- Discussed the difference between a standardized and an unstandardized CFA
- Defined and interpreted measures of global fit

We have only scratched the surface of factor analysis. Readers wanting more information about EFA can consult Rencher and Christensen (2012), Nunnally and Bernstein (1994), and Harman (1976). Likewise, readers wanting more information about CFA can consult chapters 11 and 13, as well as Acock (2013), Brown (2015), and Kline (2015).

13 Measurement invariance

Comparisons are commonplace in science. We compare treatment conditions, experimental manipulations, people with a disease to healthy controls, majority groups to minority groups, and the same people over time. Implicit in any comparison is that the outcome measure is psychometrically equivalent across groups, time, or whatever variable we compare. For example, comparing treatment groups assumes that items on the outcome measure load equally well on the latent construct across groups. If one or more items loads poorly in one of the groups, then it is impossible to say whether group differences are due to actual group difference or due to methodological differences (that is, measurement differences).

In this chapter, we examine how to use confirmatory factor analysis (CFA) to examine measurement invariance across groups and across time. In essence, this is an extension of the reliability discussion in chapter 11 because we can now evaluate how consistent a measure is for varying groups and on different occasions. In this chapter, you will learn the following:

1. The difference between construct equality and measurement equality.

2. The different levels of measurement invariance and how to set up a CFA model for each level.

3. How to evaluate measurement invariance across groups using CFA.

4. How to evaluate measurement invariance across time using CFA.

Stata commands featured in this chapter

- `sem`: for fitting CFA models
- `estimates store`: for saving the estimates and statistics from models
- `lrtest`: for conducting a likelihood-ratio test that compares the fit of two nested models
- `constraint`: for using the effects-coding method of identification

13.1 Data

This chapter includes two datasets. The first dataset is ybocs.dta, which includes responses to the first five items of the Yale–Brown Obsessive Compulsive Scale (YBOCS) (Goodman et al. 1989) for $N = 287$ participants as well as the participant's sex. The YBOCS items are rated by a clinician from 0 to 4, with 0 representing no symptom/distress and 4 representing extreme distress. The variables are

- ybocs1: Time occupied by obsessive thoughts
- ybocs2: Interference due to obsessive thoughts
- ybocs3: Distress associated with obsessive thoughts
- ybocs4: Resistance against obsessions
- ybocs5: Degree of control over obsessive thoughts
- sex: Birth sex (0 = men, 1 = women)

ybocs.dta is used to find about measurement invariance across groups (see section 13.3).

The second dataset is early_child.dta, which includes responses to four items aimed at measuring concentration in young children. This dataset is drawn from the publicly available Early Childhood Longitudinal Program that is part of the National Center for Education Statistics (https://nces.ed.gov/ecls/dataproducts.asp). The dataset for this chapter includes $N = 11986$ participants measured in kindergarten and first grade. Teachers at school rated each child on four items on a scale of 1 (extremely untrue) to 7 (extremely true). Variables with a 1 at the end are for kindergarten, and variables with a 2 are for first grade.

- concent1 and concent2: Child has strong concentration while drawing or coloring.
- involved1 and involved2: Child is very involved or works a long time when building.
- absorbed1 and absorbed2: Child is absorbed in a picture book for a long time.
- wait1 and wait2: Child can wait if asked before an activity.

early_child.dta is used to learn about measurement invariance across time (see section 13.5).

13.2 Measurement invariance

Measurement invariance analyses investigate the extent to which a measure taps the same construct in the same way across groups or time (Bollen 1989; Byrne, Shavelson, and Muthén 1989; Millsap 2011; Reise, Widaman, and Pugh 1993; Steenkamp and

Baumgartner 1998; Vandenberg and Lance 2000). There are two parts to this definition. The first part is the idea that a measure taps the same construct across groups. In the case of the YBOCS data, this is the idea that the five YBOCS items all load on a single construct that we believe represents "obsessions". The second part is the idea that a measure taps the same construct in the same way across groups. For the YBOCS data, this means that factor loadings, item means, and item residual variances are equal across groups.

Assessing measurement invariance requires several steps, each of which includes a particular CFA model. The first step starts with the least constrained, most flexible model. As the steps proceed, we introduce new constraints to test invariance across groups or time. Model fit is compared from one step to another to evaluate whether introducing constraints significantly degrades model fit. If model fit is not worse once a set of constraints is added, then we conclude that a measure is invariant for that step.

There are four steps in evaluating measurement invariance.[1]

1. **Configural invariance**: Do the groups have the same factor structure?
2. **Metric invariance**: Do the groups have the same factor loadings?
3. **Scalar invariance**: Do the groups have the same item intercepts?
4. **Residual invariance**: Do the groups have the same residual variances?

We can also assess what is called structural invariance, which is the question of whether the mean and the variance of the latent variables vary across groups. In other words, structural invariance involves whether the constructs differ once we have established that measurement invariance holds. There are two steps in evaluating structural invariance (when there is a single latent factor):

---

1. Sometimes scholars recommend a test that evaluates the equality of the covariance matrices across groups (Vandenberg and Lance 2000). The logic is that if a measure is not invariant, then the covariance matrices will be the same. The null hypothesis is that the covariance matrices do not differ. If the null is not rejected, then no further invariance tests are needed because there is no source of differences.

   This test can be conducted using the YBOCS data as follows:

   ```
   sem (ybocs1 ybocs2 ybocs3 ybocs4 ybocs5), ///
       covstructure(_Ex, unstructured) group(sex) ginvariant(none)
   estimates store omnibus_free
   sem (ybocs1 ybocs2 ybocs3 ybocs4 ybocs5), ///
       covstructure(_Ex, unstructured) group(sex) ginvariant(all)
   estimates store omnibus_const
   lrtest omnibus_free omnibus_const
   ```

   In this example, we do not reject the null hypothesis in the YBOCS data, suggesting that the additional tests are not needed. We proceed with the tests anyway because we are learning the ideas.

1. **Invariant factor variances**: Evaluate differences in the latent variance across groups.
2. **Invariant factor means**: Evaluate differences in the latent mean across groups.

In what follows, I provide details on how to test each of these models and discuss the interpretation of invariance tests as well as any implications of failure to establish invariance.

13.3 Measurement invariance across groups

13.3.1 Configural invariance

The first step to establishing measurement invariance across groups is to specify the measurement model and see whether the model fits equally well in each group. In other words, we examine configural invariance. The main goal here is to see whether the pattern of loadings is the same in all groups (Steenkamp and Baumgartner 1998). For example, in a single-factor model, do all items load well on just one factor or is there more than one factor in a group or do some items not load on any latent variable in one of the groups? If configural invariance cannot be established, then it is not useful to compare the groups on a measure, because the items are not assessing the same construct (Steenkamp and Baumgartner 1998; Vandenberg and Lance 2000).

Figure 13.1 presents a possible model for the five YBOCS obsession items. All five items load on a single latent factor labeled "Obsessions", and the factor structure is identical across groups. Note that all parameters in figure 13.1 are subscripted with either an m for men or a w for women. This indicates that all parameters will be freely estimated in each group, with the exception of any parameters needed for identification.

Figure 13.1 introduces item intercepts, denoted by τ, and latent variable intercepts, denoted by κ. As we saw in chapter 11 [(11.3)], the prediction equation in confirmatory factor analysis includes item intercepts. Previously, we ignored them, but in this chapter the intercepts will be central to the models we fit. The latent variable mean, κ, is often fixed to 0 (because the scaling of latent variables is arbitrary). However, in this chapter we will investigate whether groups have different latent means, so they become a parameter in the models. The relationship between a latent variable mean and an item mean is expressed as (Brown 2015, 225)

$$M_X = \tau_X + \lambda_X \kappa$$

Thus, the item mean is a function of the latent mean, factor loading, and item intercept. If the latent mean is fixed to 0, as it commonly is, then the item intercept is equal to the item mean.

13.3.1 Configural invariance

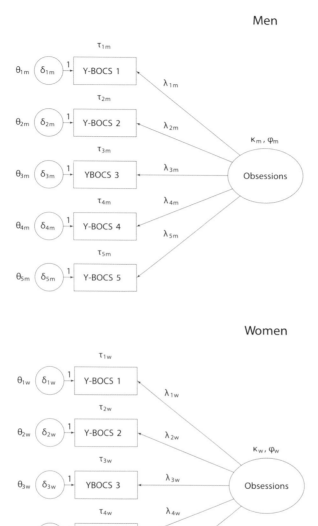

Figure 13.1. Measurement model for YBOCS across men and women

We have several options for identification (also see chapter 12), but for these models, we follow procedures that provide a straightforward way to compare the various models (Reise, Widaman, and Pugh 1993; Widaman and Reise 1997; Bontempo and Hofer 2007). Specifically, the configural model is fit as follows:

1. Fit the same model with respect to factor loadings, item intercepts, residual variances and covariances, number of factors, and covariances among the factors.

2. Select one of the groups to be a reference group (we select men in this example). Fix the mean of the latent variable to 0 and the variance of the latent variable to 1 in the reference group only. Thus, $\kappa_m = 0$ and $\phi_m = 1$.

3. Freely estimate the mean and variance of the latent variable in all other groups.

4. Choose one item that will have its factor loading and intercept constrained to be equal across groups.[2] The parameters will be estimated—the loading will not be fixed to 1 nor will the intercept be fixed to 0—and be required to be equal across groups. In the YBOCS data, we choose the first item. Thus, $\lambda_{1m} = \lambda_{1w}$ and $\tau_{1m} = \tau_{1w}$.

5. All other model parameters will be estimated and allowed to vary across groups.

To fit the configural model, we use group(*varname*) and ginvariant(*classname*). The group() option indicates which variable identifies the groups—in this example, that is women. The ginvariant() option lets us establish which parameters we want to constrain across groups. The argument for ginvariant() is what Stata calls a *classname* or a class of coefficients, which is a group of coefficients in a structural equation model. The classes we use in this chapter are

- mcoef: measurement coefficients, which are the factor loadings (λ's)
- mcons: measurement intercepts, which are the item intercepts (τ's)
- merrvar: covariances of errors, which covers covariance among the errors as well as the error variances (θ's)
- all: constrain all parameters across groups
- none: do not constrain any parameters across groups

We also manually constrain some coefficients out of necessity as well as to increase clarity of what is being constrained.

---

2. There are no hard-and-fast rules regarding which item to select to constrain. Bontempo and Hofer (2007) state the following:

> Choosing which indicator to use as the reference item can be done on a theoretical or statistical basis. The model fit is not affected by this choice. Theoretically, it might be convenient to choose an indicator that has the highest face validity with respect to the proposed latent construct. On a more statistical basis, an indicator that had the largest loading in prior EFA [exploratory factor analysis] studies could be chosen. If EFA has previously been conducted within each group, an indicator having a relatively large loading and consistent loading across groups would be desirable because these circumstances suggest that this indicator has an invariant relation with the common factor. (p. 162)

13.3.1 Configural invariance

The sem code for the configural model is

```
. use http://www.stata-press.com/data/pspus/ybocs2
. sem (0: OBSESS -> ybocs1@l1 ybocs2 ybocs3 ybocs4 ybocs5)
>     (1: OBSESS -> ybocs1@l1 ybocs2 ybocs3 ybocs4 ybocs5)
>     (0: ybocs1 <- _cons@i1)
>     (1: ybocs1 <- _cons@i1),
>     group(women) ginvariant(none)
>     var(0: OBSESS@1) var(1:OBSESS@var) means(0:OBSESS@0) means(1:OBSESS@mean)
>     nolog nocnsreport
```

Endogenous variables

Measurement: ybocs1 ybocs2 ybocs3 ybocs4 ybocs5

Exogenous variables

Latent: OBSESS

Structural equation model		Number of obs	=	287
Grouping variable = women		Number of groups	=	2
Estimation method = ml				
Log likelihood = -1733.8992				

Group   : male   Number of obs =   147

	Coef.	OIM Std. Err.	z	P>|z|	[95% Conf. Interval]	
Measurement						
ybocs1						
OBSESS	.7365645	.0774425	9.51	0.000	.5847799	.888349
_cons	2.816327	.0810754	34.74	0.000	2.657422	2.975231
ybocs2						
OBSESS	.7928292	.0766154	10.35	0.000	.6426658	.9429926
_cons	2.489796	.0807018	30.85	0.000	2.331623	2.647969
ybocs3						
OBSESS	.5565066	.060593	9.18	0.000	.4377466	.6752667
_cons	2.591837	.0629059	41.20	0.000	2.468543	2.71513
ybocs4						
OBSESS	.3872515	.0907291	4.27	0.000	.2094258	.5650772
_cons	2.217687	.0841544	26.35	0.000	2.052748	2.382627
ybocs5						
OBSESS	.3095989	.0763898	4.05	0.000	.1598777	.4593201
_cons	2.619048	.0715634	36.60	0.000	2.478786	2.759309
mean(OBSESS)	0	(constrained)				
var(e.ybocs1)	.4237369	.0720953			.3035787	.5914543
var(e.ybocs2)	.3288007	.0723199			.2136537	.5060053
var(e.ybocs3)	.2720024	.0442458			.1977472	.3741409
var(e.ybocs4)	.891084	.1086224			.7017103	1.131565
var(e.ybocs5)	.656983	.0791912			.518743	.8320626
var(OBSESS)	1	(constrained)				

```
Group                : female                    Number of obs    =       140

                           OIM
               Coef.    Std. Err.      z     P>|z|     [95% Conf. Interval]

Measurement
  ybocs1
    OBSESS    .7365645   .0774425    9.51    0.000    .5847799    .888349
    _cons     2.816327   .0810754   34.74    0.000    2.657422    2.975231

  ybocs2
    OBSESS    .7410249   .1524625    4.86    0.000    .4422039    1.039846
    _cons     2.294218   .1173951   19.54    0.000    2.064128    2.524308

  ybocs3
    OBSESS    .5870428   .1181546    4.97    0.000    .355464     .8186216
    _cons     2.496255   .0914043   27.31    0.000    2.317106    2.675404

  ybocs4
    OBSESS    .3520222   .1274159    2.76    0.006    .1022915    .6017528
    _cons     2.124927   .0956197   22.22    0.000    1.937516    2.312338

  ybocs5
    OBSESS    .3135367   .097093     3.23    0.001    .1232379    .5038356
    _cons     2.637934   .0754882   34.94    0.000    2.48998     2.785889

mean(OBSESS)  .1523898   .1523626    1.00    0.317   -.1462354    .451015

var(e.ybocs1) .4191774   .0780099                     .2910632    .6036822
var(e.ybocs2) .5178884   .0882168                     .3708883    .7231515
var(e.ybocs3) .2858424   .0511247                     .2013185    .4058536
var(e.ybocs4) .9257672   .1150527                     .7256312    1.181103
var(e.ybocs5) .5279673   .0664847                     .4124951    .6757643
var(OBSESS)   .7452013   .2471567                     .3890097    1.427535

LR test of model vs. saturated: chi2(10) =     20.57, Prob > chi2 = 0.0243
. estimates store configural
```

There are eight key aspects of the sem code.

1. The group(women) option indicates that we are fitting a multiple-group model and that one or more parameters may be allowed to vary across groups.

2. There is a leading 0: or 1: before parts of the model. The 0 and 1 refer to the level of the women variable. Syntax that begins with 0: are models for men (because men have a 0 on women), and syntax that begins with 1: are models for women.

3. The ginvariant(none) option tells Stata to not force any constraints across groups.

4. The variance of the latent variables is controlled via the var() option, and the mean of the latent variables is controlled via the means() option.

5. The variance of the latent variable is constrained to 1 for men only, as indicated by the var(0: OBSESS@1) option. The variance of the latent variable for women

13.3.1 Configural invariance

is freely estimated, as indicated by the `var(1: OBSESS@var)`. If you have used the `ginvariant(none)` option, you do not need to state `var(1: OBSESS@var)`. However, I like to be explicit in the code because it helps me keep track of what I intended and ensure that Stata fit the model I intended.

6. The mean of the latent variable is constrained to 0 for men only, as indicated by the `means(0: OBSESS@0)` option. The mean of the latent variable for women is freely estimated, as indicated by the `means(1: OBSESS@mean)`. As with the variance, the explicit option for women is not needed but is included for clarity.

7. The first two lines of code specify the loading patterns. The first loading for both men and women is constrained to be equal across groups, as indicated by `ybocs1@l1`. The `l1` is a label for the constraint, and the specific label used is arbitrary (`l1` simply stands for loading 1).

8. The third and fourth lines of code allow the intercept for the first item to be constrained to be equal across groups, as indicated by `_cons@i1` (`i1` stands for intercept 1, though again the name used is arbitrary).

The output provides parameter estimates stratified by gender if the parameter was allowed to vary. If the parameter was constrained, it is marked with a `[*]`. At the end of the output, we stored the estimated via `estimates store configural` so that we can use them at a later point (see section 13.3.5).

Table 13.1 provides the unstandardized coefficients for all invariance models. The first column represents the configural model; the loadings and the intercept for `ybocs1` are constrained to be equal across groups. All other parameters are free to vary across groups. As noted above, when evaluating configural invariance, the key issue is whether the pattern of loadings is similar across groups. The issue is not whether the loadings are equal, but rather whether the items load on the same factor across groups (Bontempo and Hofer 2007; Widaman and Reise 1997). As seen in table 13.1, this is true. One could argue that the loadings for `ybocs4` and `ybocs5` are low and may not be important. However, this is true in both groups, which means it is consistent with configural invariance.

Table 13.1. YBOCS measurement invariance results

	Configural	Metric	Scalar	Residual	Variance	Mean
YBOCS1						
λ_{men}	0.74	0.74	0.74	0.73	0.69	0.69
λ_{women}	0.74	0.74	0.74	0.73	0.69	0.69
τ_{men}	2.82	2.82	2.87	2.87	2.87	2.87
τ_{women}	2.82	2.82	2.87	2.87	2.87	2.87
YBOCS2						
λ_{men}	0.79	0.78	0.78	0.76	0.72	0.72
λ_{women}	0.74	0.78	0.78	0.76	0.72	0.72
τ_{men}	2.49	2.49	2.45	2.44	2.44	2.45
τ_{women}	2.29	2.29	2.45	2.44	2.44	2.45

Continued on next page

	Configural	Metric	Scalar	Residual	Variance	Mean
YBOCS3						
λ_{men}	0.56	0.57	0.57	0.57	0.53	0.53
λ_{women}	0.59	0.57	0.57	0.57	0.53	0.53
τ_{men}	2.59	2.59	2.59	2.59	2.59	2.59
τ_{women}	2.50	2.50	2.59	2.59	2.59	2.59
YBOCS4						
λ_{men}	0.39	0.37	0.37	0.37	0.35	0.35
λ_{women}	0.35	0.37	0.37	0.37	0.35	0.35
τ_{men}	2.22	2.22	2.20	2.20	2.20	2.20
τ_{women}	2.12	2.12	2.20	2.20	2.20	2.20
YBOCS5						
λ_{men}	0.31	0.31	0.31	0.31	0.29	0.29
λ_{women}	0.31	0.31	0.31	0.31	0.29	0.29
τ_{men}	2.62	2.62	2.65	2.65	2.65	2.65
τ_{women}	2.64	2.64	2.65	2.65	2.65	2.65
Residuals, latent mean, and latent variance						
$\theta_{1\text{men}}$	0.42	0.42	0.42	0.42	0.43	0.43
$\theta_{1\text{women}}$	0.42	0.42	0.43	0.42	0.43	0.43
$\theta_{2\text{men}}$	0.33	0.34	0.34	0.43	0.43	0.43
$\theta_{2\text{women}}$	0.52	0.50	0.51	0.43	0.43	0.43
$\theta_{3\text{men}}$	0.27	0.27	0.27	0.28	0.28	0.28
$\theta_{3\text{women}}$	0.29	0.30	0.29	0.28	0.28	0.28
$\theta_{4\text{men}}$	0.89	0.90	0.90	0.91	0.91	0.91
$\theta_{4\text{women}}$	0.93	0.92	0.92	0.91	0.91	0.91
$\theta_{5\text{men}}$	0.66	0.66	0.66	0.60	0.60	0.60
$\theta_{5\text{women}}$	0.53	0.53	0.53	0.60	0.60	0.60
κ_{men}	0.00	0.00	0.00	0.00	0.00	0.00
κ_{women}	0.15	0.15	0.01	0.01	0.01	0.00
ϕ_{men}	1.00	1.00	1.00	1.00	1.00	1.00
ϕ_{women}	0.75	0.73	0.73	0.76	1.00	1.00
N	287	287	287	287	287	287
χ^2_{diff}		0.34	3.91	4.72	1.62	0.01
DF		4	4	5	1	1
p		0.99	0.42	0.45	0.20	0.92
Adj. CFI	0.962	0.975	0.976	0.977	0.975	0.978

NOTE: The first two coefficients for every outcome are the loadings, and the second two are intercepts.

Configural invariance is a prerequisite for measurement invariance. However, it does not provide the evidence needed to begin making group comparisons on the construct because it does not help us understand whether the latent construct is on the same scale across groups. That is, even though the measure appears to be assessing the same construct across groups, differences between groups in the mean or in a correlation between the latent construct and another variable could be real or they could be due to different factor loadings or intercepts between groups. Other forms of invariance, such as metric and scalar invariance, will assist with these types of comparisons. Consequently, the configural model will serve as a baseline model for making comparisons.

13.3.2 Metric invariance

Metric invariance makes the same assumptions as configural invariance but adds the additional constraint that all loadings should be equal across groups. In this case, that means $\lambda_{1m} = \lambda_{1w}$, $\lambda_{2m} = \lambda_{2w}$, $\lambda_{3m} = \lambda_{3w}$, $\lambda_{4m} = \lambda_{4w}$, and $\lambda_{5m} = \lambda_{5w}$. This does not mean that loadings within a group are held equal. The metric invariance model syntax is

```
. sem (0: OBSESS -> ybocs1@l1 ybocs2 ybocs3 ybocs4 ybocs5)
>     (1: OBSESS -> ybocs1@l1 ybocs2 ybocs3 ybocs4 ybocs5)
>     (0: ybocs1 <- _cons@i1)
>     (1: ybocs1 <- _cons@i1),
>     group(women) ginvariant(mcoef)
>     var(0: OBSESS@1) var(1:OBSESS@var) means(0:OBSESS@0) means(1:OBSESS@mean)
  (output omitted)
. estimates store metric
```

The only change from the configural model is the addition of the `ginvariant(mcoef)` option, which forces the factor loadings to be equal across groups. The same constraints could be achieved by using the @ notation plus a name for the parameter, as we did for `ybocs1`. Even though we used `ginvariant(mcoef)`, Stata still requires that the constraint for `ybocs1` be written out (that is, the @l1 be included for both groups). If we do not use the constraint, Stata will force the factor loading for women to be 1 because we estimate the variance for women.

Table 13.1 provides the estimates for the metric model. As expected, all loadings are equal across groups, but all other parameters are free to vary. The final three rows of table 13.1 provide a χ^2-difference test [see section 11.9.1, (11.11) and (11.12)]. In the metric column, the test compares the fit of the configural model with the metric model. There are 4 degrees of freedom (DF) because the metric model estimates four fewer parameters than the configural because of the addition of the constraints on the loadings for ybocs2–ybocs5. The test is not statistically significant, which is consistent with metric invariance (that is, the assumption of equal factor loadings across groups is reasonable). Given the evidence for metric invariance, group comparisons involving variances or covariances/correlations are interpretable (Widaman and Reise 1997).

13.3.3 Scalar invariance

The scalar invariance model extends the metric invariance model by adding constraints to the item intercepts. Specifically, in addition to the constraints on loadings, we constrain $\tau_{1m} = \tau_{1w}$, $\tau_{2m} = \tau_{2w}$, $\tau_{3m} = \tau_{3w}$, $\tau_{4m} = \tau_{4w}$, and $\tau_{5m} = \tau_{5w}$. The metric invariance model in Stata is

```
. sem (0: OBSESS -> ybocs1@l1 ybocs2 ybocs3 ybocs4 ybocs5)
>     (1: OBSESS -> ybocs1@l1 ybocs2 ybocs3 ybocs4 ybocs5),
>     group(women) ginvariant(mcoef mcons)
>     var(0: OBSESS@1) var(1:OBSESS@var) means(0:OBSESS@0) means(1:OBSESS@mean)
  (output omitted)
. estimates store scalar
```

We now use `ginvariant(mcoef mcons)`, where `mcons` tells Stata to constrain the item intercepts across groups.

Table 13.1 shows that both the loadings and the intercepts are constrained in the scalar model. The χ^2-difference test compares the scalar model to the metric model. There are 4 DF because the scalar model constrains four more intercepts than the scalar model. The test is not significant, suggesting that the assumption of equal item intercepts is reasonable. This suggests that one group does not consistently score higher on the items (after accounting for the latent construct). Given the evidence for scalar invariance, group comparisons involving mean differences as well as variance/covariance differences are warranted (Widaman and Reise 1997).

13.3.4 Residual invariance

The residual invariance model extends the scalar invariance model by adding constraints to the item residual variances. In the residual model, the additional constraints are $\theta_{1m} = \theta_{1w}$, $\theta_{2m} = \theta_{2w}$, $\theta_{3m} = \theta_{3w}$, $\theta_{4m} = \theta_{4w}$, and $\theta_{5m} = \theta_{5w}$. The scalar invariance model in Stata is

```
. sem (0: OBSESS -> ybocs1@l1 ybocs2 ybocs3 ybocs4 ybocs5)
>     (1: OBSESS -> ybocs1@l1 ybocs2 ybocs3 ybocs4 ybocs5),
>     group(women) ginvariant(mcoef mcons merrvar)
>     var(0: OBSESS@1) var(1:OBSESS@var) means(0:OBSESS@0) means(1:OBSESS@mean)
```
  (output omitted)
```
. estimates score residual
```

We now use `ginvariant(mcoef mcons merrvar)`, where `merrvar` tells Stata to constrain the residual variances across groups.

Table 13.1 shows that the residual variances are constrained in the residual invariance model. The χ^2-difference test compares the residual model to the scalar model. This test has 5 DF because of the constraints on all five residual variances that were not present in the metric or scalar models. This is in contrast to the two previous difference tests, which had 4 DF. In the configural and metric models, constraints on a loading and an intercept were required for identification of the model. No constraints have been needed for the residual variances up to this point. The test comparing the scalar and the residual invariance models was not significant, suggesting that the assumption of equal residual variances is reasonable.

Residual invariance is not required to make group comparisons on means and variances; only scalar invariance is needed (Widaman and Reise 1997). The most important benefit of residual invariance is that it greatly simplifies interpretation. Widaman and Reise (1997) note that if residual invariance holds, "... then group differences in means and variances on the measured variables are a function of only group differences on the means and variances of the common factors" (p. 296).

13.3.5 Using the comparative fit index to evaluate invariance

Up to this point, we used a χ^2-difference test to determine whether adding invariance constraints significantly degraded model fit. Although none of the χ^2-difference tests in the YBOCS examples were significant, the χ^2-difference test can be too stringent of a test. For example, trivial levels of measurement invariance can lead to a significant test, especially when sample sizes are large (Cheung and Rensvold 2002; Meade, Johnson, and Braddy 2008).

An alternative is to use the change in the comparative fit index (Cheung and Rensvold 2002; Meade, Johnson, and Braddy 2008; Little 2013), which is a global fit index discussed in section 12.3.7. Simulations show that the comparative fit index (CFI) is not affected as much as the χ^2-difference test by sample size while still being sensitive to invariance. The recommended threshold for a change in the CFI is -0.01 to -0.002 (Cheung and Rensvold 2002; Meade, Johnson, and Braddy 2008). This range is fairly large, which underscores the need to not simply rely on statistical criteria as the only criteria for establishing invariance. Consider theory and previous results as context for interpreting the CFI changes.

A challenge with using the CFI is that its calculation relies on the specification of a null model [see (12.1)], which is a model that estimates all means and all variances of each item but constrains all covariances among the items to be 0. That is, the null model assumes that the items are completely unrelated. In a multiple-group setting, the null model estimates separate item means and variances across groups. Stata automatically computes the null model as part of the estimation process.

Before looking at statistics that rely on the null model, you must verify that the null model is nested in the model you are fitting. Unfortunately, when conducting invariance tests, the null model used by Stata is not the correct null model. Widaman and Thompson (2003) show that models that place constraints on the item intercepts (as is done in the scalar invariance tests) or the item residual variances (as is done in the residual invariance tests) are not nested within the standard null model. Thus, the CFI, and any changes in the CFI from one model to the next, are incorrect. Widaman and Thompson (2003) state, "If the standard null model is not nested within competing substantive models, incremental fit indices computed using the standard null model will be incorrect, have no useful interpretation, and should be disregarded" (p. 34). The difference between using the standard null model and a correct null model will not always be large. Nevertheless, I believe it is important to get things correct and precise when we can, so correctly specifying the null and computing the CFI is worthwhile.

A suitable null model for the multiple-group scenario discussed here, like the YBOCS data, is to estimate all item means and variances but constrain them to be equal across groups.[3] We can fit this model in Stata as follows:

---

3. Other null models are possible. For example, if we do not need to evaluate residual invariance, then a null model with just constrained means would be fine. If all we care about is metric invariance, then the default null model will work.

```
. sem (0: ybocs1 ybocs2 ybocs3 ybocs4 ybocs5)
>       (1: ybocs1 ybocs2 ybocs3 ybocs4 ybocs5),
>       group(women) ginvariant(meanex covex)
>       covstruct(_OEx, diagonal) nolog
Exogenous variables
Observed:   ybocs1 ybocs2 ybocs3 ybocs4 ybocs5
Structural equation model                  Number of obs    =        287
Grouping variable   = women                Number of groups =          2
Estimation method   = ml
Log likelihood      = -1879.2036
 ( 1)  [/]var(ybocs1)#0bn.women - [/]var(ybocs1)#1.women = 0
 ( 2)  [/]var(ybocs2)#0bn.women - [/]var(ybocs2)#1.women = 0
 ( 3)  [/]var(ybocs3)#0bn.women - [/]var(ybocs3)#1.women = 0
 ( 4)  [/]var(ybocs4)#0bn.women - [/]var(ybocs4)#1.women = 0
 ( 5)  [/]var(ybocs5)#0bn.women - [/]var(ybocs5)#1.women = 0
 ( 6)  [/]mean(ybocs1)#0bn.women - [/]mean(ybocs1)#1.women = 0
 ( 7)  [/]mean(ybocs2)#0bn.women - [/]mean(ybocs2)#1.women = 0
 ( 8)  [/]mean(ybocs3)#0bn.women - [/]mean(ybocs3)#1.women = 0
 ( 9)  [/]mean(ybocs4)#0bn.women - [/]mean(ybocs4)#1.women = 0
 (10)  [/]mean(ybocs5)#0bn.women - [/]mean(ybocs5)#1.women = 0
```

Group   : male   Number of obs   =   147

	Coef.	OIM Std. Err.	z	P>|z|	[95% Conf. Interval]	
mean(ybocs1)	2.87108	.0559914	51.28	0.000	2.761339	2.980821
mean(ybocs2)	2.449477	.057361	42.70	0.000	2.337052	2.561903
mean(ybocs3)	2.58885	.0442775	58.47	0.000	2.502068	2.675633
mean(ybocs4)	2.198606	.059914	36.70	0.000	2.081177	2.316036
mean(ybocs5)	2.651568	.0486754	54.47	0.000	2.556166	2.74697
var(ybocs1)	.899756	.0751102			.763955	1.059697
var(ybocs2)	.9443116	.0788296			.8017858	1.112173
var(ybocs3)	.5626631	.0469702			.4777399	.6626823
var(ybocs4)	1.030242	.0860029			.8747467	1.213378
var(ybocs5)	.6799888	.0567644			.5773575	.8008639

Group   : female   Number of obs   =   140

	Coef.	OIM Std. Err.	z	P>|z|	[95% Conf. Interval]	
mean(ybocs1)	2.87108	.0559914	51.28	0.000	2.761339	2.980821
mean(ybocs2)	2.449477	.057361	42.70	0.000	2.337052	2.561903
mean(ybocs3)	2.58885	.0442775	58.47	0.000	2.502068	2.675633
mean(ybocs4)	2.198606	.059914	36.70	0.000	2.081177	2.316036
mean(ybocs5)	2.651568	.0486754	54.47	0.000	2.556166	2.74697
var(ybocs1)	.899756	.0751102			.763955	1.059697
var(ybocs2)	.9443116	.0788296			.8017858	1.112173
var(ybocs3)	.5626631	.0469702			.4777399	.6626823
var(ybocs4)	1.030242	.0860029			.8747467	1.213378
var(ybocs5)	.6799888	.0567644			.5773575	.8008639

LR test of model vs. saturated: chi2(30) =   311.17, Prob > chi2 = 0.0000

```
. estimates store null
```

13.3.5 Using the comparative fit index to evaluate invariance

The first two lines tell Stata to estimate only the means and variances for items (that is, no latent variables or regression relationships among variables). `group(women)` specifies a multiple group model, and `ginvariant(meanex covex)` tells Stata to constrain the means and variances to be equivalent across men and women. If we stopped the model here, Stata would also estimate all covariances among items (and set them to be equal across groups). The `covstruct(_OEx, diagonal)` option forces the covariances among the items to be 0. The `covstruct()` option is a convenience function for applying a structure to covariances. The first argument, here _OEx, indicates what covariances you want to apply the structure to. _OEx refers to all observed, exogenous variables. Exogenous variables are variables that are not predicted by any other variable in the model, which in this case is all the items. The second argument, `diagonal`, specifies a diagonal structure to the covariance matrix among the items. A diagonal structure means that the covariance matrix will have nonzero values on the diagonal, which in this case is the variance of each item, and zeros on the off-diagonal, which in this case is all the covariances.

We can use (12.1) to compute the adjusted CFI based on the correct null model. I have written a program, called `adjfit`,[4] that will do it for you. This program takes two arguments, stored estimates for the null model followed by stored estimates from the proposed model (that is, configural, metric, etc.). This is why we stored the model results for each model fit previously. The correct CFI for the configural model is

```
. adjfit null configural
The null model is null
The proposed model is configural
The Adjusted CFI is .962
```

Table 13.1 includes the adjusted CFI values for each of the models. The CFI is used to compare the measurement invariance models, and the χ^2-difference test is used for structural invariance tests (Little 2013). The changes in the CFI, moving from one column to the next (that is, metric to scalar and scalar to residual), all meet the -0.002 criterion, consistent with measurement invariance. Most of the constrained models actually show better fit than the more general models, especially when moving from the configural model to the metric model. This can occur when the χ^2 does not change much from model to model, which it did not in this case. If the χ^2 does not change much from the configural model, then the first term in the numerator of (12.1) will be functionally constant across models. Adding constraints to a model will increase the DF, which means the second term will be bigger in the more constrained model. Consequently, the denominator will be smaller in the more constrained model, which means the CFI will be bigger in the more constrained model. Therefore, improvements in the CFI should be consistent with invariance.

---

4. The code is available in a do-file, which can be downloaded from the Stata Press website.

13.4 Structural invariance

The final two types of invariance tests involve placing constraints on the latent factor variances and means. These are tests about group differences on the latent variable itself. These are typically of substantive interest, and often there is not an expectation that they will be equal.

13.4.1 Invariant factor variances

The first structural invariance test involves fixing all the latent variances to 1; that is, $\phi_{\text{men}} = 1$ and $\phi_{\text{women}} = 1$.[5] Although one might expect that a latent construct is more variable in one group than another, equal variances (that is, homoskedasticity) across groups is important from a measurement perspective. Specifically, if the variance in one group differs markedly from another, then reliability of the measure can be reduced—for example, if there is restriction of range in one group—which can negatively affect correlation and regression models (Bontempo and Hofer 2007). The invariant factor-variances model in Stata is

```
. sem (0: OBSESS -> ybocs1@l1 ybocs2 ybocs3 ybocs4 ybocs5)
>     (1: OBSESS -> ybocs1@l1 ybocs2 ybocs3 ybocs4 ybocs5),
>     group(women) ginvariant(mcoef mcons merrvar)
>     var(0:OBSESS@1) var(1:OBSESS@1) means(0:OBSESS@0) means(1:OBSESS@mean)
  (output omitted)
. estimates store struct_var
```

The only change from the residual invariance model is that the variance for OBSESS is fixed at 1 for women.[6]

Table 13.1 includes the estimates for the invariant factor-variances model. The χ^2-difference test compares this model with the residual invariance model. There is a single degree of freedom because the invariant factor-variances model adds just one additional constraint over the residual invariance model. The test was not statistically significant, suggesting that the assumption that the variability in obsessions does not significantly differ across men and women is reasonable.

13.4.2 Invariant factor means

The second structural invariance model involves fixing the latent means to 0; that is, $\kappa_{\text{men}} = 0$ and $\kappa_{\text{women}} = 0$. This invariance test is the structural equation modeling equivalent of an independent-groups t test (Bontempo and Hofer 2007). Thus, this tests whether groups are equal on the latent construct—do men and women have equal levels of obsessions? The invariant factor-means model in Stata is

---

5. If the CFA includes multiple latent variables, we can also test whether the covariances among the latent variables are equal (Bontempo and Hofer 2007; Widaman and Reise 1997).
6. Constraining factor variances to be equal across groups is only interpretable if metric invariance has been satisfied (Widaman and Reise 1997).

13.5.1 Configural invariance

```
. sem (0: OBSESS -> ybocs1@l1 ybocs2 ybocs3 ybocs4 ybocs5)
>     (1: OBSESS -> ybocs1@l1 ybocs2 ybocs3 ybocs4 ybocs5),
>     group(women) ginvariant(mcoef mcons merrvar)
>     var(0:OBSESS@1) var(1:OBSESS@1) means(0:OBSESS@0) means(1:OBSESS@0)
  (output omitted)
. estimates store struct_mean_var
```

The only change from the invariant factor-variances model is that the mean for OBSESS is fixed at 0 for women.

Table 13.1 includes the estimates for the invariant factor-means model. The χ^2-difference test compares this model with the invariant factor-variances model. There is a single degree of freedom because the invariant factor-means model adds just one additional constraint over the invariant factor-variances model. The test was not statistically significant, suggesting that men and women do not differ in their average level of obsessions.

13.5 Measurement invariance across time

In addition to comparing groups on a construct, often we compare people on a construct over time. This occurs in developmental psychology to study identity development, psychopathology to study the course of disorders, psychotherapy to study change, and many other areas. Comparing people over time requires that the measures we use are invariant over time. Remember, this does not mean that construct level cannot change over time (in fact, we often believe that the construct will change over time); rather, it means that the psychometric properties of the construct will not change over time or repeated measurements.

The process for examining longitudinal invariance is similar to group invariance. However, we will no longer be able to use the group() option in Stata because longitudinal invariance does not use multiple-group models. Multiple-group analyses are appropriate when comparing models across independent groups where the observations in one model are independent from the observations in the other. This is not the case with longitudinal data given that the participants are repeatedly observed over time. Consequently, the longitudinal invariance model is fit as a single model. Unfortunately, this means that we no longer have access to the ginvariant() option for imposing constraints; we must instead introduce constraints by hand. Manually constraining models is not too difficult and can be useful when learning the material to ensure that one understands the concepts. Given that we are no longer fitting a multiple-group model, the process of identification is also different from the multiple-group scenario. Consequently, in this section we explore a different method of identifying a CFA model.

13.5.1 Configural invariance

As with group invariance, examining longitudinal invariance begins with establishing that the pattern of loadings is similar across time (Little 2013). With two timepoints, the configural model is fit as follows (see figure 13.2):

1. Fit a model with two factors, each one representing the latent construct at a given time. The latent factors load on the items for the appropriate timepoint. In the concentration data, the four items from a given timepoint load on a factor for that timepoint.

2. Include a covariance between the latent variables to allow for the construct likely being correlated over time. This covariance is represented by the double-headed arrow between the latent variables in figure 13.2 and is denoted by ϕ_{12}.

3. Include a covariance between the residuals for a given item across the timepoints. The observations are from repeated observations on the same person, which means that the observations are most likely correlated. For example, why might INVOLVED1 be related to ABSORB2? Figure 13.2 shows that the reason for the correlation among these items is that each loads on one of two latent factors, which themselves are related. That is, items are correlated because a) they are indicators of concentration and b) a given participant's concentration at time 1 is correlated with their concentration at time 2. This explains why different items (for example, INVOLVED1 and ABSORB2) are related over time but does not fully account for why an item could be related with itself over time (for example, INVOLVED1 with INVOLVED2). There may be systematic variability in the same item that is above and beyond the construct variance due to method effects (for example, the same item stem or the same response options). A covariance between the residuals (that is, the part of the item not related because of the latent construct) allows systematic unique variance in an item across time. These covariances are represented by the double-headed arrows between the residuals in figure 13.2 and are denoted by θ_{15}, θ_{26}, θ_{37}, and θ_{48}.

4. Except for any constraints needed to identify the model, estimate all loadings, intercepts, residuals, means, and variances freely across groups.

If configural invariance holds, then the pattern of loadings should be equivalent across time.

I cover an identification method called effects coding. Other possibilities include the method used previously in this chapter as well as the identification methods discussed in chapter 12 (Little 2013, compare chap. 5). All methods will produce equivalent models, including equivalent fit. The fixed-factor and effects-coding methods have some interpretational benefits when it comes to the parameters themselves. Thus, if the invariance test is the initial step in a larger analysis, the fixed-factor or effects-coding method is recommended.

13.5.1 Configural invariance

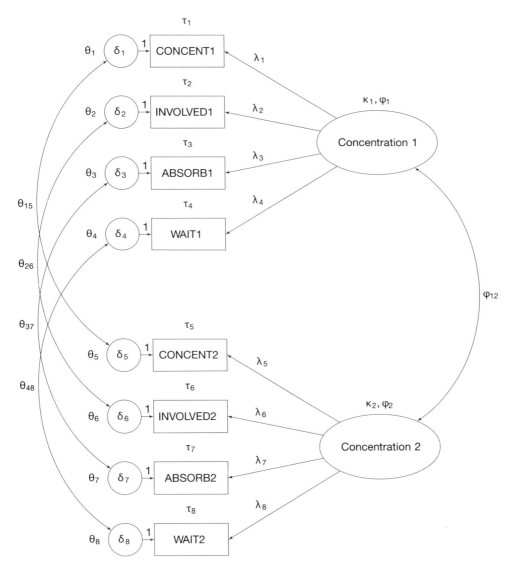

Figure 13.2. Measurement model for concentration items across two timepoints

Effects coding for identification

The effects-coding method is so named because of its relationship to the contrasts in analysis of variance (Little, Slegers, and Card 2006). The loadings are constrained so that they average to 1, and the intercepts are constrained so that they average to 0. This is done by constraining the loadings to sum to the number of items and constraining the intercepts to sum to 0 (Little, Slegers, and Card 2006, eq. 4):

$$\sum_{i=1}^{I} \lambda_i = I$$

$$\sum_{i=1}^{I} \tau_i = 0$$

where I is the total number of items loading on a factor. Thus, if loadings sum to the number of items, then dividing by the number of items produces 1. Likewise, if the intercepts sum to 0, dividing by the number of items produces 0.

In the case of time 1 in figure 13.2, $I = 4$. Thus, the sum of the loadings should be 4.

$$\lambda_1 + \lambda_2 + \lambda_3 + \lambda_4 = 4$$

One of the loadings needs to be constrained. We can do this by setting one of the loadings equal to 4 minus each of the other loadings.

$$\lambda_1 = 4 - \lambda_2 - \lambda_3 - \lambda_4 \tag{13.1}$$

With estimates of λ_2, λ_3, and λ_4, λ_1 is constrained to be a specific value. In other words, Stata will estimate three loadings and determine the fourth via (13.1) (Little, Slegers, and Card 2006). It does not matter which loading is placed on the left-hand side of this equation—the answer will be the same.[7] The process for the intercepts is the same, except this time we replace 4 with 0:

$$\tau_1 = 0 - \tau_2 - \tau_3 - \tau_4 \tag{13.2}$$

Why bother with this? The primary benefit is interpretation. Little, Slegers, and Card (2006) summarize the benefits:

> This method uses the effects' constraints to provide an optimal balance across the possible indicators to establish the scale for the estimated parameters, where the average intercept is zero, but no individual manifest intercept is fixed to be zero. Similarly, the loading parameters are estimated as an optimal balance around 1.0, but no individual loading is necessarily constrained to be 1.0. This method results in estimates of the latent variances that are the average of the indicators' variances accounted by the

---

7. I leave it to the readers to verify this for themselves.

13.5.1 Configural invariance

construct, and the latent means are estimates as optimally weighted averages of the set of indicator means for a given construct. In other words, the estimated latent variances and latent means reflect the observed metric of the indicators, optimally weighted by the degree to which each indicator represents the underlying latent construct. (p. 63)

In short, the estimated loadings compactly show how much information each item contributes to the latent construct. Loadings higher than 1 provide more than the average amount of information, and loadings lower than 1 provide less than the average. The estimated intercepts show where on the construct each item provides the most information. Intercepts above 0 are above the average on the construct, and intercepts below 0 are below the average on the construct. Finally, the latent means and variances are scaled by averaging across items rather than being linked to just a single item (marker-variable method) or being constrained (Little, Slegers, and Card 2006; Little 2013).

Effects-coding constraints in Stata

One method for applying the necessary constraints for effects coding is to use the `constraint` command to define the constraint and then pass the constraint to `sem`.[8]

To use `constraint`, we need to know the coefficient names. Begin by fitting the configural model with `sem`'s default constraints but add the `coeflegend` option. At this point, ignore the covariances among the residuals because we are not placing any constraints on them.

---

8. You can also implement the constraints directly in `sem` by using @ notation. For example, the configural model can be estimated as follows:

```
// Using @ for constraints.
sem (CONCEN1 -> concent1@(4-a-b-c) involved1@a absorb1@b wait1@c) ///
    (_cons -> concent1@(0-d-e-f) involved1@d absorb1@e wait1@f) ///
    (CONCEN2 -> concent2@(4-g-h-i) involved2@g absorb2@h wait2@i) ///
    (_cons -> concent2@(0-j-k-l) involved2@j absorb2@k wait2@l), ///
    mean(CONCEN1@mean1) mean(CONCEN2@mean2) ///
    var(CONCEN1@var1) var(CONCEN2@var2) ///
    cov(e.concent1*e.concent2) ///
    cov(e.involved1*e.involved2) ///
    cov(e.absorb1*e.absorb2) cov(e.wait1*e.wait2)
```

We can estimate metric, scalar, and residual invariance models by changing the appropriate labels.

```
. sem (CONCEN1 -> concent1 involved1 absorb1 wait1)
>     (CONCEN2 -> concent2 involved2 absorb2 wait2), coeflegend nolog
(6188 observations with missing values excluded)

Endogenous variables

Measurement:  concent1 involved1 absorb1 wait1 concent2 involved2 absorb2
              wait2

Exogenous variables

Latent:       CONCEN1 CONCEN2

Structural equation model                       Number of obs    =     11,986
Estimation method  = ml
Log likelihood     = -155703.99

 ( 1)  [concent1]CONCEN1 = 1
 ( 2)  [concent2]CONCEN2 = 1
```

	Coef.	Legend
Measurement		
concent1		
CONCEN1	1	_b[concent1:CONCEN1]
_cons	4.974971	_b[concent1:_cons]
involved1		
CONCEN1	.8761141	_b[involved1:CONCEN1]
_cons	5.031871	_b[involved1:_cons]
absorb1		
CONCEN1	.7330925	_b[absorb1:CONCEN1]
_cons	4.340981	_b[absorb1:_cons]
wait1		
CONCEN1	.6076173	_b[wait1:CONCEN1]
_cons	5.259219	_b[wait1:_cons]
concent2		
CONCEN2	1	_b[concent2:CONCEN2]
_cons	5.220674	_b[concent2:_cons]
involved2		
CONCEN2	.8784984	_b[involved2:CONCEN2]
_cons	5.314617	_b[involved2:_cons]
absorb2		
CONCEN2	.8168365	_b[absorb2:CONCEN2]
_cons	4.633322	_b[absorb2:_cons]
wait2		
CONCEN2	.6020706	_b[wait2:CONCEN2]
_cons	5.450943	_b[wait2:_cons]

13.5.1 Configural invariance

```
      var(e.conce~1)    .8181718   _b[/var(e.concent1)]
      var(e.invol~1)    .7073948   _b[/var(e.involved1)]
      var(e.absorb1)   1.276973    _b[/var(e.absorb1)]
       var(e.wait1)    1.535082    _b[/var(e.wait1)]
      var(e.conce~2)    .7867425   _b[/var(e.concent2)]
      var(e.invol~2)    .6583896   _b[/var(e.involved2)]
      var(e.absorb2)   1.296537    _b[/var(e.absorb2)]
       var(e.wait2)    1.551405    _b[/var(e.wait2)]
       var(CONCEN1)    1.597688    _b[/var(CONCEN1)]
       var(CONCEN2)    1.533908    _b[/var(CONCEN2)]
    ─────────────────────────────────────────────────────
         cov(CONCEN1,
             CONCEN2)   1.097948   _b[/cov(CONCEN1,CONCEN2)]

  LR test of model vs. saturated: chi2(19)   =    3776.79, Prob > chi2 = 0.0000
```

We care about the coefficient names for the loadings and the intercepts. We need to specify four constraints: two loadings constraints (one for each factor) and two intercept constraints (again, one for each factor). The constraints take the form of (13.1) and (13.2).

```
. // intercepts for time 1
. constraint 1  _b[concent1:_cons] = 0 - _b[involved1:_cons]
>  - _b[absorb1:_cons] - _b[wait1:_cons]
. // loadings for time 1
. constraint 2  _b[concent1:CONCEN1] = 4 - _b[involved1:CONCEN1]
>  - _b[absorb1:CONCEN1] - _b[wait1:CONCEN1]
. // intercepts for time 2
. constraint 3 _b[concent2:_cons] = 0 - _b[involved2:_cons] - _b[absorb2:_cons]
>  - _b[wait2:_cons]
. // loadings for time 2
. constraint 4  _b[concent2:CONCEN2] = 4 - _b[involved2:CONCEN2]
>  - _b[absorb2:CONCEN2] - _b[wait2:CONCEN2]
```

The number following constraint labels the constraint.

The constraints are passed to sem.

```
. sem (CONCEN1 -> concent1 involved1 absorb1 wait1)
>     (CONCEN2 -> concent2 involved2 absorb2 wait2),
>     mean(CONCEN1@mean1) mean(CONCEN2@mean2)
>     var(CONCEN1@var1) var(CONCEN2@var2)
>     cov(e.concent1*e.concent2) cov(e.involved1*e.involved2)
>     cov(e.absorb1*e.absorb2) cov(e.wait1*e.wait2)
>     constraints(1/4)
  (output omitted)
. estimates store configural
```

Adding `constraints(1/4)` as an option tells Stata to impose the four constraints we just defined. The `sem` syntax also includes the `mean()` and `var()` options to have Stata estimate the latent mean and variances. Stata will automatically estimate the latent variances, but you must include `mean()` to estimate the latent means. We included both here for transparency and clarity. Lastly, we have added residual covariances with the `cov()` option. The `e.` before the variable name indicates that we are working with residuals, and the `*` indicates that we want a covariance between two parts of the model.

Table 13.2 includes the estimates for the configural model. Given that we used effects coding for identification, the loadings for a given latent variable average to 1. Thus, for the `CONCEN1` factor, `concent1` and `involved1` have higher than average loadings (that is, they provide an above average amount of information about the latent construct), and `absorb1` and `wait1` have below average loadings. The same general pattern of loadings is present at time 2 for `CONCEN2`, which is consistent with configural invariance. The intercepts for a given latent variable average to 0. Thus, the intercepts for `concent1`, `involved1`, and `absorb1` all have below average intercepts, and `wait1` has an above average intercept. This means that participants tended to score a bit lower on the 1 to 7 scale for `concent1`, `involved1`, and `absorb1` than they did for `wait1`. The pattern was the same at time 2.

The mean of the latent variables, κ_1 and κ_2, indicates the average score on the indicators of the latent variable. For `CONCEN1`, $\kappa_1 = 4.9$, indicating that participants averaged 4.9 (on the 1 to 7 scale) on `concent1`, `involved1`, `absorb1`, and `wait1`. For `CONCEN2`, $\kappa_2 = 5.15$, indicating an average of 5.15 on `concent2`, `involved2`, `absorb2`, and `wait2` (the invariance test for the time 1 to time 2 difference in latent means can test whether the difference is statistically significant). Finally, the latent variances, ϕ_1 and ϕ_2, indicate the average amount of construct variance that each item provides for the latent factor (Little, Slegers, and Card 2006; Little 2013).

13.5.1 Configural invariance

Table 13.2. Concentration longitudinal invariance results

	(1) Configural	(2) Metric	(3) Scalar	(4) Residual	(5) Variance	(6) Mean
concent1						
λ_1	1.25	1.23	1.22	1.22	1.22	1.23
τ_1	−1.15	−1.08	−1.06	−1.06	−1.06	−1.09
involved1						
λ_2	1.13	1.11	1.11	1.11	1.11	1.11
τ_2	−0.48	−0.42	−0.43	−0.43	−0.43	−0.43
absorb1						
λ_3	0.90	0.94	0.95	0.95	0.95	0.95
τ_3	−0.09	−0.28	−0.29	−0.29	−0.30	−0.28
wait1						
λ_4	0.72	0.71	0.71	0.71	0.71	0.71
τ_4	1.73	1.78	1.78	1.78	1.78	1.79
concent2						
λ_5	1.22	1.23	1.22	1.22	1.22	1.23
τ_5	−1.06	−1.14	−1.06	−1.06	−1.06	−1.09
involved2						
λ_6	1.10	1.11	1.11	1.11	1.11	1.11
τ_6	−0.35	−0.42	−0.43	−0.43	−0.43	−0.43
absorb2						
λ_7	0.98	0.94	0.95	0.95	0.95	0.95
τ_7	−0.44	−0.23	−0.29	−0.29	−0.30	−0.28
wait2						
λ_8	0.70	0.71	0.71	0.71	0.71	0.71
τ_8	1.86	1.79	1.78	1.78	1.78	1.79
mean(CONCEN1)	4.90	4.90	4.91	4.91	4.91	5.03
mean(CONCEN2)	5.15	5.15	5.15	5.15	5.15	5.03
var(e.concent1)	0.84	0.85	0.86	0.83	0.83	0.82
var(e.involved1)	0.66	0.67	0.66	0.64	0.64	0.64
var(e.absorb1)	1.29	1.27	1.27	1.30	1.30	1.31

Continued on next page

	(1) Configural	(2) Metric	(3) Scalar	(4) Residual	(5) Variance	(6) Mean
var(e.wait1)	1.57	1.57	1.57	1.57	1.57	1.57
var(e.concent2)	0.81	0.80	0.81	0.83	0.83	0.82
var(e.involved2)	0.62	0.61	0.61	0.64	0.64	0.64
var(e.absorb2)	1.31	1.33	1.33	1.30	1.30	1.31
var(e.wait2)	1.58	1.58	1.58	1.57	1.57	1.57
var(CONCEN1)	1.01	1.02	1.02	1.03	1.02	1.03
var(CONCEN2)	1.02	1.01	1.02	1.01	1.02	1.03
cov(e.concent1,e.concent2)	0.23	0.23	0.23	0.23	0.23	0.23
cov(e.involved1,e.involved2)	0.12	0.11	0.11	0.11	0.11	0.11
cov(e.absorb1,e.absorb2)	0.41	0.41	0.41	0.41	0.41	0.40
cov(e.wait1,e.wait2)	0.54	0.54	0.54	0.54	0.54	0.54
cov(CONCEN1,CONCEN2)	0.66	0.66	0.66	0.66	0.66	0.64
N	11986	11986	11986	11986	11986	11986
χ^2_{diff}		44.11	57.75	24.15	0.70	664.75
DF		3	3	4	1	1
p		0.000	0.000	0.000	0.403	0.000
Adj. CFI	0.987	0.985	0.984	0.983	0.983	0.966

NOTE: The first two coefficients for every outcome are the loadings, and the second two are intercepts.

13.5.1 Configural invariance

Finally, we can examine the covariances between the latent variables and the covariances among the residuals. Unfortunately, interpreting covariances is difficult because the scaling is arbitrary. To overcome this problem, we use the standardized estimates to turn the covariances into correlations. The simplest way to do this for the configural model is to "replay" the results but this time ask for the standardized coefficients.

```
. sem, standardized
Structural equation model                       Number of obs     =     11,986
Estimation method = ml
Log likelihood    = -154081.07

 ( 1)  [concent1]CONCEN1 + [involved1]CONCEN1 + [absorb1]CONCEN1 +
       [wait1]CONCEN1 = 4
 ( 2)  [concent2]CONCEN2 + [involved2]CONCEN2 + [absorb2]CONCEN2 +
       [wait2]CONCEN2 = 4
 ( 3)  [concent1]_cons + [involved1]_cons + [absorb1]_cons + [wait1]_cons = 0
 ( 4)  [concent2]_cons + [involved2]_cons + [absorb2]_cons + [wait2]_cons = 0
```

		OIM				
Standardized	Coef.	Std. Err.	z	P>|z|	[95% Conf. Interval]	
Measurement						
concent1						
CONCEN1	.8082784	.0050065	161.45	0.000	.7984658	.818091
_cons	-.7413086	.0291102	-25.47	0.000	-.7983635	-.6842537
involved1						
CONCEN1	.8115949	.0049684	163.35	0.000	.801857	.8213327
_cons	-.3465819	.0314114	-11.03	0.000	-.4081472	-.2850166
absorb1						
CONCEN1	.6246588	.0066284	94.24	0.000	.6116675	.6376502
_cons	-.061511	.0333259	-1.85	0.065	-.1268285	.0038064
wait1						
CONCEN1	.5014039	.0077935	64.34	0.000	.486129	.5166788
_cons	1.19377	.0409019	29.19	0.000	1.113604	1.273937
concent2						
CONCEN2	.8079042	.0049195	164.23	0.000	.7982623	.8175462
_cons	-.6975022	.0302402	-23.07	0.000	-.756772	-.6382325
involved2						
CONCEN2	.8164217	.0048453	168.50	0.000	.806925	.8259183
_cons	-.2600096	.0328346	-7.92	0.000	-.3243642	-.195655
absorb2						
CONCEN2	.6555182	.0062658	104.62	0.000	.6432375	.6677988
_cons	-.2887342	.0326655	-8.84	0.000	-.3527574	-.2247111
wait2						
CONCEN2	.4890862	.0078501	62.30	0.000	.4737002	.5044722
_cons	1.287281	.0431063	29.86	0.000	1.202795	1.371768
mean(CONCEN1)	4.873494	.0412967	118.01	0.000	4.792554	4.954434
mean(CONCEN2)	5.101769	.0429285	118.84	0.000	5.017631	5.185908

var(e.conce~1)	.346686	.0080933			.3311808	.3629172
var(e.invol~1)	.3413138	.0080646			.3258678	.3574919
var(e.absorb1)	.6098014	.0082809			.5937851	.6262496
var(e.wait1)	.7485941	.0078153			.733432	.7640697
var(e.conce~2)	.3472908	.0079489			.3320555	.3632251
var(e.invol~2)	.3334557	.0079117			.3183041	.3493284
var(e.absorb2)	.5702959	.0082146			.5544207	.5866257
var(e.wait2)	.7607947	.0076788			.7458924	.7759947
var(CONCEN1)	1	.			.	.
var(CONCEN2)	1	.			.	.
cov(e.conce~1, e.concent2)	.2757806	.0128981	21.38	0.000	.2505007	.3010605
cov(e.invol~1, e.involved2)	.1803238	.0138239	13.04	0.000	.1532295	.2074182
cov(e.absorb1, e.absorb2)	.3152403	.0093336	33.77	0.000	.2969468	.3335337
cov(e.wait1, e.wait2)	.3436101	.0085764	40.06	0.000	.3268006	.3604196
cov(CONCEN1, CONCEN2)	.6477918	.0071212	90.97	0.000	.6338346	.661749

```
LR test of model vs. saturated: chi2(15) =    530.95, Prob > chi2 = 0.0000
```

The correlation between the latent factors is 0.65, indicating a moderate to strong relationship between the participants' latent concentration at time 1 and time 2. The correlations among the residuals are sometimes interpreted as "residual stability coefficients", meaning that they provide an estimate of how stable the scores on an item are from one time to the next after controlling for the latent variable. The most stable items were `absorb` and `wait`, and the least stable item was `involved`.

13.5.2 Metric invariance

As before, metric invariance means that the model will constrain factor loadings to be equivalent across time. This is done in Stata as follows:

```
. sem (CONCEN1 -> concent1@l1 involved1@l2 absorb1@l3 wait1@l4)
>     (CONCEN2 -> concent2@l1 involved2@l2 absorb2@l3 wait2@l4),
>     mean(CONCEN1@mean1) mean(CONCEN2@mean2)
>     var(CONCEN1@var1) var(CONCEN2@var2)
>     cov(e.concent1*e.concent2) cov(e.involved1*e.involved2)
>     cov(e.absorb1*e.absorb2) cov(e.wait1*e.wait2)
>     constraints(1 2 3)
  (output omitted)
. estimates store metric
```

The addition of the constraints (for example, @l1) on the loadings force the loadings for a given item to be the same across time. Note also that we can drop constraint #4 from the `constraints()` option. Holding the loadings constant across time means that we no longer explicitly have to state the effects-coding constraint at time 2 because the constraint is part of the identification for time 1.

13.5.3 Scalar invariance

The results are in table 13.2. The χ^2-difference test is statistically significant, but given the large sample size, it may not be meaningful invariance. Table 13.2 also presents the CFI value for each model computed using the correct null model. The change from the configural model to the metric model falls within the -0.01 to -0.002 criterion, consistent with invariance. This suggests we can move on to comparing the metric and scalar models.

The correct null model is a longitudinal model that estimates the mean and variance for each item but constrains the mean and variance for a given item to be equal across time. Further, all covariances among items are fixed to 0. This can be done in Stata as follows:

```
. sem (concent1 involved1 absorb1 wait1 concent2 involved2 absorb2 wait2),
>       mean(concent1@m1 concent2@m1 involved1@m2 involved2@m2
>            absorb1@m3 absorb2@m3 wait1@m4 wait2@m4)
>       covstruct(_OEx, diagonal)
>       var(concent1@v1 concent2@v1 involved1@v2 involved2@v2
>            absorb1@v3 absorb2@v3 wait1@v4 wait2@v4)
  (output omitted)
. estimates store null
```

As with the multiple-group model in section 13.3.5, we specify a model with no latent variables and no relationships among the items. The `mean()` and `var()` options allow us to constrain item means and variances across time. The `covstruct(_OEx, diagonal)` option allows us to specify a diagonal structure to the covariance matrix of the residuals (see section 13.3.5). Note that `covstruct()` needs to come before `var()` in this example or Stata will not correctly constrain the item variances to be equal across time.

13.5.3 Scalar invariance

Scalar invariance means that we constrain the item intercepts to be equal across time. In Stata, this is done as

```
. sem (CONCEN1 -> concent1@l1 involved1@l2 absorb1@l3 wait1@l4)
>      (CONCEN2 -> concent2@l1 involved2@l2 absorb2@l3 wait2@l4)
>      (concent1 concent2 <- _cons@m1)
>      (involved1 involved2 <- _cons@m2)
>      (absorb1 absorb2 <- _cons@m3)
>      (wait1 wait2 <- _cons@m4),
>      mean(CONCEN1@mean1) mean(CONCEN2@mean2)
>      var(CONCEN1@var1) var(CONCEN2@var2)
>      cov(e.concent1*e.concent2) cov(e.involved1*e.involved2)
>      cov(e.absorb1*e.absorb2) cov(e.wait1*e.wait2)
>      constraints(1 2)
  (output omitted)
. estimates store scalar
```

This code has two primary changes from the metric invariance code. First, the code for constraining the intercepts was added. The intercepts are referred to as constants, hence the term `_cons`. Placing the variables for a given item to the left of `<-` is a convenient way to constrain the intercepts for a specific item. The other change is that the identification constraint for the means at time 2 (that is, constraint #3) is no longer needed because the over-time constraints produce that constraint.

Table 13.2 includes the estimates from the metric model. The χ^2-difference test is statistically significant; however, the difference in the CFI between the metric and scalar models falls within the -0.01 to -0.002 criterion, consistent with invariance. This suggests we can move on to comparing the scalar and residual models.

13.5.4 Residual invariance

Testing residual invariance involves constraining the residual variances to be equal across time. This is done in Stata as

```
. sem (CONCEN1 -> concent1@l1 involved1@l2 absorb1@l3 wait1@l4)
>     (CONCEN2 -> concent2@l1 involved2@l2 absorb2@l3 wait2@l4)
>     (concent1 concent2 <- _cons@m1)
>     (involved1 involved2 <- _cons@m2)
>     (absorb1 absorb2 <- _cons@m3)
>     (wait1 wait2 <- _cons@m4),
>     mean(CONCEN1@mean1) mean(CONCEN2@mean2)
>     var(CONCEN1@var1) var(CONCEN2@var2)
>     cov(e.concent1*e.concent2) cov(e.involved1*e.involved2)
>     cov(e.absorb1*e.absorb2) cov(e.wait1*e.wait2)
>     var(e.concent1@e1) var(e.concent2@e1)
>     var(e.involved1@e2) var(e.involved2@e2)
>     var(e.absorb1@e3) var(e.absorb2@e3)
>     var(e.wait1@e4) var(e.wait2@e4)
>     constraints(1 2)
```
  (output omitted)
```
. estimates store residual
```

The only change from the metric invariance model is the addition of more instances of the var() option, which is used to constrain the residual variances.

Table 13.2 includes the estimates from the residual model. The χ^2-difference test is statistically significant; however, the difference in the CFI between the metric and scalar models falls within the -0.01 to -0.002 criterion, consistent with invariance. The final tests we can perform are structural invariance tests.

13.6 Structural invariance

Testing for structural invariance in longitudinal data follows the same principles as it did with multiple groups. The code for invariant factor variances is

```
. sem (CONCEN1 -> concent1@l1 involved1@l2 absorb1@l3 wait1@l4)
>       (CONCEN2 -> concent2@l1 involved2@l2 absorb2@l3 wait2@l4)
>       (concent1 concent2 <- _cons@m1)
>       (involved1 involved2 <- _cons@m2)
>       (absorb1 absorb2 <- _cons@m3)
>       (wait1 wait2 <- _cons@m4),
>       mean(CONCEN1@mean1) mean(CONCEN2@mean2)
>       var(CONCEN1@var1) var(CONCEN2@var1)
>       cov(e.concent1*e.concent2) cov(e.involved1*e.involved2)
>       cov(e.absorb1*e.absorb2) cov(e.wait1*e.wait2)
>       var(e.concent1@e1) var(e.concent2@e1)
>       var(e.involved1@e2) var(e.involved2@e2)
>       var(e.absorb1@e3) var(e.absorb2@e3)
>       var(e.wait1@e4) var(e.wait2@e4)
>       constraints(1 2)
  (output omitted)
. estimates store struct_var
```

The code for invariant factor means is

```
. sem (CONCEN1 -> concent1@l1 involved1@l2 absorb1@l3 wait1@l4)
>       (CONCEN2 -> concent2@l1 involved2@l2 absorb2@l3 wait2@l4)
>       (concent1 concent2 <- _cons@m1)
>       (involved1 involved2 <- _cons@m2)
>       (absorb1 absorb2 <- _cons@m3)
>       (wait1 wait2 <- _cons@m4),
>       mean(CONCEN1@mean1) mean(CONCEN2@mean1)
>       var(CONCEN1@var1) var(CONCEN2@var1)
>       cov(e.concent1*e.concent2) cov(e.involved1*e.involved2)
>       cov(e.absorb1*e.absorb2) cov(e.wait1*e.wait2)
>       var(e.concent1@e1) var(e.concent2@e1)
>       var(e.involved1@e2) var(e.involved2@e2)
>       var(e.absorb1@e3) var(e.absorb2@e3)
>       var(e.wait1@e4) var(e.wait2@e4)
>       constraints(1 2)
  (output omitted)
. estimates store struct_mean_var
```

Table 13.2 includes the estimates from both structural invariance models. For the invariant factor-variances model, the χ^2-difference test between it and the residual invariance model is not statistically significant, suggesting that the variance in latent concentration does not differ across time. Likewise, the CFI does not change much when the latent variances are constrained. However, introducing constraints on the latent means produces both a statistically significant χ^2-difference test and substantial change in the CFI. This suggests that the mean latent concentration differed across time. This difference, coupled with the fact that the invariance tests were largely supported (via the CFI criterion), suggests that we can directly interpret this significant difference as a true difference in the latent variable rather than as a function of psychometric differences across time.

13.7 Summary

The primary goal of this chapter was to introduce the measurement invariance and illustrate how to use CFA to statistically evaluate invariance. In this chapter, we

- defined measurement invariance for multiple-group and longitudinal data structures
- discussed the different levels of measurement invariance, including configural, metric, scalar, and residual invariance
- discussed structural invariance
- discussed how to identify models for invariance tests
- learned how to use sem to impose constraints in either a multiple group or longitudinal dataset
- learned how to compare models to test invariance
- discussed the issue of null models when computing fit indices and how to specify an appropriate null model for multiple-group and longitudinal data

Along the way, we covered options in sem that simplify multiple-group comparisons and we learned how to specify constraints external to sem and then add them with the constraint() option.

References

Acock, A. C. 2013. *Discovering Structural Equation Modeling Using Stata*. Revised ed. College Station, TX: Stata Press.

Aiken, L. S., and S. G. West. 1991. *Multiple Regression: Testing and Interpreting Interactions*. Thousand Oaks, CA: Sage.

American Psychiatric Association. 2013. *Diagnostic and Statistical Manual of Mental Disorders: DSM-5*. 5th ed. Washington, DC: American Psychiatric Association.

Baldwin, A. S., A. J. Rothman, M. W. Vander Weg, and A. J. Christensen. 2013. Examining causal components and a mediating process underlying self-generated health arguments for exercise and smoking cessation. *Health Psychology* 32: 1209–1217.

Baldwin, S. A. 2017. Improving the rigor of psychophysiology research. *International Journal of Psychophysiology* 111: 5–16.

Baldwin, S. A., D. J. Bauer, E. Stice, and P. Rohde. 2011. Evaluating models for partially clustered designs. *Psychological Methods* 16: 149–165.

Baldwin, S. A., A. Berkeljon, D. C. Atkins, J. A. Olsen, and S. L. Nielsen. 2009. Rates of change in naturalistic psychotherapy: Contrasting dose-effect and good-enough level models of change. *Journal of Consulting and Clinical Psychology* 77: 203–211.

Baldwin, S. A., and G. W. Fellingham. 2013. Bayesian methods for the analysis of small sample multilevel data with a complex variance structure. *Psychological Methods* 18: 151–164.

Baldwin, S. A., and Z. E. Imel. 2013. Therapist effects: Findings and methods. In *Bergin and Garfield's Handbook of Psychotherapy and Behavior Change*, ed. M. J. Lambert, 6th ed., 258–297. Hoboken, NJ: Wiley.

Baldwin, S. A., Z. E. Imel, S. R. Braithwaite, and D. C. Atkins. 2014. Analyzing multiple outcomes in clinical research using multivariate multilevel models. *Journal of Consulting and Clinical Psychology* 82: 920–930.

Baldwin, S. A., D. M. Murray, and W. R. Shadish. 2005. Empirically supported treatments or type I errors? Problems with the analysis of data from group-administered treatments. *Journal of Consulting and Clinical Psychology* 73: 924–935.

Baldwin, S. A., B. E. Wampold, and Z. E. Imel. 2007. Untangling the alliance-outcome correlation: Exploring the relative importance of therapist and patient variability in the alliance. *Journal of Consulting and Clinical Psychology* 75: 842–852.

Baum, C. F. 2016. *An Introduction to Stata Programming*. 2nd ed. College Station, TX: Stata Press.

Bem, D. J. 2004. Writing the empirical journal article. In *The Compleat Academic: A Career Guide*, ed. J. M. Darley, M. P. Zanna, and H. L. Roediger, III, 2nd ed., 185–219. Washington, DC: American Psychological Association.

Ben-Porath, Y. S. 2012. *Interpreting the MMPI-2-RF*. Minneapolis, MN: University of Minnesota Press.

Beutler, L. E., P. P. P. Machado, and S. A. Neufeldt. 1994. Therapist variables. In *Handbook of Psychotherapy and Behavior Change*, ed. A. E. Bergin and S. L. Garfield, 4th ed., 229–269. New York: Wiley.

Bollen, K. A. 1989. *Structural Equations with Latent Variables*. New York: Wiley.

Bontempo, D. E., and S. M. Hofer. 2007. Assessing factorial invariance in cross-sectional and longitudinal studies. In *Oxford Handbook of Methods in Positive Psychology*, ed. A. D. Ong and M. H. M. van Dulmen, 153–175. New York: Oxford University Press.

Brown, B. L., S. B. Hendrix, D. W. Hedges, and T. B. Smith. 2011. *Multivariate Analysis for the Biobehavioral and Social Sciences: A Graphical Approach*. Hoboken, NJ: Wiley.

Brown, T. A. 2015. *Confirmatory Factor Analysis for Applied Research*. 2nd ed. New York: Guilford Press.

Button, K. S., J. P. A. Ioannidis, C. Mokrysz, B. A. Nosek, J. Flint, E. S. J. Robinson, and M. R. Munafò. 2013. Power failure: Why small sample size undermines the reliability of neuroscience. *Nature Reviews Neuroscience* 14: 365–376.

Byrne, B. M., R. J. Shavelson, and B. O. Muthén. 1989. Testing for the equivalence of factor covariance and mean structures: The issue of partial measurement invariance. *Psychological Bulletin* 105: 456–466.

Carsey, T. M., and J. J. Harden. 2014. *Monte Carlo Simulation and Resampling Methods for Social Science*. Thousand Oaks, CA: Sage.

Cattell, R. B. 1966. The scree test for the number of factors. *Multivariate Behavioral Research* 1: 245–276.

Cheung, G. W., and R. B. Rensvold. 2002. Evaluating goodness-of-fit indexes for testing measurement invariance. *Structural Equation Modeling: A Multidisciplinary Journal* 9: 233–255.

Cohen, J. 1962. The statistical power of abnormal-social psychological research: A review. *Journal of Abnormal and Social Psychology* 65: 145–153.

———. 1988. *Statistical Power Analysis for the Behavioral Sciences*. 2nd ed. Hillsdale, NJ: Lawrence Erlbaum.

———. 1994. The earth is round (p < .05). *American Psychologist* 49: 997–1003.

Cohen, J., P. Cohen, S. G. West, and L. S. Aiken. 2003. *Applied Multiple Regression/Correlation Analysis for the Behavioral Sciences*. 3rd ed. Mahwah, NJ: Lawrence Erlbaum.

Cox, N. J. 2000. egenmore: Stata modules to extend the generate function. Statistical Software Components S386401, Department of Economics, Boston College. https://ideas.repec.org/c/boc/bocode/s386401.html.

———. 2004. Speaking Stata: Graphing categorical and compositional data. *Stata Journal* 4: 190–215.

———. 2015. A short history of Stata on its 30th anniversary. In *Thirty Years with Stata: A Retrospective*, ed. E. Pinzon, 135–147. College Station, TX: Stata Press.

Crits-Christoph, P., K. Baranackie, J. Kurcias, A. Beck, K. Carroll, K. Perry, L. Luborsky, A. McLellan, G. Woody, L. Thompson, D. Gallagher, and C. Zitrin. 1991. Meta-analysis of therapist effects in psychotherapy outcome studies. *Psychotherapy Research* 1: 81–91.

Crits-Christoph, P., and J. Mintz. 1991. Implications of therapist effects for the design and analysis of comparative studies of psychotherapies. *Journal of Consulting and Clinical Psychology* 59: 20–26.

Cronbach, L. J. 1951. Coefficient alpha and the internal structure of tests. *Psychometrika* 16: 297–334.

Cronbach, L. J., and P. E. Meehl. 1955. Construct Validity in Psychological Tests. *Psychological Bulletin* 52: 281–302.

Curran, P. J. 2003. Have multilevel models been structural equation models all along? *Multivariate Behavioral Research* 38: 529–569.

Darby, B. W., and D. Jeffers. 1988. The effects of defendant and juror attractiveness on simulated courtroom trial decisions. *Social Behavior and Personality: An International Journal* 16: 39–50.

Decety, J., J. M. Cowell, K. Lee, R. Mahasneh, S. Malcolm-Smith, B. Selcuk, and X. Zhou. 2015. The negative association between religiousness and children's altruism across the world. *Current Biology* 25: 2951–2955.

DeRubeis, R. J., M. A. Brotman, and C. J. Gibbons. 2005. A conceptual and methodological analysis of the nonspecifics argument. *Clinical Psychology: Science and Practice* 12: 174–183.

Dimidjian, S., S. D. Hollon, K. S. Dobson, K. B. Schmaling, R. J. Kohlenberg, M. E. Addis, R. Gallop, J. B. McGlinchey, D. K. Markley, J. K. Gollan, D. C. Atkins, D. L. Dunner, and N. S. Jacobson. 2006. Randomized trial of behavioral activation, cognitive therapy, and antidepressant medication in the acute treatment of adults with major depression. *Journal of Consulting and Clinical Psychology* 74: 658–670.

Dinno, A. 2009. Implementing Horn's parallel analysis for principal component analysis and factor analysis. *Stata Journal* 9: 291–298.

Ebbesen, E. B., and V. J. Konecni. 1975. Decision making and information integration in the courts: The setting of bail. *Journal of Personality and Social Psychology* 32: 805–821.

Efran, M. G. 1974. The effect of physical appearance on the judgment of guilt, interpersonal attraction, and severity of recommended punishment in a simulated jury task. *Journal of Research in Personality* 8: 45–54.

Efron, B., and D. V. Hinkley. 1978. Assessing the accuracy of the maximum likelihood estimator: Observed versus expected Fisher information. *Biometrika* 65: 457–482.

Enders, C. K. 2010. *Applied Missing Data Analysis*. New York: Guilford Press.

Faes, C., G. Molenberghs, M. Aerts, G. Verbeke, and M. G. Kenward. 2009. The effective sample size and an alternative small-sample degrees-of-freedom method. *American Statistician* 63: 389–399.

Fanelli, D. 2012. Negative results are disappearing from most disciplines and countries. *Scientometrics* 90: 891–904.

Feaster, D. J., F. L. Newman, and C. Rice. 2003. Longitudinal analysis when the experimenter does not determine when treatment ends: What is dose–response? *Clinical Psychology & Psychotherapy* 10: 352–360.

Fitzmaurice, G. M., N. M. Laird, and J. H. Ware. 2011. *Applied Longitudinal Analysis*. 2nd ed. Hoboken, NJ: Wiley.

Flake, J. K., J. Pek, and E. Hehman. 2017. Construct validation in social and personality research: Current practice and recommendations. *Social Psychological and Personality Science* 8: 370–378.

Fox, J. 2008. *Applied Regression Analysis and Generalized Linear Models*. 2nd ed. Thousand Oaks, CA: Sage.

Gelman, A. 2011. Why tables are really much better than graphs. *Journal of Computational and Graphical Statistics* 20: 3–7.

Gelman, A., and J. Carlin. 2014. Beyond power calculations: Assessing type S (sign) and type M (magnitude) errors. *Perspectives on Psychological Science* 9: 641–651.

Gelman, A., and J. Hill. 2007. *Data Analysis Using Regression and Multilevel/Hierarchical Models*. New York: Cambridge University Press.

Gelman, A., and E. Loken. 2014. The statistical crisis in science. *American Scientist* 102: 460–465.

Gelman, A., C. Pasarica, and R. Dodhia. 2002. Let's practice what we preach: Turning tables into graphs. *American Statistician* 56: 121–130.

Gelman, A., and H. Stern. 2006. The difference between "significant" and "not significant" is not itself statistically significant. *American Statistician* 60: 328–331.

Goodman, W. K., L. H. Price, S. A. Rasmussen, C. Mazure, R. L. Fleischmann, C. L. Hill, G. R. Heninger, and D. S. Charney. 1989. The Yale–Brown Obsessive Compulsive Scale: I. Development, use, and reliability. *Archives of General Psychiatry* 46: 1006–1011.

Gottfredson, N. C., D. J. Bauer, and S. A. Baldwin. 2014. Modeling change in the presence of non-randomly missing data: Evaluating a shared parameter mixture model. *Structural Equation Modeling: A Multidisciplinary Journal* 21: 196–209.

Gottfredson, N. C., D. J. Bauer, S. A. Baldwin, and J. C. Okiishi. 2014. Using a shared parameter mixture model to estimate change during treatment when termination is related to recovery speed. *Journal of Consulting and Clinical Psychology* 82: 813–827.

Gould, W. 2018. *The Mata Book: A Book for Serious Programmers and Those Who Want to Be*. College Station, TX: Stata Press.

Greenwald, A. G. 1975. Consequences of prejudice against the null hypothesis. *Psychological Bulletin* 82: 1–20.

Guttman, L. 1945. A basis for analyzing test-retest reliability. *Psychometrika* 10: 255–282.

———. 1954. Some necessary conditions for common-factor analysis. *Psychometrika* 19: 149–161.

Harman, H. H. 1976. *Modern Factor Analysis*. Rev. 3rd ed. Chicago: University of Chicago Press.

Horn, J. L. 1969. On the Internal Consistency Reliability of Factors. *Multivariate Behavioral Research* 4: 115–125.

Ioannidis, J. P. A. 2005. Why most published research findings are false. *PLOS Medicine* 2: e124.

———. 2008. Why most discovered true associations are inflated. *Epidemiology* 19: 640–648.

———. 2012. Why science is not necessarily self-correcting. *Perspectives on Psychological Science* 7: 645–654.

———. 2013. Implausible results in human nutrition research. *British Medical Journal* 347: f6698.

———. 2014. How to make more published research true. *PLOS Medicine* 11: e1001747.

Judd, C. M., J. Westfall, and D. A. Kenny. 2012. Treating stimuli as a random factor in social psychology: A new and comprehensive solution to a pervasive but largely ignored problem. *Journal of Personality and Social Psychology* 103: 54–69.

Kaiser, H. F. 1960. The application of electronic computers to factor analysis. *Educational and Psychological Measurement* 20: 141–151.

Kenny, D. A., and C. M. Judd. 1986. Consequences of violating the independence assumption in analysis of variance. *Psychological Bulletin* 99: 422–431.

Kenny, D. A., L. Mannetti, A. Pierro, S. Livi, and D. A. Kashy. 2002. The statistical analysis of data from small groups. *Journal of Personality and Social Psychology* 83: 126–137.

Kenward, M. G., and J. H. Roger. 2009. An improved approximation to the precision of fixed effects from restricted maximum likelihood. *Computational Statistics and Data Analysis* 53: 2583–2595.

Keppel, G., and T. D. Wickens. 2004. *Design and Analysis: A Researcher's Handbook.* 4th ed. Upper Saddle River, NJ: Prentice Hall.

Kline, R. B. 2011. *Principles and Practice of Structural Equation Modeling.* 3rd ed. New York: Guilford Press.

———. 2015. *Principles and Practice of Structural Equation Modeling.* 4th ed. New York: Guilford Press.

Larson, M. J., and K. A. Carbine. 2017. Sample size calculations in human electrophysiology (EEG and ERP) studies: A systematic review and recommendations for increased rigor. *International Journal of Psychophysiology* 111: 33–41.

Littell, R. C., G. A. Milliken, W. W. Stroup, R. D. Wolfinger, and O. Schabenberger. 2007. *SAS for Mixed Models.* 2nd ed. Cary, NC: SAS Institute.

Littell, R. C., W. W. Stroup, and R. J. Freund. 2002. *SAS for Linear Models.* 4th ed. Cary, NC: SAS Institute.

Little, T. D. 2013. *Longitudinal Structural Equation Modeling.* New York: Guilford Press.

Little, T. D., D. W. Slegers, and N. A. Card. 2006. A non-arbitrary method of identifying and scaling latent variables in SEM and MACS models. *Structural Equation Modeling: A Multidisciplinary Journal* 13: 59–72.

Long, J. S. 2009. *The Workflow of Data Analysis Using Stata.* College Station, TX: Stata Press.

Lord, C. G. 2004. A guide to PhD graduate school: How they keep score in the big leagues. In *The Compleat Academic: A Career Guide*, ed. J. M. Darley, M. P. Zanna, and H. L. Roediger, III, 2nd ed., 3–15. Washington, DC: American Psychological Association.

MacCallum, R. C., and J. T. Austin. 2000. Applications of structural equation modeling in psychological research. *Annual Review of Psychology* 51: 201–226.

MacCallum, R. C., S. Zhang, K. J. Preacher, and D. D. Rucker. 2002. On the practice of dichotomization of quantitative variables. *Psychological Methods* 7: 19–40.

Maxwell, S. E., H. D. Delaney, and K. Kelley. 2018. *Designing Experiments and Analyzing Data: A Model Comparison Perspective*. 3rd ed. New York: Routledge.

McConnell, A. R., C. M. Brown, T. M. Shoda, L. E. Stayton, and C. E. Martin. 2011. Friends with benefits: On the positive consequences of pet ownership. *Journal of Personality and Social Psychology* 101: 1239–1252.

McDonald, R. P. 1999. *Test Theory: A Unified Treatment*. Mahwah, NJ: Lawrence Erlbaum.

McElreath, R. 2016. *Statistical Rethinking: A Bayesian Course with Examples in R and Stan*. Boca Raton, FL: CRC Press.

Meade, A. W., E. C. Johnson, and P. W. Braddy. 2008. Power and sensitivity of alternative fit indices in tests of measurement invariance. *Journal of Applied Psychology* 93: 568–592.

Meehl, P. E. 1978. Theoretical risks and tabular asterisks: Sir Karl, Sir Ronald, and the slow progress of soft psychology. *Journal of Consulting and Clinical Psychology* 46: 806–834.

Millsap, R. E. 2011. *Statistical Approaches to Measurement Invariance*. New York: Routledge.

Mitchell, M. N. 2012a. *Interpreting and Visualizing Regression Models Using Stata*. College Station, TX: Stata Press.

———. 2012b. *A Visual Guide to Stata Graphics*. 3rd ed. College Station, TX: Stata Press.

———. 2015. *Stata for the Behavioral Sciences*. College Station, TX: Stata Press.

Mooney, C. Z., and R. D. Duval. 1993. *Bootstrapping: A Nonparametric Approach to Statistical Inference*. Newbury Park, CA: Sage.

Morey, R. D., R. Hoekstra, J. N. Rouder, M. D. Lee, and E.-J. Wagenmakers. 2016. The fallacy of placing confidence in confidence intervals. *Psychonomic Bulletin & Review* 23: 103–123.

Muthén, L. K., and B. O. Muthén. 2002. How to use a Monte Carlo study to decide on sample size and determine power. *Structural Equation Modeling: A Multidisciplinary Journal* 9: 599–620.

Nelson, L. D., J. P. Simmons, and U. Simonsohn. 2012. Let's publish fewer papers. *Psychological Inquiry* 23: 291–293.

Nieuwenhuis, S., B. U. Forstmann, and E.-J. Wagenmakers. 2011. Erroneous analyses of interactions in neuroscience: A problem of significance. *Nature Neuroscience* 14: 1105–1107.

Nunnally, J. C., and I. H. Bernstein. 1994. *Psychometric Theory*. 3rd ed. New York: McGraw–Hill.

Open Science Collaboration. 2015. Estimating the reproducibility of psychological science. *Science* 349: aac4716.

Palmer, T. M., and J. A. C. Sterne, ed. 2016. *Meta-Analysis in Stata: An Updated Collection from the Stata Journal*. 2nd ed. College Station, TX: Stata Press.

Pashler, H., and C. R. Harris. 2012. Is the replicability crisis overblown? Three arguments examined. *Perspectives on Psychological Science* 7: 531–536.

Poi, B. P. 2004. From the help desk: Some bootstrapping techniques. *Stata Journal* 4: 312–328.

Preacher, K. J., M. J. Zyphur, and Z. Zhang. 2010. A general multilevel SEM framework for assessing multilevel mediation. *Psychological Methods* 15: 209–233.

Prochaska, J. O., C. C. DiClemente, and J. C. Norcross. 1992. In search of how people change. Applications to addictive behaviors. *American Psychologist* 47: 1102–1114.

Rabe-Hesketh, S., and A. Skrondal. 2012a. *Multilevel and Longitudinal Modeling Using Stata. Volume I: Continuous Responses*. 3rd ed. College Station, TX: Stata Press.

———. 2012b. *Multilevel and Longitudinal Modeling Using Stata. Volume II: Categorical Responses, Counts, and Survival*. 3rd ed. College Station, TX: Stata Press.

Raudenbush, S. W., and A. S. Bryk. 2001. *Hierarchical Linear Models: Applications and Data Analysis Methods*. 2nd ed. Thousand Oaks, CA: Sage.

Raykov, T. 1997. Scale reliability, Cronbach's coefficient alpha, and violations of essential tau-equivalence with fixed congeneric components. *Multivariate Behavioral Research* 32: 329–353.

———. 1998. A method for obtaining standard errors and confidence intervals of composite reliability for congeneric items. *Applied Psychological Measurement* 22: 369–374.

———. 2001. Estimation of congeneric scale reliability using covariance structure analysis with nonlinear constraints. *British Journal of Mathematical and Statistical Psychology* 54: 315–323.

Raykov, T., and D. Grayson. 2003. A test for change of composite reliability in scale development. *Multivariate Behavioral Research* 38: 143–159.

Raykov, T., and P. E. Shrout. 2002. Reliability of scales with general structure: Point and interval estimation using a structural equation modeling approach. *Structural Equation Modeling: A Multidisciplinary Journal* 9: 195–212.

Reise, S. P., K. F. Widaman, and R. H. Pugh. 1993. Confirmatory factor analysis and item response theory: Two approaches for exploring measurement invariance. *Psychological Bulletin* 114: 552–566.

Rencher, A. C., and W. F. Christensen. 2012. *Methods of Multivariate Analysis*. 3rd ed. Hoboken, NJ: Wiley.

Rencher, A. C., and G. B. Schaalje. 2008. *Linear Models in Statistics*. 2nd ed. New York: Wiley.

Rogosa, D., D. Brandt, and M. Zimowski. 1982. A growth curve approach to the measurement of change. *Psychological Bulletin* 92: 726–748.

Roman, D. D., G. E. Edwall, R. J. Buchanan, and J. H. Patton. 1991. Extended norms for the paced auditory serial addition task. *Clinical Neuropsychologist* 5: 33–40.

Satterthwaite, F. E. 1946. An approximate distribution of estimates of variance components. *Biometrics Bulletin* 2: 110–114.

Savage, S. L. 2009. *The Flaw of Averages: Why We Underestimate Risk in the Face of Uncertainty*. Hoboken, NJ: Wiley.

Shariff, A. F., A. K. Willard, M. Muthukrishna, S. R. Kramer, and J. Henrich. 2016. What is the association between religious affiliation and children's altruism? *Current Biology* 26: R699–R700.

Simmons, J. P., L. D. Nelson, and U. Simonsohn. 2011. False-positive psychology: Undisclosed flexibility in data collection and analysis allows presenting anything as significant. *Psychological Science* 22: 1359–1366.

Singer, J. D., and J. B. Willet. 2003. *Applied Longitudinal Data Analysis: Modeling Change and Event Occurrence*. New York: Oxford University Press.

Snijders, T. A. B., and R. J. Bosker. 2011. *Multilevel Analysis: An Introduction to Basic and Advance Multilevel Modeling*. 2nd ed. London: Sage.

Spielberger, C. D., R. L. Gorsuch, R. Lushene, P. R. Vagg, and G. A. Jacobs. 1983. *Manual for the State-Trait Anxiety Inventory*. Palo Alto, CA: Consulting Psychologists Press.

StataCorp. 2017a. *Stata 15 Base Reference Manual*. College Station, TX: Stata Press.

———. 2017b. *Stata 15 Bayesian Analysis Reference Manual*. College Station, TX: Stata Press.

———. 2017c. *Stata 15 Graphics Reference Manual*. College Station, TX: Stata Press.

———. 2017d. *Stata 15 Mata Reference Manual*. College Station, TX: Stata Press.

———. 2017e. *Stata 15 Multilevel Mixed-Effects Reference Manual*. College Station, TX: Stata Press.

———. 2017f. *Stata 15 Multivariate Statistics Reference Manual*. College Station, TX: Stata Press.

———. 2017g. *Stata 15 Power and Sample-Size Reference Manual*. College Station, TX: Stata Press.

———. 2017h. *Stata 15 Programming Reference Manual*. College Station, TX: Stata Press.

———. 2017i. *Stata 15 Structural Equation Modeling Reference Manual*. College Station, TX: Stata Press.

Steenkamp, J.-B. E. M., and H. Baumgartner. 1998. Assessing measurement invariance in cross–national consumer research. *Journal of Consumer Research* 25: 78–107.

Steiger, J. H. 2002. When constraints interact: A caution about reference variables, identification constraints, and scale dependencies in structural equation modeling. *Psychological Methods* 7: 210–227.

Stewart, J. E., II. 1985. Appearance and punishment: The attraction-leniency effect in the courtroom. *Journal of Social Psychology* 125: 373–378.

Vandenberg, R. J., and C. E. Lance. 2000. A review and synthesis of the measurement invariance literature: Suggestions, practices, and recommendations for organizational research. *Organizational Research Methods* 3: 4–70.

Wampold, B. E., B. Davis, and R. H. Good, III. 1990. Hypothesis validity of clinical research. *Journal of Consulting and Clinical Psychology* 58: 360–367.

Wampold, B. E., and Z. E. Imel. 2015. *The Great Psychotherapy Debate: The Evidence for What Makes Psychotherapy Work*. 2nd ed. New York: Routledge.

Wampold, B. E., and R. C. Serlin. 2000. The consequence of ignoring a nested factor on measures of effect size in analysis of variance. *Psychological Methods* 5: 425–433.

Warner, R. M. 2013. *Applied Statistics: From Bivariate through Multivariate Techniques: From Bivariate through Multivariate Techniques*. 2nd ed. Thousand Oaks, CA: Sage.

Webb, C. A., R. J. DeRubeis, and J. P. Barber. 2010. Therapist adherence/competence and treatment outcome: A meta-analytic review. *Journal of Consulting and Clinical Psychology* 78: 200–211.

Westfall, J., C. M. Judd, and D. A. Kenny. 2015. Replicating studies in which samples of participants respond to samples of Stimuli. *Perspectives on Psychological Science* 10: 390–399.

Widaman, K. F., and S. P. Reise. 1997. Exploring the measurement invariance of psychological instruments: Applications in the substance use domain. In *The Science of Prevention: Methodological Advances from Alcohol and Substance Abuse Research*, ed. K. J. Bryant, M. Windle, and S. G. West, 281–324. Washington, DC: American Psychological Association.

Widaman, K. F., and J. S. Thompson. 2003. On specifying the null model for incremental fit indices in structural equation modeling. *Psychological Methods* 8: 16–37.

Winer, B. J., D. R. Brown, and K. M. Michels. 1991. *Statistical Principles in Experimental Design*. 3rd ed. New York: McGraw–Hill.

Winter, N., and A. Nichols. 2008. vioplot: Stata module to produce violin plots with current graphics. Statistical Software Components S456902, Department of Economics, Boston College. https://ideas.repec.org/c/boc/bocode/s456902.html.

Wuensch, K. L., R. C. Chia, W. A. Castellow, C.-J. Chuang, and B.-S. Cheng. 1993. Effects of physical attractiveness, sex, and type of crime on mock juror decisions: A replication with Chinese students. *Journal of Cross-Cultural Psychology* 24: 414–427.

Author index

A
Acock, Alan C. 395, 396
Addis, M. E. 112, 116, 121
Aerts, M. 180, 249
Aiken, L. S. . . 57, 59, 61, 74, 96, 99, 110
American Psychiatric Association . . 315
Atkins, D. C. 112, 116, 121, 284, 310
Austin, J. T. 392, 395

B
Baldwin, A. S. 131
Baldwin, S. A. 6, 174, 190, 197, 209,
       239, 242, 244, 250, 267, 273,
       284, 310
Baranackie, K. 244
Barber, J. P. 220
Bauer, D. J. 174, 190, 197, 209, 310
Baum, C. F. 33, 40
Baumgartner, H. 399, 400
Beck, A. 244
Bem, D. J. 7
Ben-Porath, Y. S. 356
Berkeljon, A. 310
Bernstein, I. H. 315, 361, 366, 367,
       388, 396
Beutler, L. E. 244
Bollen, K. A. 321, 335, 344, 398
Bontempo, D. E. 401, 402, 405, 412
Bosker, R. J. 237, 279, 284
Braddy, P. W. 409
Braithwaite, S. R. 284
Brandt, D. 290
Brotman, M. A. 264, 265
Brown, B. L. 361, 370
Brown, C. M. 131
Brown, D. R. . . . 117, 121–123, 126, 180

Brown, T. A. . . . 330, 346, 348, 360, 371,
       382–384, 388, 393–396, 400
Bryk, A. S. 237, 242, 255, 273, 284
Buchanan, R. J. 169
Button, K. S. 201, 230, 235
Byrne, B. M. 398

C
Carbine, K. A. 201
Card, N. A. 416, 417, 420
Carlin, J. 64, 230–234
Carroll, K. 244
Carsey, T. M. 59
Castellow, W. A. 132, 133, 152
Cattell, R. B. 367
Charney, D. S. 398
Cheng, B.-S. 132, 133, 152
Cheung, G. W. 409
Chia, R. C. 132, 133, 152
Christensen, A. J. 131
Christensen, W. F. . . . 361, 363, 365, 396
Chuang, C.-J. 132, 133, 152
Cohen, J. . . . 57, 59, 61, 64, 74, 110, 201,
       223, 226
Cohen, P. 57, 59, 61, 74, 110
Cowell, J. M. 85
Cox, N J 252
Crits-Christoph, P. 244, 246
Cronbach, L. J. 344, 360
Curran, P. J. 324

D
Darby, B. W. 132
Davis, B. 120
Decety, J. 85
Delaney, H. D. 129, 138, 166
DeRubeis, R. J. 220, 264, 265

DiClemente, C. C. 273
Dimidjian, S. 112, 116, 121
Dinno, A. 368
Dobson, K. S. 112, 116, 121
Dodhia, R. 5
Dunner, D. L. 112, 116, 121
Duval, R. D. 343

E
Ebbesen, E. B. 131
Edwall, G. E. 169
Efran, M. G. 132
Efron, B. 190
Enders, C. K. 168

F
Faes, C. 180, 249
Fanelli, D. 233
Feaster, D. J. 310
Fellingham, G. W. 250
Fitzmaurice, G. M. 310
Flake, J. K. 360
Fleischmann, R. L. 398
Flint, J. 201, 230, 235
Forstmann, B. U. 271
Fox, J. 105, 110, 389
Freund, R. J. 167, 168, 190, 200

G
Gallagher, D. 244
Gallop, R. 112, 116, 121
Gelman, A. 5, 64, 65, 72, 230–234,
    246, 254, 255, 258, 271
Gibbons, C. J. 264, 265
Gollan, J. K. 112, 116, 121
Good, R. H., III 120
Goodman, W. K. 398
Gorsuch, R. L. 317, 357
Gottfredson, N. C. 310
Gould, W. 363
Grayson, D. 340
Greenwald, A. G. 229, 233
Guttman, L. 344, 366

H
Harden, J. J. 59
Harman, H. H. 361, 396
Harris, C. R. 6, 64
Hedges, D. W. 361, 370
Hehman, E. 360
Hendrix, S. B. 361, 370
Heninger, G. R. 398
Henrich, J. 85
Hill, C. L. 398
Hill, J. 65, 72, 246, 254, 255, 258
Hinkley, D. V. 190
Hoekstra, R. 59
Hofer, S. M. 401, 402, 405, 412
Hollon, S. D. 112, 116, 121
Horn, J. L. 368

I
Imel, Z. E. 217, 237, 239, 244, 250,
    267, 273, 284
Ioannidis, J. P. A. ... 5, 6, 64, 201, 230,
    235

J
Jacobs, G. A. 317, 357
Jacobson, N. S. 112, 116, 121
Jeffers, D. 132
Johnson, E. C. 409
Judd, C. M. 143, 242, 246

K
Kaiser, H. F. 366
Kashy, D. A. 242
Kelley, K. 129, 138, 166
Kenny, D. A. 143, 242, 246
Kenward, M. G. 180, 190, 249
Keppel, G. 126, 127, 129, 131, 141–
    143, 150, 154, 155, 166, 167
Kline, R. B. ... 348, 383, 384, 386, 388,
    391, 393–396
Kohlenberg, R. J. 112, 116, 121
Konecni, V. J. 131
Kramer, S. R. 85
Kurcias, J. 244

L

Laird, N. M. 310
Lance, C. E. 399, 400
Larson, M. J. 201
Lee, K. 85
Lee, M. D. 59
Littell, R. C. . . . 167, 168, 178, 190, 200
Little, T. D. . . . 409, 411, 413, 414, 416, 417, 420
Livi, S. 242
Loken, E. 7
Long, J. S. 5, 9, 31
Lord, C. G. 6
Luborsky, L. 244
Lushene, R. 317, 357

M

MacCallum, R. C. 107, 392, 395
Machado, P. P. P. 244
Mahasneh, R. 85
Malcolm-Smith, S. 85
Mannetti, L. 242
Markley, D. K. 112, 116, 121
Martin, C. E. 131
Maxwell, S. E. 129, 138, 166
Mazure, C. 398
McConnell, A. R. 131
McDonald, R. P. 315, 317–319, 340, 344, 349
McElreath, R. 7, 255
McGlinchey, J. B. 112, 116, 121
McLellan, A. 244
Meade, A. W. 409
Meehl, P. E. 7, 64, 229, 360
Michels, K. M. . . . 117, 121–123, 126, 180
Milliken, G. A. 167, 178, 190, 200
Millsap, R. E. 398
Mintz, J. 246
Mitchell, M. N. 22, 23, 110, 160, 166
Mokrysz, C. 201, 230, 235
Molenberghs, G. 180, 249
Mooney, C. Z. 343
Morey, R. D. 59
Munafò, M. R. 201, 230, 235
Murray, D. M. 242

Muthén, B. O. 209, 235, 398
Muthén, L. K. 209, 235
Muthukrishna, M. 85

N

Nelson, L. D. 6, 7, 62, 160
Neufeldt, S. A. 244
Newman, F. L. 310
Nichols, A. 113
Nielsen, S. L. 310
Nieuwenhuis, S. 271
Norcross, J. C. 273
Nosek, B. A. 201, 230, 235
Nunnally, J. C. 315, 361, 366, 367, 388, 396

O

Okiishi, J. C. 310
Olsen, J. A. 310
Open Science Collaboration 5

P

Pasarica, C. 5
Pashler, H. 6, 64
Patton, J. H. 169
Pek, J. 360
Perry, K. 244
Pierro, A. 242
Poi, B. P. 343
Preacher, K. J. 107, 284
Price, L. H. 398
Prochaska, J. O. 273
Pugh, R. H. 398, 401

R

Rabe-Hesketh, S. 173, 200, 238, 240, 244, 246, 248, 249, 255, 267, 279, 280, 282, 284, 298, 299, 310, 311
Rasmussen, S. A. 398
Raudenbush, S. W. 237, 242, 255, 273, 284
Raykov, T. 340, 341, 346, 353
Reise, S. P. 398, 401, 405, 407, 408, 412

Rencher, A. C. . . 49, 361, 363, 365, 396
Rensvold, R. B. 409
Rice, C. 310
Robinson, E. S. J. 201, 230, 235
Roger, J. H. 190
Rogosa, D. 290
Rohde, P. 174, 190, 197, 209
Roman, D. D. 169
Rothman, A. J. 131
Rouder, J. N. 59
Rucker, D. D. 107

S

Satterthwaite, F. E. 178
Savage, S. L. 329
Schaalje, G. B. 49
Schabenberger, O. . . . 167, 178, 190, 200
Schmaling, K. B. 112, 116, 121
Selcuk, B. 85
Serlin, R. C. 244, 246
Shadish, W. R. 242
Shariff, A. F. 85
Shavelson, R. J. 398
Shoda, T. M. 131
Shrout, P. E. 340
Simmons, J. P. 6, 7, 62, 160
Simonsohn, U. 6, 7, 62, 160
Singer, J. D. 237, 290, 307, 310
Skrondal, A. . . . 173, 200, 238, 240, 244,
246, 248, 249, 255, 267, 279,
280, 282, 284, 298, 299, 310,
311
Slegers, D. W. 416, 417, 420
Smith, T. B. 361, 370
Snijders, T. A. B. 237, 279, 284
Spielberger, C. D. 317, 357
StataCorp. . 23, 25, 30, 33, 74, 179, 214,
250, 311, 331, 337, 363, 366
Stayton, L. E. 131
Steenkamp, J.-B. E. M. 399, 400
Steiger, J. H. 384
Stern, H. 271
Stewart, J. E., II 132
Stice, E. 174, 190, 197, 209
Stroup, W. W. . . 167, 168, 178, 190, 200

T

Thompson, J. S. 394, 409
Thompson, L. 244

V

Vagg, P. R. 317, 357
Vandenberg, R. J. 399, 400
Vander Weg, M. W. 131
Verbeke, G. 180, 249

W

Wagenmakers, E.-J. 59, 271
Wampold, B. E. 120, 217, 237, 239,
244, 246, 267, 273
Ware, J. H. 310
Warner, R. M. 149
Webb, C. A. 220
West, S. G. . . 57, 59, 61, 74, 96, 99, 110
Westfall, J. 143, 246
Wickens, T. D. . . . .
. . 126, 127, 129, 131, 141–143,
150, 154, 155, 166, 167
Widaman, K. F. 394, 398, 401, 405,
407–409, 412
Willard, A. K. 85
Willet, J. B. 237, 290, 307, 310
Winer, B. J. 117, 121–123, 126, 180
Winter, N. 113
Wolfinger, R. D. 167, 178, 190, 200
Woody, G. 244
Wuensch, K. L. 132, 133, 152

Z

Zhang, S. 107
Zhang, Z. 284
Zhou, X. 85
Zimowski, M. 290
Zitrin, C. 244
Zyphur, M. J. 284

Subject index

A
adding commands/programs........33
adjust for multiple comparisons...126–129
    experimentwise error rate.....126
    familywise error rate..........126
    Scheffé....................126–127
    Tukey's HSD...............126–129
anova command....................119
    factorial ANOVA..........144–146, 159–160
    one-way ANOVA...............119

B
benefits of Stata....................4–5
bootstrap command..............342
    computing the confidence interval 343
by processing......................18
bysort prefix.....................268

C
categorical predictors............77–86
    dummy coding..............78–86
        changing the base category...84
        factor variables...........83–84
        generate and replace commands.................79
        generate command only.....80
        interpretation of coefficients....
       81–83
        null hypothesis...............83
        rules........................78
        tabulate command..........80
        using with regress command..
       81
    incorrect implementation.......78
cd command........................12
conditional variance 53
    formula.......................53
    root mean squared error........53
confidence intervals..................58
    coverage rate 59–60
    formula.......................58
    interpretation...............59–60
confirmatory factor analysis...377–396
    χ^2-difference test 346–349
    causal models.................321
    computing reliability (ω)......340
    connection to regression.......323
    correlated residuals 350
    reliability...................353
    estimating in Stata........see sem command
    global fit 392–396
    goodness of fit 328
        χ^2 test.................330–334
        χ^2....................392–393
        CFI........................394
        printing in Stata..... see estat gof command
        RMSEA 393
        SRMR..................394–395
        TLI........................394
    identification 383, 401–402
        constraints..............384–386
        just-identified...............386
        overidentified 386
        underidentified..............386
    intercepts.....................400
    introduction 320
    latent variables................321
    mean structure..........381, 383

confirmatory factor analysis, *continued*
    model-implied covariance 323, 382
    nested models 346
    parallel items 349
    path model 321
    parts 321
    predicted covariance matrix ($\widehat{\Sigma}$) 326, 329, 334
    prediction equation 323–326
    R^2 for an item 335
        estimating in Stata ... *see* `estat eqgof` command
        formula 335
    reliability
        ω versus α 343
        impact of weak loadings 343
        internal consistency 343
        tau-equivalence 344
    sample versus asymptotic variance 336
    saturated model 331
    shared variance across all items 338
    standardized solution 389
    tau-equivalent models 346
    unstructured model 321
    variance of an item 382
    variance structure 381
    versus exploratory factor analysis 377
`contrast` command
    factorial ANOVA 147–149, 164–165
    `mcompare()` option 148, 165
    `nowald` option 165
    `pveffects` option 165
    repeated measures 176, 196
    `small` option 176, 196
    small samples 190
`corr` command 322, 359
    `covariance` option 329
cumulative normal distribution function (Φ) 211

D
Data Editor 15
`describe` command 14
do-files .... *see* reproducible analysis
dummy coding .... *see* categorical predictors, dummy coding

E
Early Childhood Longitudinal Program data 398
effect size
    η^2 149–150
    ω^2 150–151
    f 223
    Cohen's d 116
        versus Hedges's g 116
    estimating in Stata ..... *see* `esize` command
    factorial ANOVA 149–151, 165
    partial-η^2 149
    partial-ω^2 150
`egen` command 17
    `mean()` option 252, 268
    `tag` option 268
`egenmore` command
    `semean` option 252
`esize` command 116
`esize twosample` command 116
`estat bootstrap` command 343
`estat eqgof` command 335
`estat esize` command
    factorial ANOVA 150–151
    `omega` option 151
`estat framework` command
    `fitted` option 329, 334, 383
`estat gof` command 330, 391, 392
`estat ic` command
    following `mixed` 179
`estat icc` command 250
`estat wcorrelation` command ... 178, 182, 184, 187, 196
`estimates store` command 348
exploratory factor analysis 358–377
    common factor model 359
    communality 364

exploratory factor analysis, *continued*
    data reduction 359
    eigenvalues 363
    equation 360
    estimating in Stata .... *see* `factor` command
    extracting factors 360
        choosing the number 366
        eigenvalues-greater-than-one rule 366
        scree plot 367
    extraction methods 360
        principal-component factor .. 360
    interpreting loadings 362
    notation 360
    orthogonal factors 362
    parallel analysis 368
    promax rotation 374
    rotation
        `estat common` command 376
        `estat structure` command 376
        `rotate` command 372, 374
    simple structure 370
    uniqueness 364
    varimax rotation 370
    versus confirmatory factor analysis 358
    versus principal-component analysis 365
extending syntax over multiple lines ... 25

F
`factor` command 362
    parallel analysis .... *see* `paran` command
    `pcf` option 362
    scree plot .... *see* `screeplot` command
factor variables .... *see* categorical predictors, dummy coding, factor variables

factorial ANOVA
    benefits 131
    degrees of freedom 142–143
    effect size .... *see* effect size
    estimating in Stata ..... *see* `anova` command
    interactions 139–140, 156–158
        null hypothesis 140, 156
    main effects 138–139, 155–158
        null hypothesis 139
        versus first-order effects 139–140
    marginal means 138–139, 154–155
        one-way marginal means 155
        two-way marginal means 154
    notation 133–134
    partitioning the variance .. 140–142
        SS between 141
        SS interaction 141–142
        SS main effects 141
    simple effects 146–149, 163–165
        estimating in Stata .... *see* `contrast` command
        null hypothesis 147
    source table 142–143
    three-factor design 151–166
    three-way interactions 156, 158–159
        null hypothesis 158–159
    two-factor design 134–151
    visualize data ... 134–138, 152–153
file paths 12

G
General Social Survey 38
`generate` command 17
    `_n` function 288
    handling missing data 18–19
    getting help 32–33
`graph box` command 24, 287
    `by()` option 153
    factorial ANOVA 135–136, 153
    `over()` option 153
    repeated measures 171, 192

graph dot command 261
graphics introduction 22–27

H
histogram command 23, 281, 290

I
in qualifier 15, 16
interactions 90–109
    categorical by continuous .. 91–107
    continuous by continuous 107–109
        interpretation of coefficients ....
        108
        margins command 108–109
        marginsplot command 108–109
        visualizing relationships 108–109
    dichotomous by continuous 91–101
    factor-variable notation 94
    first-order coefficients 96
    interpreting coefficients 95–101
    multiple intercept, multiple slope model 99–101
        factor notation 100
        lincom command 100
    polytomous by continuous 101–107
        interpretation of coefficients ....
        102–103
        margins command 104
        marginsplot command 104
        using factor notation to prevent errors 102
        visualizing the interaction .. 104–107
    probing meaning 96–101
    problems with dichotomization ...
    107–109
    product between variables 94

interactions, *continued*
    simple regression equation .. 96–101
        lincom command 98–99
        margins command 97–98
        marginsplot command ... 97–98
        plotting 97–98
interocular trauma test 329

L
label values command 19
label variable command 19
labels 19
    value labels 19
    variable labels 19
lincom command 73, 305
    between-clusters and within-cluster relationships 271
    contrast after the regress command 73–74
    planned comparisons 123–125
linear combinations .... *see* lincom command
lines 43
    intercept 43
    slope 43
list command 15
    combined with if and in qualifiers 17
    repeated measures 173
lrtest command ... 279, 298, 348, 350, 353, 389
    repeated measures 199

M
margins command 50–51, 299, 302, 305, 309
    expected values 65
    following the anova command ....
    .... 145–146, 160
    inferential uncertainty 65–70
marginsplot command 66–70, 299, 302, 305, 309
    by option 309
    factorial ANOVA 136–138

marginsplot command, *continued*
    following the anova command ....
      ...... 145–146, 160–163
    methods for improving......67–70
    noci option...................309
    xdimension() option 162
mata command 363, 364
mean................................40
measurement invariance.......398–427
    across groups.............400–413
      configural invariance ... 400–406
      invariant factor means..412–413
      invariant factor variances ...412
      metric invariance 407
      residual invariance..........408
      scalar invariance 407–408
      sem command..........402–405
      structural invariance ... 412–413
    across time................413–427
      configural invariance ... 413–424
      identification............414–424
      metric invariance.......424–425
      residual invariance..........426
      scalar invariance 425–426
      sem command..........417–424
      structural invariance........427
    steps.....................399–400
    using the CFI to evaluate..409–411
mixed command
    between-clusters and within-cluster
      relationships...............270
    cov(unstructured) option ...278
    covariance(unstructured)
      option....................297
    cross-level interaction 304
    dfmethod() option.......175, 190
    dfmethod(satterthwaite) option
      ....178, 190
    display residual correlation matrix
      .... *see* estat wcorrelation
      command
    fit indices ....*see* estat ic
      command
    longitudinal data..............297
    noconstant option.......175, 306

mixed command, *continued*
    nofetable option.............249
    noheader option..............249
    random slopes 278
    random-effects specification ...175
    random-intercept model ..248–252
    reml option...................175
    repeated measures........175, 195
    residuals() option
      by() option.................198
    residuals(ar 1) option......181
    residuals(exchangeable) option
     178
    residuals(independent)
      option...............175, 176
    residuals(toeplitz) option ....
     183
    residuals(unstructured) option
      ....186, 195
    separate intercepts, separate slopes
      model....................306
    stddeviations option...178, 184,
      249, 279
    syntax basics.............175–176
    t() option for residuals()..181,
      183, 186
    time-invariant covariate 301
    time-varying covariate.........308
multilevel models
    adding a predictor........262–264
    atomistic fallacy 267
    basic model...................244
    Bayes's theorem...............255
    between-clusters variance.....244,
      245
    between-clusters versus within-
      cluster effects........264–266
    between-clusters versus within-
      cluster relationships..267–273
    caterpillar plot 253–254
    centering.....................292
    complete pooling..............254
    conditional independence......246
    contextual effect 273
    correlated random effects......276

multilevel models, *continued*
    definition of random effects....246
    ecological fallacy..............267
    fitting them in Stata....*see* `mixed` command
    interpreting random slopes...280–283
    intraclass correlation.....250–252, 280
        fitting them in Stata..*see* `estat icc` command
    issues with clustered data....239–243
    level-specific relationships.....266
    longitudinal data..............296
        need for growth model......290
        random slope...............296
    longitudinal data introduction....287
    no pooling....................254
    nonindependence.........239–242
        sources................242–243
        substantive benefits....242–243
    partial pooling...........255–259
    partitioning variance.....244–246
    predicting cluster means..252–261
    random effects
        compared to fixed effects....247
        examples..................246
    random intercepts........246–249
    random slopes............273–283
    separate intercepts, separate slopes model...................306
    time-varying covariate.........307
    total relationships.............266
    total residual.................244
    versus repeated-measures models..243
    visualizing partitioning........244
    within-cluster variance...244, 245

multiple comparisons..........120–129
    adjust for multiple comparisons...*see* adjust for multiple comparisons
    adjusting in Stata..*see* `pwcompare` command
    α_{joint}....................121–122
    protecting against α inflation.....122–129
multiple regression..............86–90
    centering predictors.........87–88
    interpreting coefficients.....87–90
    model fit.....................86
    R^2.........................87
    root mean squared error.....87
    partial slopes..................88
    `regress` command.............86
    relationship among predictors..89–90
    rescaling predictors.........88–89

N

`nlcom` command.....338, 340, 341, 353
`normal()` function................211
null hypothesis significance testing..60–63, 202
    alternative hypothesis..........60
    criticisms..................63–64
    null hypothesis................60
    p-value.......................63
    `regress` output..............63
    steps.....................61–63

O

one-way ANOVA...............116–129
    alternative hypothesis.........117
    between versus within variance...117–119
    degrees of freedom............118
    estimating in Stata....*see* `oneway` command or `anova` command
    follow-up tests............119–129
    mean squares............118–119
    multiple comparisons..........*see* multiple comparisons

one-way ANOVA, *continued*
    null hypothesis 117
    ratio of variances 119
    sum of squares 117–118
oneway command 119

P

paran command 368
    pcf option 368
partitioning variance
    regression 51–53
planned comparisons 122–125
    contrast weights 122
    degrees of freedom 123
    F test 123
    linear combinations 122–125
    Stata estimation . . . .
        . . . . . . *see* test command and
        lincom command
    sum of squares 123
point-and-click 9
    problems 9
    versus command language 10
population parameters versus sample-
    based estimates 44
power . . . . . *see also* simulate command
    definition 202
    estimating in Stata . . . . . *see* power
        command
    null and alternative distribution . .
        202–210
    simulation 204–210
    type M error 230, 233–235
    type S error 230–233
    z test 210–214
    z test formula 211
power command 214–229
    correlation 220–223
        detectable correlation . . 222–223
        sample-size calculation 221–
        222
    diff() option 218
    factorial ANOVA 226–229
    graph() option 216
    knownsds option 215

power command, *continued*
    one-way ANOVA 223–226
        f 223
        sample-size estimates 225
    onecorrelation 221–223
    oneway 224–226
        parallel option 225–226
    sample-size estimates 229
    t test 215–220
        detectable difference 216–217
        sample-size calculation 218
    twomeans 214–215
    twoway 227–229
        parallel option 229
        varcolumn() option 227
        varrow() option 227
        varrowcolumn() option 227
    varying multiple parameters . . . 219
    z test 214–215
predict command 49, 70–73
    residual option 49
    fitted option 299
    inferential uncertainty 70–71
    plotting the results 71–73
    predictive uncertainty 71–73
    stdf option 72
    stdp option 70
predictive uncertainty . . . . *see*
    predict command, predictive
    uncertainty
program command 204
    bootstrapping 342
program define command 40
Project Manager . . . . . . *see* reproducible
    analysis
pwcompare command 191
    effects option 191
    mcompare() option 191
    Scheffé 126–127
    small option 191
    small samples 190
    Tukey's HSD 127–129

R

R^2 52
    formula 52
    interpretation 53
    Stata output 52
regress command 45
    output 45
regression
    centering predictors 47
    conceptual introduction 43
    equation 43
    expected values 44
    general bivariate equation 44
    intercept computation 45
    intercept interpretation 47–48
    margins command ... see margins command
    predict command ... see predict command
    predicted values 49–51
    reason for error term 43
    residuals 48–49
        definition 48
    slope computation 45
    slope interpretation 46–47
    Stata command ...... see regress command
reliability 315
    definition 319
    factor analysis .. see factor analysis
    partitioning variability in items ... 319
    reasons for factor analysis 320
    shared variance across items .. 318–320
    variability in item scores 318
repeated measures
    Σ 174
    benefits of mixed models .. 167–168
    block diagonal matrix 189
    covariance among residuals 174

repeated measures, *continued*
    covariance structures 176–189
        autoregressive 180–182
        compound symmetry ... 177–180
        independent 176–177
        Toeplitz 183–185
        unstructured 186–189
    degrees of freedom 189–190
    denominator degrees of freedom 179
    estimating in Stata ..... see mixed command
    heteroskedastic residuals .. 197–200
    maximum likelihood versus least squares 167
    model formulation 172–173
    multiple factors 192–197
    nonindependence 173
replace command 17
replication 5–8
reproducible analysis 5, 27–31
    log files 30
        text versus SMCL 30
    Project Manager 30–31
    role of do-files 28
        annotating 29
        versus point-and-click 29
    suggestions for 28
    workflow 31
        recommendations 31
reshape command 135, 171–172
    i() option 172
    j() option 172
    naming recommendations 172
reshape long command 172
reshape wide command 172
residual variance .... see conditional variance
residuals .... see regression residuals
root mean squared error .... see conditional variance

S

sampling distribution 56, 202
 mean of 56
 shape of 56
 simulating .... see simulation
 standard deviation of 56
 standard error .. see standard error
screeplot command 367
sem command 380
 constraining latent correlation ....
  388
 constraining latent variables ... 386
 constraints 328, 345
 equation goodness of fit .... see
  estat gof command
 making sense of the output 328
 model builder versus syntax ... 326
 one-factor model 327
 options
  coeflegend 338, 340
  covariance() 351, 381
  covstructure 349
  ginvariant 402, 407, 408
  group(varname) 402
  nm1 336
 parallel items constraints 349
 standardized option 389
 syntax basics 327
 tau-equivalent constraints 347
 tau-equivalent model 344
simdemo command 40
simple effects ...... see factorial ANOVA
simulate command 40–41
 arguments 41
 options 41
 power 205–209
simulation 40
 sampling distribution 54–56, 58
standard errors 56
 regression 56–58
Stata interface 10
 Command window 10
 Properties window 11
 Results window 11

Stata interface, *continued*
 Review window 11
 Variable window 11
statsby command 275, 292–293
sum of squares 52
 model 52
 residual 52
 total 52
summarize command 20–21
 assisting with centering 47
 repeated measures 170
 returned values 47

T

t test 114–116
 effect size 115–116
 formula and procedures 114
 null hypothesis 114
table command 21–22
 by() option 134, 152
 contents() option 134
 factorial ANOVA 134–135, 152
 format() option 134
 marginal means 138
tabstat command 113
 stat option 336
tabulate command 21–22
test command
 planned comparisons 123–125
ttest command 114–115
twoway
 function command 260
 lfit command 269, 292, 294
 rcap command 276, 293
 scatter command 25–27, 260,
  269, 276, 291–294, 369
type I error rate 203
type II error 203

U

use command 12
 file paths 12

V

validity 315
variable types
      date and time 14
      numeric 14
      string 15
violin plots 113
      `vioplot` command 113

W

wide versus long datasets 169–170
working directory 12
      benefits of setting 13

Y

Yale–Brown Obsessive Compulsive
      Scale 398